Lecture Notes in Computer Science 14988

Founding Editors

Gerhard Goos
Juris Hartmanis

Editorial Board Members

Elisa Bertino, *Purdue University, West Lafayette, IN, USA*
Wen Gao, *Peking University, Beijing, China*
Bernhard Steffen, *TU Dortmund University, Dortmund, Germany*
Moti Yung, *Columbia University, New York, NY, USA*

The series Lecture Notes in Computer Science (LNCS), including its subseries Lecture Notes in Artificial Intelligence (LNAI) and Lecture Notes in Bioinformatics (LNBI), has established itself as a medium for the publication of new developments in computer science and information technology research, teaching, and education.

LNCS enjoys close cooperation with the computer science R & D community, the series counts many renowned academics among its volume editors and paper authors, and collaborates with prestigious societies. Its mission is to serve this international community by providing an invaluable service, mainly focused on the publication of conference and workshop proceedings and postproceedings. LNCS commenced publication in 1973.

Andrea Ceccarelli · Mario Trapp ·
Andrea Bondavalli · Friedemann Bitsch
Editors

Computer Safety, Reliability, and Security

43rd International Conference, SAFECOMP 2024
Florence, Italy, September 18–20, 2024
Proceedings

Editors
Andrea Ceccarelli
University of Florence
Florence, Firenze, Italy

Andrea Bondavalli
Department of Mathematics
University of Florence
Florence, Italy

Mario Trapp
Fraunhofer IKS
Munich, Germany

Friedemann Bitsch
GTS Deutschland GmbH
Ditzingen, Germany

ISSN 0302-9743   ISSN 1611-3349 (electronic)
Lecture Notes in Computer Science
ISBN 978-3-031-68605-4   ISBN 978-3-031-68606-1 (eBook)
https://doi.org/10.1007/978-3-031-68606-1

© The Editor(s) (if applicable) and The Author(s), under exclusive license
to Springer Nature Switzerland AG 2024

This work is subject to copyright. All rights are solely and exclusively licensed by the Publisher, whether the whole or part of the material is concerned, specifically the rights of translation, reprinting, reuse of illustrations, recitation, broadcasting, reproduction on microfilms or in any other physical way, and transmission or information storage and retrieval, electronic adaptation, computer software, or by similar or dissimilar methodology now known or hereafter developed.
The use of general descriptive names, registered names, trademarks, service marks, etc. in this publication does not imply, even in the absence of a specific statement, that such names are exempt from the relevant protective laws and regulations and therefore free for general use.
The publisher, the authors and the editors are safe to assume that the advice and information in this book are believed to be true and accurate at the date of publication. Neither the publisher nor the authors or the editors give a warranty, expressed or implied, with respect to the material contained herein or for any errors or omissions that may have been made. The publisher remains neutral with regard to jurisdictional claims in published maps and institutional affiliations.

This Springer imprint is published by the registered company Springer Nature Switzerland AG
The registered company address is: Gewerbestrasse 11, 6330 Cham, Switzerland

If disposing of this product, please recycle the paper.

Preface

The SAFECOMP conference series was initiated in 1979 by EWICS TC7, the Technical Committee on Reliability, Safety and Security of the European Workshop on Industrial Computer Systems, with the aim of offering a regular platform for knowledge and technology transfer across academia, industry, research, and licensing institutions. Since 1985, the International Conference on Safety, Reliability and Security of Computer-based Systems (SAFECOMP) has taken place annually. This volume contains the papers presented at the 43rd SAFECOMP edition, held in Florence, Italy, in September 2024.

This year, we received 80 high-quality submissions from 13 European and 11 extra-European countries, with evidence of considerable cooperative effort across geographical and institutional boundaries. Each submitted article was single-blind reviewed by at least three independent reviewers; the decision on the conference program was jointly taken during the International Program Committee meeting in April 2024. In total, 19 articles were finally accepted for publication within the present proceedings volume as well as for presentation in September 2024 during the conference.

The program was enriched by three keynotes given by renowned speakers: "Combinatorial testing for safety-critical systems" by W. Eric Wong; "The automotive industry is transitioning to the Software Defined Car (SDV) - How to get there safely?" by Stefan Poledna; and "Autonomous driving - safety, security, and liability", by Isabella Ferrari.

As in previous years, the conference was organized as a single-track conference, allowing intensive networking during breaks and social events, and participation in all presentations and discussions.

For this edition, we had 4 high-quality workshops in parallel the day before the main conference: DECSoS, SASSUR, TOASTS, and WAISE - These workshops differed according to the topic, goals, and organizing group(s), and are published in separate Springer LNCS SAFECOMP Workshop Proceedings.

We would like to express our gratitude and thanks to all those who contributed to making this conference possible: the authors of submitted papers and the invited speakers; the International Program Committee members and the external reviewers; EWICS and the supporting organizations; last but not least, the Local Organization Committee who took care of the local arrangements.

We hope that the reader will find these proceedings interesting and stimulating.

September 2024

Andrea Bondavalli
Andrea Ceccarelli
Mario Trapp

Organization

EWICS TC7 Chair

Mario Trapp — Technical University of Munich, Germany

General Co-chairs

Andrea Bondavalli — University of Florence, Italy
Andrea Ceccarelli — University of Florence, Italy

Conference Program Co-chairs

Andrea Ceccarelli — University of Florence, Italy
Mario Trapp — Technical University of Munich, Germany

General Workshop and Special Session Co-chairs

Barbara Gallina — Mälardalen University, Sweden
Erwin Schoitsch — AIT Austrian Institute of Technology, Austria
Elena Troubitsyna — KTH Royal Institute of Technology, Sweden

Publication Chair

Friedemann Bitsch — Hitachi Rail GTS Deutschland GmbH, Germany

Industrial Contacts and Publicity Co-chairs

António Casimiro — Universidade de Lisboa, Portugal
Ilir Gashi — City, University of London, UK
Wilfried Steiner — TTTech Computertechnik AG, Austria

Web Chair

Francesco Mariotti — University of Florence, Italy

Local Organization Committee

Muhammad Atif — University of Florence, Italy
Francesco Mariotti — University of Florence, Italy
Tommaso Puccetti — University of Florence, Italy
Marzieh Kordi — University of Florence, Italy

International Program Committee

Magnus Albert — SICK AG, Germany
Uwe Becker — Drägerwerk AG & Co KGaA, Germany
Alessandro Biondi — Scuola Superiore Sant'Anna, Italy
Peter G. Bishop — Adelard, UK
Friedemann Bitsch — Hitachi Rail GTS Deutschland GmbH, Germany
Andrea Bondavalli — University of Florence, Italy
Jeroen Boydens — Katholieke Universiteit Leuven, Belgium
Simon Burton — University of York, UK
António Casimiro — Universidade de Lisboa, Portugal
Marsha Chechik — University of Toronto, Canada
Peter Daniel — EWICS TC7, UK
Ewen Denney — KBR/NASA Ames Research Center, USA
Felicita Di Giandomenico — ISTI-CNR, Italy
Wolfgang Ehrenberger — Fulda University of Applied Sciences, Germany
John Favaro — Intecs, Italy
Francesco Flammini — Linnaeus University, Sweden
Barbara Gallina — Mälardalen University, Sweden
Janusz Górski — Gdańsk University of Technology, Poland
Lars Grunske — Humboldt University Berlin, Germany
Jérémie Guiochet — LAAS-CNRS, France
Ibrahim Habli — University of York, UK
Maritta Heisel — University of Duisburg-Essen, Germany
Andreas Heyl — Robert Bosch GmbH, Germany
Yan Jia — University of York, UK
Phil Koopman — Carnegie Mellon University, USA
Paolo Lollini — University of Florence, Italy
Francesca Lonetti — ISTI-CNR, Italy

John McDermid	University of York, UK
Zoltan Micskei	BME - Budapest University of Technology and Economics, Hungary
Leonardo Montecchi	NTNU - Norwegian University of Science and Technology, Norway
Ganesh Pai	KBR/NASA Ames Research Center, USA
Philippe Palanque	ICS-IRIT, University of Toulouse 3, France
Yiannis Papadopoulos	University of Hull, UK
Peter Popov	City, University of London, UK
Andrew Rae	Griffith University, Australia
Alexander Romanovsky	Newcastle University, UK
Matteo Rossi	Politecnico di Milano, Italy
Martin Rothfelder	Siemens AG, Germany
Juan Carlos Ruiz Garcia	Universitat Politècnica de València, Spain
John Rushby	SRI International, USA
Francesca Saglietti	University of Erlangen-Nuremberg, Germany
Behrooz Sangchoolie	RISE Research Institutes of Sweden, Sweden
Erwin Schoitsch	AIT Austrian Institute of Technology, Austria
Christel Seguin	Office National d'Études et de Recherches Aérospatiales, France
Evgenia Smirni	College of William and Mary, USA
Wilfried Steiner	TTTech Computertechnik AG, Austria
Mark Alexander Sujan	Loughborough University, UK
Kenji Taguchi	National Institute of Informatics, Japan
Stefano Tonetta	FBK Fondazione Bruno Kessler, Italy
Elena Troubitsyna	KTH Royal Institute of Technology, Sweden
Martin Törngren	KTH Royal Institute of Technology, Sweden
Marcus Völp	University of Luxemburg, Luxembourg
Tommaso Zoppi	University of Trento, Italy

Sub-reviewers

Federico Aromolo	Scuola Superiore Sant'Anna, Italy
Koorosh Aslansefat	University of Hull, UK
Muhammad Atif	University of Florence, Italy
Pierre Bieber	Office National d'Études et de Recherches Aérospatiales, France
Laure Buysse	Katholieke Universiteit Leuven, Belgium
Kester Clegg	University of York, UK
Franca Corradini	Polo universitario Lugano, Switzerland
Brent De Blaere	Katholieke Universiteit Leuven, Belgium

x   Organization

Kevin Delmas   Office National d'Études et de Recherches
                                  Aérospatiales, France
Allan Espindola   University of Lisbon, Portugal
Camille Fayollas   ICS-IRIT, University of Toulouse, France
Gianfilippo Fornaro   KTH Royal Institute of Technology, Sweden
José Manuel Gaspar Sánchez   KTH Royal Institute of Technology, Sweden
Magnus Gyllenhammar   KTH Royal Institute of Technology, Sweden
Fajar Haifani   Fondazione Bruno Kessler, Italy
Victoria Hodge   University of York, UK
Boyue Caroline Hu   University of Toronto, Canada
Kaushik Madala   UL Solutions, Japan
Sam Michiels   Katholieke Universiteit Leuven, Belgium
Samaneh Mohammadi   Mälardalen University, Sweden
Logan Murphy   University of Toronto, Canada
Federico Nesti   Scuola Superiore Sant'Anna, Italy
Philippa Ryan Conmy   University of York, UK
Muhammad Rusyadi Ramli   KTH Royal Institute of Technology, Sweden
Gianluca Redondi   Fondazione Bruno Kessler, Italy
Giulio Rossolini   Scuola Superiore Sant'Anna, Italy
Septavera Sharvia   University of Hull, UK
Kaige Tan   KTH Royal Institute of Technology, Sweden
Jens Vankeirsbilck   Katholieke Universiteit Leuven, Belgium
Torin Viger   University of Toronto, Canada

Main Sponsors

IMAGINARY

SATODEV
Safety Tools Development

Organization xi

SICK AG
The SICK logo is a registered trademark of SICK AG
in Germany and other countries.

SafeComp Sponsors

Critical Systems Labs Inc.

ResilTech S.R.L.

Supporting Institutions

European Workshop on
Industrial Computer Systems
Technical Committee 7 on
Reliability, Safety and Security

Università Degli Studi Firenze
Dipartimento Di Matematica E
Informatica "Ulisse Dini"

Logos-RI
Research and Innovation

Technical University of Munich

Austrian Institute of Technology

Mälardalen University (MDU)

Hitachi Rail
GTS Deutschland GmbH

Lecture Notes in Computer Science (LNCS),
Springer Science + Business Media

ERCIM
European Research Consortium
for Informatics and Mathematics

European Research Consortium for
Informatics and Mathematics

European
Network of
Clubs for
REliability and
Safety of
Software

Technical Group ENCRESS
in GI and ITG

xiv Organization

Gesellschaft für Informatik (GI)

Informationstechnische Gesellschaft
(ITG) im VDE

Austrian Computer Society

Austrian Software Innovation Association

Contents

Fault Injection and Tolerance

In-Memory Zero-Space Floating-Point-Based CNN Protection Using
Non-significant and *Invariant* Bits 3
   *Juan Carlos Ruiz, David de Andrés, Luis-J. Saiz-Adalid,
and Joaquín Gracia-Morán*

A Failure Model Library for Simulation-Based Validation of Functional
Safety 18
   Tiziano Munaro, Irina Muntean, and Alexander Pretschner

Strategic Resilience Evaluation of Neural Networks Within Autonomous
Vehicle Software 33
   *Anna Schmedding, Philip Schowitz, Xugui Zhou, Yiyang Lu,
Lishan Yang, Homa Alemzadeh, and Evgenia Smirni*

System and Software Safety Assurance

Reconciling Safety Measurement and Dynamic Assurance 51
   Ewen Denney and Ganesh Pai

Safety Invariant Engineering for Interlocking Verification 68
   *Alexei Iliasov, Dominic Taylor, Linas Laibinis,
and Alexander Romanovsky*

Assurance Case Synthesis from a Curated Semantic Triplestore 84
   Saswata Paul, Baoluo Meng, Kit Siu, Abha Moitra, and Michael Durling

CyberDS: Auditable Monitoring in the Cloud 100
   Lev Sorokin and Ulrich Schoepp

Automated Driving Systems

Anatomy of a Robotaxi Crash: Lessons from the Cruise Pedestrian
Dragging Mishap 119
   Philip Koopman

Comprehensive Change Impact Analysis Applied to Advanced Automotive Systems 134
Nicholas Annable, Mehrnoosh Askarpour, Thomas Chiang, Sahar Kokaly, Mark Lawford, Richard F. Paige, Ramesh Sethu, and Alan Wassyng

A Case Study of Continuous Assurance Argument for Level 4 Automated Driving 150
Hideaki Kodama, Yutaka Matsuno, Toshinori Takai, Hiroshi Ota, Manabu Okada, and Tomoyuki Tsuchiya

Security of Safety-Critical Systems

TitanSSL: Towards Accelerating OpenSSL in a Full RISC-V Architecture Using OpenTitan Root-of-Trust 169
Alberto Musa, Franco Volante, Emanuele Parisi, Luca Barbierato, Edoardo Patti, Andrea Bartolini, Andrea Acquaviva, and Francesco Barchi

A Lightweight and Responsive On-Line IDS Towards Intelligent Connected Vehicles System 184
Jia Liu, Wenjun Fan, Yifan Dai, Eng Gee Lim, and Alexei Lisitsa

Evaluating the Vulnerability Detection Efficacy of Smart Contracts Analysis Tools 200
Silvia Bonomi, Stefano Cappai, and Emilio Coppa

Safety-Security Analysis via Attack-Fault-Defense Trees: Semantics and Cut Set Metrics 218
Reza Soltani, Milan Lopuhaä-Zwakenberg, and Mariëlle Stoelinga

Safety Verification

Coyan: Fault Tree Analysis – Exact and Scalable 235
Nazareno Garagiola, Holger Hermanns, and Pedro R. D'Argenio

Safety Argumentation for Machinery Assembly Control Software 251
Julieth Patricia Castellanos-Ardila, Sasikumar Punnekkat, Hans Hansson, and Peter Backeman

Sound Non-interference Analysis for C/C++ 267
Daniel Kästner, Laurent Mauborgne, Sebastian Hahn, Stephan Wilhelm, Jörg Herter, Christoph Cullmann, and Christian Ferdinand

Autonomous Systems

A Dynamic Assurance Framework for an Autonomous Survey Drone 285
  *Philippa Ryan, Sepeedeh Shahbeigi, Jie Zou, Ioannis Stefanakos,
  and John Molloy*

Redefining Safety for Autonomous Vehicles 300
  Philip Koopman and William Widen

Author Index 315

Fault Injection and Tolerance

In-Memory Zero-Space Floating-Point-Based CNN Protection Using *Non-significant* and *Invariant* Bits

Juan Carlos Ruiz[✉], David de Andrés, Luis-J. Saiz-Adalid, and Joaquín Gracia-Morán

ITACA - Universitat Politècnica de València (UPV), Camino de Vera s/n, Valencia, Spain
{jcruizg,ddandres}@disca.upv.es, {ljsaiz,jgracia}@itaca.upv.es

Abstract. Convolutional Neural Networks (CNNs) have accomplished significant success in various domains, including transportation, health care and banking. Millions of weights, loaded from main memory into the internal buffers of CNN accelerators, are repeatedly used in the inference process. Accidental and malicious bit-flips targeting these buffers may negatively impact the CNN's accuracy. This paper proposes a methodology to tolerate the effect of (multiple) bit-flips on floating-point-based CNNs using the *non-significant* and the *invariant* bits of CNN parameters. The former, determined after fault injection, do not significantly affect the accuracy of the inference process regardless of their value. The latter, determined after analyzing the network parameters, have the same value for all of them. Slight modifications can be applied to carefully selected parameters to increase the number of *invariant* bits. Since *non-significant* and *invariant* bits do not require protection against faults, they are employed to store the parity bits of error control codes. The methodology preserves the CNN accuracy, keeps its memory footprint, and does not require any retraining. Its usefulness is exemplified through the FP32 and BFloat16 versions of the LeNet-5 and GoogleNet CNNs.

Keywords: Floating-point-based CNN · ECC · *invariant* bits · *non-significant* bits.

1 Introduction

Convolutional neural networks (CNNs) are currently of great interest in critical domains demanding image classification to support advanced safety-oriented features [20], such as those deployed in autonomous driving, medical image analysis or worker activity monitoring systems [4]. Providing high degrees of accuracy in object recognition comes with a high computational cost that requires the support of specific hardware accelerators.

Grant PID2020-120271RB-I00 funded by MCIN/AEI/10.13039/501100011033.

© The Author(s), under exclusive license to Springer Nature Switzerland AG 2024
A. Ceccarelli et al. (Eds.): SAFECOMP 2024, LNCS 14988, pp. 3–17, 2024.
https://doi.org/10.1007/978-3-031-68606-1_1

These accelerators support the execution of a series of computation layers in CNNs. Each layer takes the output of the preceding layer as its input. During inference, these inputs are scaled using a large set of weights determined during training. The evaluation of every image requires reusing these weights. Modern CNNs exchange hundreds of megabytes of weights between the on-chip processing elements and the off-chip accelerator memory. This leads to a severe bandwidth bottleneck limiting performance that is usually leveraged by caching weights on-chip to speed up the inference process [7].

Reducing the precision of weights without significantly altering the accuracy of a CNN is also a trend that enables the allocation of more complex models in existing hardware accelerators. This means changing the weights encoding from the conventional IEEE-754 32-bit single-precision floating-point (FP32) format to reduced-precision FP formats, such as the IEEE-754 16-bit FP format (FP16) or the Google Brain 16-bit FP format (BF16) [16]. Qualcomm Cloud AI, the AMD AI Engine Technology, Cerebras WS-2 [11], or Groq Chip [3], to name a few, support today inference using both 32-bit and 16-bit FP arithmetic. The use of non-FP CNNs falls out of the considerations discussed in this contribution.

Once loaded into accelerators' buffers, existing memory protection mechanisms do not protect CNN weights. Thus, during inference, they are exposed to faults that may lead to object misclassifications, provoking unsafe situations [17]. On the one hand, the multiplicity of bits that can be potentially flipped by single event upsets increases with larger technology scales [5]. On the other hand, as CNNs interconnect to other systems, they become further exposed to malicious faults (attacks) that may crush their inference process by simply flipping a small number of vulnerable weight bits [22]. Guaranteeing protection against accidental and malicious multiple bit-flips is crucial today for safety-critical systems.

This paper defines a step forward in the provision of multiple bit-flip protection for FP-based CNNs First, fault injection is used to determine which bits from all weights can be simultaneously altered in all network tensors without affecting the CNN's accuracy. Once identified, these *non-significant* bits can be used to hold the parity bits of error correction codes (ECC). This proposal is in line with the research presented in [25]. However, that approach did not address what to do when insufficient *non-significant* bits are available to deploy the desired ECC. The current contribution leverages that limitation by additionally locating and considering *invariant* bits for ECC deployment. These additional bits can be *pure invariant* bits or bits that are *forced to become invariant* by applying slight modifications that do not significantly alter the accuracy of the considered network. The feasibility of the methodology is exemplified through FP32- and BF16-based versions of LeNet-5 and GoogLeNet CNNs, requiring the protection of 45,379 and 6,624,904 parameters (weights and biases), respectively. Conversely to other alternatives, the proposal has the merit of enabling the *in-parameter* deployment of ECCs while preserving the accuracy of the protected CNN and without i) requiring retraining, ii) increasing the size in bytes of protected network parameters, and iii) inducing a significant temporal overhead in the protected inference process.

The rest of this paper is organised as follows. Section 2 details the research context, Sect. 3 describes the proposed methodology and Sect. 4 reports the results of experiments carried out. The pros and cons of the proposal are discussed in Sect. 5, and Sect. 6 concludes the paper.

2 Background

Despite their intrinsic robustness [17], FP-based CNNs still require fault tolerance to ensure a safe operation. Their robustness in the presence of multiple bit-flips is essential when considering the evolution of CMOS integrated circuits (ICs) and the impact of *bit-flip attacks (BFA)* on the CNN behaviour.

Typically, the effect of ionising particles on CMOS ICs is modelled using the single bit-flip fault model. However, recent studies [5] have shown that the close proximity of semiconductor regions in 3D manufacturing technologies, such as FinFET, results in multiple regions collecting charge after an ion strike. In other words, when ionizing particles strike FiFET-based ICs, they manifest as multiple (adjacent) bit-flips. It must be noted that this conclusion only applies to storage elements, such as the buffers used to save CNN parameters, where transistors are placed closer to each other compared to other IC components.

The multiple bit-flip fault model is also useful to mimic the effect of modifications that attackers can apply on parameters to alter the CNN behaviour. BFAs differ from accidental parameters alterations in the number of bits that can be simultaneously affected by the fault and the location of such bits, that may not be adjacent in the case of an attack. Research has shown that manipulating a few CNN parameters may significantly perturb the network inference process [26] or enable the discovery of existing network vulnerabilities [23]. In [22], for instance, authors showed that altering only 13 out of 93 million bits of a ResNet18 CNN was enough to completely crash the network.

Triplicating network kernels (weights) to tolerate the occurrence of bit-flips on one of the replicas is the solution proposed in [9]. However, this also triplicates the amount of memory required to hold and process the kernels, which can be prohibitive under certain circumstances and/or for certain computing devices.

An alternative is to force the parameters of a network to follow specific patterns and make them less sensitive to faults. For instance, Schron [26] normalises weights to equalize their criticality, but the CNN must be retrained to compensate for the accuracy loss. Other proposals involve the injection of faults during training to enable the CNN to compensate via adaptation [13]. Both protection mechanisms require network retraining, which is computationally and timing expensive.

It is also possible to take advantage of the *non-significant* bits existing in parameters to store an ECC, so no extra bits are required for parity bit allocation. The proposal in [10] combines interleaved single-error correction (SEC) and triple repetition codes. However, this is an ad hoc solution for a specific CNN rather than a generic methodology that could be applied to any network. Burel [6] proposes randomly selecting a parity bit and setting the parameter to 0 upon detecting a fault to mitigate its impact. This approach increases the robustness of FP-based CNNs. Nevertheless, having only one parity bit exposes parameters to the effect of multiple even bit-flips that can be part of a BFA.

Invariant bits in parameters can be another potential source for ECC parity bit allocation. For instance, in [12], authors found that bits 30 and 29 of the FP32 weights of AlexNet were all 0 and 1, respectively. Based on this observation, they defined bit 30 as the "most unreliable bit," and masked it to avoid changes from 0 to 1. A different approach was presented in [18] where they found that 99.5% of the FP32 weights of GoogLeNet, AlexNet, and ResNet50 have bits (30:28) set to "011". In that work, researchers proposed two fault tolerance mechanisms. In the first one, the three parity bits of a Hamming SEC (7, 4) are stored in bits (30:28), to protect bits 31, 27, 26, and 25. After the error correction process, the value of bits (30:28) is set to "011". In the second proposal, the four parity bits of a Hamming SEC (13, 9) used to protect the sign and exponent bits are stored in the four LSbs of the parameters. Finally, in [10], and after studying the nine CNNs supported by Darknet [24], investigators found that 99.97% of the exponents of FP32 parameters have the value "011" in their MSbs. Accordingly, they propose to protect these parameters using Triple Modular Redundancy (TMR) and the rest using an SEC-Double Error Detection (DED). It is worth mentioning that this technique also required two extra bits (stored in bits 0 and 1 of the parameters) to indicate the fault tolerance method that was used for each parameter.

As can be seen, although the protection of CNN parameters through ECCs has been largely explored, some works require using additional bits to store parity bits, whereas others make limited use of *invariant* bits. To the best of the authors' knowledge, this is the first contribution proposing a method to protect 32-bit and 16-bit FP-based CNNs against multiple bit-flips by simultaneously exploiting the information redundancy existing in *non-significant* and *invariant* weight bits.

3 Proposed Methodology

The training process of CNNs commonly uses the IEEE FP32 format to compute its parameters. This 32-bit format stores the sign (S) of the represented value in bit 31, the exponent (E) is represented in excess of 127 in bits 30 to 23, and the mantissa (the fractional bits with an implicit leading 1) is stored in bits 22 to 0.

This format allows the representation of a wide range of values (computed by Eq. 1) with high precision, as required for general-purpose computers. For instance, the smallest and largest positive real numbers that can be represented are 2^{-149} ($\sim 1.4 \times 10^{-45}$) and $(2 - 2^{-23}) \times 2^{127}$ ($\sim 3.4 \times 10^{38}$), respectively, and the least significant bit (LSb) value is a 2^{-23} ($\sim 1.2 \times 10^{-7}$) fraction of the value of the leading bit.

$$value = (-1)^S 2^{E-127} \left(1.0 + \sum_{i=1}^{23} b_{23-i} 2^{-i}\right) \qquad (1)$$

However, CNN parameters do not usually need all this range, as weights are in the range [-1.0, +1.0] and, frequently, do not require high precision. That is why some systems, deep-learning accelerators, and libraries (for instance [1]) employ the Google BF16 format [28]. This 16-bit format keeps the FP32 format but removes its 16 LSbs, thus saving memory. Accordingly, it maintains the range of FP32 (as the exponent has the same range) but reduces its precision (only the seven most significant bits (MSbs) of the mantissa remain).

The replacement of FP32 parameters by their BF16 counterparts relies on these changes not altering the output of a CNN (probabilities to make a decision), or even if this output changes, the resulting classification still being correct (the likeliest category remains the same). The over-precision provided by the FP32 format and, to a lesser extent, the BF16 format enables the location of *non-significant* bits (can take any value without changing the classification). In addition, the subset of exponents employed to represent the parameters of CNNs creates *invariant* bits (those with the same value for all the CNN parameters). Hence, these bits could store parity bits for adapted ECCs without any memory overhead. The next sections detail how to proceed to locate these bits.

3.1 Identification of *non-Significant* bits

Our goal is to provide a systematic procedure to determine the set of bits within the CNN parameters that, even if changed, do not significantly affect the accuracy of the network. Fault injection is a suitable approach [17] to tamper with all parameters of the CNN simultaneously: first, the LSb (bit 0) of all parameters; then, the two LSbs (bits 1:0); later, the three LSBs (bits 2:0), and so on. Comparing the accuracy of the CNN in the absence and the presence of injected faults enables determining whether the set of injected bits significantly impacts the network or not.

This proposal improves the approach to find *non-significant* bits already introduced in [25]. Two different fault injection campaigns were carried out in that approach to change the parameters to their lowest (stuck-at-0 faults) and highest (stuck-at-1 faults) possible values. However, some parameters may present minimal changes or be completely unaffected (stuck-at faults only affect bits with the opposite value). Although the injection of bit-flip faults may be considered to deal with this problem, 0 to 1 transitions in some bits may compensate for 1 to 0 transitions in others, resulting in minor differences.

Our new proposal tries to maximise the difference between the original and the injected values for all parameters in each fault injection campaign. To accomplish this objective, the MSb of the set of considered bits is flipped to the opposite value (bit-flip fault); then, the rest of these bits are set (stuck-at fault) to the same value as the MSb. Table 1 compares this proposal with the other alternatives for the value 1.6875 represented using a 4-bit mantissa ("1011").

As Table 1 shows, the approach based on stuck-at faults sometimes offers the maximum difference between the original and the injected parameters, but not always. The bit-flip approach gets the maximum difference only when all the bits in the considered set have the same value. The new fault injection proposal

Table 1. Original versus injected values for alternative fault injection (FI) approaches.

Bits	Stuck-at-0			Stuck-at-1			Bit-flip			New FI approach		
	Mantissa	Value	Difference	Mantissa	Value	Difference	Mantissa	Value	Difference	Mantissa	Value	Difference
(0:0)	1010	1.625	0.0625	1011	1.6875	0.0000	1010	1.625	0.0625	1010	1.6250	0.0625
(1:0)	1000	1.500	0.1875	1011	1.6875	0.0000	1000	1.500	0.1875	1000	1.5000	0.1875
(2:0)	1000	1.500	0.1875	1111	1.9375	0.2500	1100	1.750	0.0625	1111	1.9375	0.2500
(3:0)	0000	1.000	0.6875	1111	1.9375	0.2500	0100	1.250	0.4375	0000	1.0000	0.6875

consistently achieves the highest differences and will be more effective in the search for *non-significant* bits.

It is worth mentioning that although identified non-significant bits can be removed to save memory space, our proposal is to promote their use to store the parity bits of ECCs without increasing memory usage. This is why our approach can be defined as an *in-memory zero-space* protection mechanism.

3.2 Location of *invariant* bits

Invariant bits (those with the same value in all the network parameters) can be found by simply analysing the network parameters. For instance, if all parameters were positive numbers, the sign (bit 31) would be 0. Thus, this bit could be removed to save memory space or, as proposed in this contribution, it could be used to store more parity bits and enhance the fault tolerance capabilities of deployed ECCs without requiring additional memory.

Empirical evidence shows that the exponent of the parameters very infrequently is greater than 0, represented as "01111111" in FP32 format. As the exponent is represented in excess of 127, relatively small negative exponent values will all have their MSbs set to "011...", like -1 ("01111110"), -2 ("01111101"), and so on. This is why it is possible to locate *invariant* bits in the exponent.

Even though the use of *invariant* bits has been previously explored [10,12,18], we extend their application to allow higher error coverage by combining them with *non-significant* bits and by *forcing* the occurrence of *invariant* bits.

After analysing a CNN for *invariant* bits, additional *invariant* bits could be found if relatively few parameters had a different exponent. Our proposal to force new *invariant* bits consists in setting the parameter to the smallest value that could be represented with the desired higher exponent or the largest value that could be represented with the desired lower exponent.

For instance, let us consider that the 5 MSbs of the exponent for all the parameters in a given CNN are set to "01111" except for some parameters with the value -8 as exponent ("01110111"). This limits the available *invariant* bits to four. However, by changing the values with exponent -8 (in the range [0.0039, 0.0078[) to the smallest value that can be represented with exponent -7 (0.0078), we could get five *invariant* bits with limited differences between the original and the changed values (at most 0.0039 in our example).

After forcing *invariant* bits, the CNN should be tested to ensure that those changes do not significantly affect its accuracy. When the number of changed parameters is limited, the effect on the CNN inference process is usually minimal. Experimental results presented in this work show the validity of this approach.

4 Case Study: LeNet-5 and GoogLeNet

Two different CNN architectures have been selected as a case study: LeNet-5 and GoogLeNet. In both cases, their parameters are encoded using FP32 and BF16 formats, thus providing two different versions for each case study.

LeNet-5 is a simple CNN architecture defined by LeCun [14] to classify images from the MINIST database [15] (10,000 greyscale pictures of 28 × 28 pixels representing handwritten digits). This architecture is relatively simple by today's standards (4 layers with 45,379 parameters), achieving an accuracy of 98.23% (117 misses out of 10,000 images). Still, it will ease its integration with the proposed ECCs to assess the resulting overhead after being implemented on a field-programmable gate array (FPGA).

GoogLeNet is a 22-layer deep CNN, with 6,624,904 parameters, proposed by Szegedy [27] from Google that won the ImageNet Large-Scale Visual Recognition Challenge (ILSVRC) in 2014. It classifies images from the ImageNet dataset [2], with an accuracy of 69.772% (15,114 misses out of 50,000 images), consisting of 50,000 RGB images resized to 256 × 256 pixels from 1,000 different categories, including objects and several types of animals. This CNN will show how the proposal could be applied to more complex architectures with much lower accuracy and, thus, potentially less capacity to deal with errors.

It is worth noting that, even though both CNNs natively encode their parameters using the FP32 format, their BF16-based counterparts can be easily obtained by simply truncating the lower half (15:0) of all parameters and rounding bit 16 if necessary (when bit 15 was 1).

4.1 Locating Non-Significant Bits

The PyTorch [19] models of both CNNs have been targeted by fault injection campaigns, as described in Sect. 3.1, to locate those bits that are not significant and can be used to store parity bits to protect those that are significant.

Table 2 lists the results obtained after modifying the LSbs of all the weights and biases (100% and 99.367% of the LeNet-5 and GoogLeNet parameters, respectively) to maximise the distance with their original value. This approach tries to mimic the worst possible effect that random (yet unknown) parity bits may have on the inference process. To help visualise the results, the cells listing the accuracy difference in percentage points (p.p.) have been colour-coded: green if they are identical, grey if the difference is below 0.5 p.p., orange if it is between 0.5 and 1.0 p.p., and red if it is higher than 1.0 p.p. The table also lists the number of images that were initially misclassified and are correctly identified after the introduced modifications (*corrected misses*) and the number of images that were originally correctly classified and are now misclassified (*unexpected misses*).

As can be seen, tampering with the 22 LSbs of LeNet-5 has a negligible effect on its accuracy, whereas it drops sharply for GoogLeNet when bit 19 is considered. Accordingly, if we look for ECCs that could be applied to both CNNs, we can only consider non-significant bits in the range (18:0).

4.2 Locating Invariant Bits

In the case of using an FP32 format, it could be possible to deploy powerful ECCs without taking additional memory space, as up to 19 bits are available to store the required parity bits. However, when using a BF16 format, only the three LSbs would be non-significant; thus, no ECC could be deployed to protect

Table 2. Difference in accuracy after maximising the distance to their original value for all the weights and biases for a range of LSBs.

		FP32-based LeNet-5				FP32-based GoogLeNet				
Bits	Misses	Accuracy	Corrected misses	Unexpected misses	Difference (in p.p.)	Misses	Accuracy	Corrected misses	Unexpected misses	Difference (in p.p.)
(0:0)	177	98.23%	0	0	0	15,114	69.772%	0	0	0
(1:0)	177	98.23%	0	0	0	15,114	69.772%	0	0	0
(2:0)	177	98.23%	0	0	0	15,114	69.772%	0	0	0
(3:0)	177	98.23%	0	0	0	15,114	69.772%	0	0	0
(4:0)	177	98.23%	0	0	0	15,114	69.772%	0	0	0
(5:0)	177	98.23%	0	0	0	15,114	69.772%	0	0	0
(6:0)	177	98.23%	0	0	0	15,114	69.772%	0	0	0
(7:0)	177	98.23%	0	0	0	15,112	69.776%	2	0	-0.004
(8:0)	177	98.23%	0	0	0	15,113	69.774%	1	0	-0.002
(9:0)	177	98.23%	0	0	0	15,111	69.778%	4	1	-0.006
(10:0)	177	98.23%	0	0	0	15,112	69.776%	5	3	-0.004
(11:0)	177	98.23%	0	0	0	15,115	69.770%	11	12	0.002
(12:0)	178	98.22%	0	1	0.01	15,116	69.768%	18	20	0.004
(13:0)	178	98.22%	0	1	0.01	15,122	69.756%	38	46	0.016
(14:0)	179	98.21%	0	2	0.02	15,097	69.806%	85	68	-0.034
(15:0)	178	98.22%	1	2	0.01	15,097	69.806%	130	113	-0.034
(16:0)	179	98.21%	2	4	0.02	150,87	69.826%	320	293	-0.054
(17:0)	176	98.24%	6	5	-0.01	15,355	69.290%	549	790	0.482
(18:0)	183	98.17%	4	10	0.06	15,546	68.908%	927	1,359	0.864
(19:0)	191	98.09%	17	31	0.14	17,513	64.974%	1,537	3,936	4.798
(20:0)	212	97.88%	29	64	0.35	24,524	50.952%	1,569	10,979	18.82
(21:0)	197	98.03%	42	62	0.2	49,100	1.800%	116	34,102	67.972
(22:0)	1,180	88.20%	33	0	10.03	49,932	0.136%	21	34,839	69.636
(23:0)	3,306	66.94%	30	3,159	31.29	49,956	0.088%	8	34,850	69.684
(24:0)	3,584	64.16%	35	3,442	34.07	49,947	0.106%	10	34,843	69.666
(25:0)	7,637	23.63%	12	7,472	74.6	49,946	0.108%	25	34,857	69.664
(26:0)	9,183	8.17%	18	9,024	90.06	49,948	0.104%	10	34,844	69.668
(27:0)	9,020	9.80%	8	8,851	88.43	49,950	0.100%	7	34,843	69.672
(28:0)	9,020	9.80%	8	8,851	88.43	49,950	0.100%	18	34,854	69.672
(29:0)	9,020	9.80%	8	8,851	88.43	49,950	0.100%	3	34,839	69.672
(30:0)	9,020	9.80%	8	8,851	88.43	49,950	0.100%	3	34,839	69.672
(31:0)	9,020	9.80%	8	8,851	88.43	49,950	0.100%	3	34,839	69.672

the rest of the bits. Hence, it is necessary to look for other sources of bits that could be used to store the parity bits.

As described in Sect. 3.2, the PyTorch dictionary of both models has been analysed to locate those bits with invariant values for all the weights and biases. Results, listed in Table 3, show that bits in the range (30:28), highlighted in green, have the constant value "011" for both CNNs. This means these bits could be used to store more parity bits, and the corresponding decoder could set them to their proper value. Hence, three more bits can be used to store parity bits, raising their number to six for the BF16 format. Nevertheless, more bits are required to protect the parameters beyond the standard single and double error correction (SEC and DEC) codes.

Table 3. Distribution of the value of the 16 MSbs for all CNN parameters. Results for the 16 LSbs of the FP32 CNNs are all close to a 50%/50% distribution.

	LeNet-5		GoogLeNet	
Bit	0s (%)	1s (%)	0s (%)	1s (%)
16	50.02%	49.98%	50.11%	49.89%
17	50.20%	49.80%	50.30%	49.70%
18	50.54%	49.46%	50.54%	49.46%
19	51.56%	48.44%	51.13%	48.87%
20	52.12%	47.88%	52.27%	47.73%
21	54.38%	45.62%	54.41%	45.59%
22	58.63%	41.37%	58.50%	41.50%
23	48.16%	51.84%	50.14%	49.86%
24	35.91%	64.09%	49.75%	50.25%
25	84.49%	15.51%	79.38%	20.62%
26	9.49%	90.51%	20.31%	79.69%
27	0.02%	99.98%	0.08%	99.92%
28	0.00%	100.00%	0.00%	100.00%
29	0.00%	100.00%	0.00%	100.00%
30	100.00%	0.00%	100.00%	0.00%
31	50.20%	49.80%	45.19%	54.81%

We could focus on bits in the range (27:25), highlighted in orange in the table, to try to obtain more invariant bits, as the rest are close to a 50%/50% distribution. Bit 27 is 1 in more than 99.9% of the parameters for both models, so it is most likely to become an invariant bit, whereas bits 26 and 25 are 1 and 0, respectively, for roughly 80% parameters. Table 4 lists the results obtained after rounding the rest of the parameters to the nearest number that forces the occurrence of invariants in these bits.

As can be seen, forcing a fourth or fifth invariant barely impacts the accuracy of the CNNs (highlighted in green). However, forcing the sixth invariant (bit

Table 4. Difference in accuracy after forcing bits to be invariant.

Bits	Value	Format	Misses	Accuracy	Corrected misses	Unexpected misses	Difference (in p.p.)	Misses	Accuracy	Corrected misses	Unexpected misses	Difference (in p.p.)
					LeNet					GoogLeNet		
(30:27)	0111	FP32	177	98.23%	0	0		15,114	69.77%	0	0	0.00%
		BF16	178	98.22%	0	1	0.01%	15,077	69.85%	155	118	-0.07%
(30:26)	01111	FP32	176	98.20%	3	4	0.02%	15,198	69.60%	652	568	0.17%
		BF16	176	98.20%	3	4	0.02%	15,189	69.62%	594	669	0.15%
(30:25)	011110	FP32	413	95.87%	54	290	2.36%	49,950	0.10%	18	34,854	69.77%
		BF16	416	95.84%	54	293	2.39%	49,950	0.10%	18	34,854	69.67%

25) drops the accuracy by more than 2 p.p. for LeNet-5 and down to 0.1% for GoogLeNet (in red in the table). Thus, we can safely use these two bits as storage for parity bits to deploy ECCs with greater protection capabilities.

4.3 Proposed Error Correcting Codes

Considering the non-significant and invariant bits discovered in the previous sections, a different ECC has been proposed for each CNN version (FP32 and BF16) to make use of the bare minimum resources required to provide single error correction (SEC) and double adjacent error correction (DAEC). Likewise, a different ECC has also been proposed for each data-formatted CNN to maximise the protection offered with the available storage space. The parity check matrices of the proposed ECCs are depicted in Fig. 1. The notation used in the Figure for proposed codes is (X,Y)+Z, where X reflects the bits of the *codeword* and Y indicates the bits in the *dataword*. Thus, the number of bits available to hold ECC parity bits is $X - Y$, among which Z are invariant bits. For instance, a (16,10)+4 ECC indicates a codeword of 16 bits, where a total of 6 (16 − 10) bits can be used to allocate ECC parity bits. Among such 6 bits, 4 are invariant bits and the rest (2 in this case) non-significant bits.

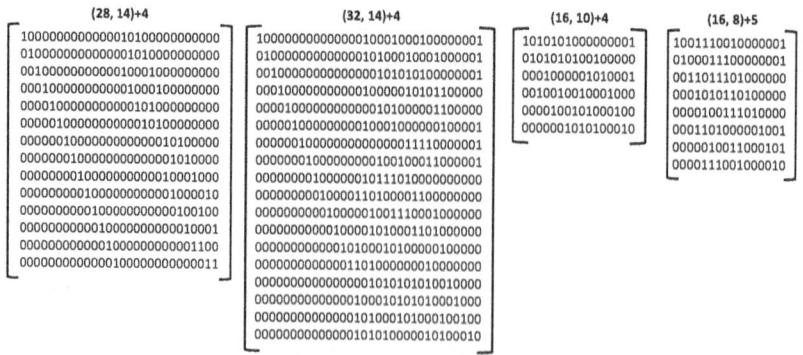

Fig. 1. Parity check matrices for the proposed ECCs. The fault tolerance capabilities of each ECC are detailed in Table 5.

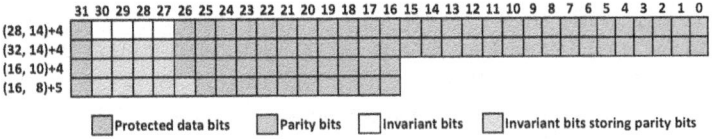

Fig. 2. Distribution of protected, parity and invariant bits for the proposed ECCs.

The (28,14)+4 code (see Fig. 2) only uses 28 bits out of the 32 bits available (FP32 parameters). In other words, the code is not considering the 4 invariant bits available. This reflects a situation where invariant bits are not protected, since it is not necessary, and they are not used to hold any parity bit. As a result the protection offered by this code (single error correction and double adjacent error correction, known as SEC-DAEC) is more modest than the one that could be deployed integrating such 4 invariant bits in the codeword. This is what the (32, 24)+4 code does, providing a massive protection of SEC, double error correction (DEC), triple error correction (TEC), and 4- and 5-bit burst error correction (4bBEC and 5bBEC) in all bits of the codeword (32 bits in this case). In the case of the BF16 format, the simplest (16, 10)+4 code is required to use the invariants as storage for parity bits to be able to provide SEC and DAEC protection. To increase this protection to SEC, DEC, triple and 4-bit adjacent error correction (TAEC and 4AEC), the (16, 8)+5 code requires two additional parity bits: one non-significant (although it decreases the accuracy a little bit) and one obtained by forcing the fifth invariant bit.

The level of protection each code provides against errors and their impact on the accuracy of the CNN is listed in Table 5. As can be seen, there is a gain in fault tolerance capabilities with negligible impact on accuracy, although to obtain 4/5 invariants bits, it is necessary to adjust a total of 11/4,316 parameters, respectively, for LeNet-5 and 5,505/1,347,751 parameters, for GoogLeNet.

4.4 Overhead Estimation of the Proposed ECCs

To obtain an estimation of the overhead induced by the proposed ECCs in custom hardware accelerators for CNNs, the LeNet-5 architecture has been synthesised

Table 5. Level of protection provided by ECCs and their impact on accuracy.

		LeNet-5			GoogLeNet		
ECC	Fault tolerance capabilities	Misses	Accuracy	Difference (in p.p)	Misses	Accuracy	Difference (in p.p)
None	None	177	98.23%		15,114	69.77%	
(28, 14)+4	SEC-DAEC	178	98.22%	0.01%	15,109	69.78%	-0.01%
(32, 14)+4	SEC-DEC-TEC-4bBEC-5bBEC	177	98.23%	0.00%	15,072	69.86%	-0.09%
(16, 10)+4	SEC-DAEC	176	98.24%	-0.01%	15,166	69.67%	0.10%
(16, 8)+5	SEC-DEC-TAEC-4AEC	179	98.21%	0.02%	15,433	69.13%	0.64%

using the AMD Vitis High-Level Synthesis (HLS) tool and, after that, implemented on an AMD Zynq UltraScale+ XCZU7EV-2FFVC1156 MPSoC device using the AMD Vivado ML Editions tool. Due to the size of the available prototype board, it was not possible to implement the GoogLeNet CNN, but the obtained results could be used as a reference.

Table 6. Overhead in terms of resources, latency, and power consumption for each ECC with respect to the unprotected LeNet-5 CNN.

CNN	LUT	FF	BRAM	DSP	Decoders	Decoders size (LUTs)	Latency (clock cycles)	Energy (mW/image)
LeNet-5	85,655	107,478	156	500	0	0	7,443	0.231
(28, 14)+4	−0.57%	3.44%	16.03%	0.00%	237	353	0.04%	5.90%
(32, 14)+4	52.32%	−49.24%	6.41%	−80.20%	39	5817	391.15%	297.40%
(16, 10)+4	15.15%	5.93%	−78.53%	0.00%	237	240	0.04%	0.97%
(16, 8)+5	19.93%	4.84%	−78.53%	0.00%	237	638	0.04%	−2.90%

The synthesis process has been configured to minimise the latency of the CNN (parallel execution of as many operations as possible) within the limits of the available FPGA resources. Table 6 lists the obtained results for the unprotected LeNet-5 CNN and the overhead after deploying the proposed ECCs.

In addition to the look-up tables (LUTs) and flip-flops (FFs) required for implementing combinational and sequential logic, respectively, the LeNet-5 CNN makes use of 156 small memory blocks (BRAM) to ease the parallel access to its parameters, and 500 multiply-and-accumulate (DSP) hard cores to accelerate the computation. The inference process for each image takes 7,443 clock cycles and consumes 0.231 mW.

It must be noted that BRAMs will need a decoder in each output port (they can be configured as dual-port memories). As the decoders of the (32, 14)+4 code are so large, it was necessary to reduce the level of parallelism to save resources for its implementation. So, it is five times slower than the rest of the versions and consumes up to three times more energy per image processed. This is not surprising due to its ambitious fault tolerance capabilities.

The latency is barely affected for the rest of the proposed codes, and as the achieved clock frequency does not change much, variations in the energy consumption are attributed to the different amounts of resources used.

The (28, 14)+4 code uses 16% more BRAMs to store the code, increasing the energy consumed per image by 6%. The codes developed for the BF16 format trade BRAMs for LUTs, as less storage is required, barely impact and even improve the energy consumed.

5 Discussion

Using ECCs for fault tolerance in CNNs may seem inappropriate, as they may induce unaffordable temporal overheads over protected networks [17]. Although this is true for software implementations, results show these overheads could be managed when decoders are directly deployed on hardware accelerators.

It is also true that ECCs, although effective against BFAs, can be easily bypassed using alternative attack methods [21]. However, it is important to understand that this happens because these alternative attacks fall out from the fault hypothesis under consideration when designing the protection codes.

Although reported results are particular for considered case studies, the proposed methodology for locating non-significant and invariant bits and forcing more bits to become invariant is general and applicable to any CNN. Obviously, the approach must be repeated for each CNN, each dataset, and after retraining the network, although minimum differences are expected in this latter case.

The proposed ECCs are also particular and adapted to the bits available in the case studies. However, different types of layers (convolution, batch normalisation, fully connected, etc.) and/or kinds of parameters (weights, biases, means, variances, etc.) may present a distinct number of *non-significant* and *invariant* bits and may require different types of ECCs for protection. This analysis can be done in general (treating all the parameters as a whole), specifically for each kind of parameter on each layer, or grouping layers or types of parameters. These different possibilities will allow tuning ECCs attending to the particular target selected for protection and the set of fault hypotheses that can be covered with the *non-significant* and *invariant* bits available.

Beyond the use of classical ECCs, it is also possible to discriminate *significant bits* attending to their level of criticality for the global inference process deployed by the considered CNN. Accordingly, not all significant bits may require equal protection against considered faults. In this context, unequal error control codes (UECs) or unequal error protection codes (UEPs) can be attractive.

The usefulness of the current proposal will be limited for quantised CNNs, as their parameters are usually represented using 8-bit integers without much space left to store parity bits. This is why protecting quantised networks against multiple bit-flips is so challenging, requiring network retraining even for single bit-flip protection [8]. Further research is required to marry the performance and memory footprint benefits of quantised CNNs with their robustness needs.

6 Conclusions

In this work, we have presented a new methodology to find *non-significant* and *invariant* bits that have been used to protect CNNs using ECCs. The feasibility of the proposal has been exemplified with two CNNs: LeNet-5 and GoogLeNet.

Results show how far we can arrive with the protection of a CNN when the number of non-significant and invariant bits is enough. For instance, in the case of the considered FP32 CNNs, it was possible to use up to 18 bits for protection, which enabled the deployment of a (32,14)+4 ECC with impressive (SEC-DEC-TEC-4bBED-5bBEC) fault tolerance capabilities. Increasing the number of invariants by carefully forcing the values of selected parameters has also enabled, in the case of the considered BF16 CNNs, the deployment of a (16,8)+5 ECC with powerful SEC-DEC-TAEC-4AEC fault tolerance capabilities.

This work opens two main lines of future research. On the one hand, the analysis of the parameters can be carried out attending to various levels of criticality that can be specified considering the impact of each parameter bit (or set of bits) on the inference carried out by the network. As a result, more critical parameters (or parameter bits) can be more protected than less critical ones. On the other hand, further research is required to protect low precision (8-bit FP or INT) parameters using ECCs against multiple single bit-flips.

References

1. Abadi, M., et al.: TensorFlow: Large-scale machine learning on heterogeneous systems (2015). https://www.tensorflow.org/, software available from tensorflow.org
2. Howard, A., Eunbyung Park, W.K.: ImageNet object localization challenge (2018). https://kaggle.com/competitions/imagenet-object-localization-challenge
3. Ahmed, I., et al.: Answer fast: accelerating BERT on the tensor streaming processor. In: 33rd IEEE International Conference Application-specific Systems, Architectures and Processors, pp. 80–87. IEEE (2022)
4. Akter, M., Ansary, S., Khan, M.A.M., Kim, D.: Human activity recognition using attention-mechanism-based deep learning feature combination. Sensors **23**(12), 5715 (2023)
5. Bhuva, B.L., et al.: Multi-cell soft errors at advanced technology nodes. IEEE Trans. Nucl. Sci. **62**(6), 2585–2591 (2015)
6. Burel, S., Evans, A., Anghel, L.: Zero-overhead protection for CNN weights. In: 2021 IEEE International Symposium on Defect and Fault Tolerance in VLSI and Nanotechnology Systems (DFT), pp. 1–6 (2021)
7. Chen, Y., Xie, Y., Song, L., Chen, F., Tang, T.: A survey of accelerator architectures for deep neural networks. Engineering **6**(3), 264–274 (2020)
8. Guan, H., Ning, L., Lin, Z., Shen, X., Zhou, H., Lim, S.H.: In-Place Zero-Space Memory Protection for CNN. Curran Associates Inc., Red Hook, NY, USA (2019)
9. Ibrahim, Y., et al.: Soft error resilience of deep residual networks for object recognition. IEEE Access **8**, 19490–19503 (2020)
10. Jang, M., Jeongkyu, H.: MATE: memory- and retraining-free error correction for convolutional neural network weights. J. Inf. Commun. Convergence Eng. **19**, 22–28 (2021). https://api.semanticscholar.org/CorpusID:245755260
11. Khairy, M.: TPU vs GPU vs Cerebras vs Graphcore: A fair comparison between ML hardware (2020). https://khairy2011.medium.com/tpu-vs-gpu-vs-cerebras-vs-graphcore-a-fair-comparison-between-ml-hardware-3f5a19d89e38. Accessed 06 Jan 2024
12. Kim, J.S., Yang, J.S.: DRIS-3: deep neural network reliability improvement scheme in 3D die-stacked memory based on fault analysis. In: 2019 56th ACM/IEEE Design Automation Conference (DAC), pp. 1–6 (2019)
13. Kim, S., Howe, P., Moreau, T., Alaghi, A., Ceze, L., Sathe, V.: MATIC: learning around errors for efficient low-voltage neural network accelerators. In: Design, Automation & Test in Europe Conference & Exhibition, pp. 1–6 (2018)
14. Lecun, Y., Bottou, L., Bengio, Y., Haffner, P.: Gradient-based learning applied to document recognition. Proc. IEEE **86**(11), 2278–2324 (1998)
15. LeCun, Y., Cortes, C., Burges, C.J.: MNIST handwritten digit database. ATT Labs **2** (2010). http://yann.lecun.com/exdb/mnist

16. Lee, J., et al.: Resource-efficient convolutional networks: a survey on model-, arithmetic-, and implementation-level techniques. ACM Comput. Surv. **55**(13s), 1–36 (2023)
17. Li, G., et al.: Understanding error propagation in deep learning neural network (DNN) accelerators and applications. In: SC17: International Conference for High Performance Computing, Networking, Storage and Analysis, pp. 1–12 (2017)
18. Nguyen, D.T., Ho, N.M., Chang, I.J.: St-DRC: stretchable dram refresh controller with no parity-overhead error correction scheme for energy-efficient DNNs. In: 2019 56th ACM/IEEE Design Automation Conference (DAC), pp. 1–6 (2019)
19. Paszke, A., et al.: PyTorch: an imperative style, high-performance deep learning library. In: Advances in Neural Information Processing Systems, vol. 32. Curran Associates, Inc. (2019)
20. Perez-Cerrolaza, J., et al.: Artificial intelligence for safety-critical systems in industrial and transportation domains: a survey. ACM Comput. Surv. **56**(7), 1–40 (2023). just Accepted
21. Qian, C., Zhang, M., Nie, Y., Lu, S., Cao, H.: A survey of bit-flip attacks on deep neural network and corresponding defense methods. Electronics **12**(4), 853 (2023)
22. Rakin, A.S., He, Z., Fan, D.: Bit-flip attack: crushing neural network with progressive bit search. In: 2019 IEEE/CVF International Conference on Computer Vision (ICCV), pp. 1211–1220 (2019)
23. Rakin, A.S., He, Z., Li, J., Yao, F., Chakrabarti, C., Fan, D.: T-BFA: targeted bit-flip adversarial weight attack. IEEE Trans. Pattern Anal. Mach. Intell. **44**(11), 7928–7939 (2022)
24. Redmon, J.: Darknet: Open source neural networks in C (2013–2016). http://pjreddie.com/darknet/
25. Ruiz, J.C., de Andrés, D., Saiz-Adalid, L.J., Gracia-Morán, J.: Zero-space in-weight and in-bias protection for floating-point-based CNNs. In: 19th European Dependable Computing Conference (EDCC), Accepted (2024)
26. Schorn, C., Guntoro, A., Ascheid, G.: An efficient bit-flip resilience optimization method for deep neural networks. In: Design, Automation & Test in Europe Conference & Exhibition (2019)
27. Szegedy, C., et al.: Going deeper with convolutions (2014). CoRR abs/1409.4842, http://arxiv.org/abs/1409.4842
28. Wang, S., Kanwar, P.: BFloat16: The secret to high performance on cloud TPUs. AI & Machine Learning (2019). https://cloud.google.com/blog/products/ai-machine-learning/bfloat16-the-secret-to-high-performance-on-cloud-tpus

A Failure Model Library for Simulation-Based Validation of Functional Safety

Tiziano Munaro[1]($\boxtimes$), Irina Muntean[1], and Alexander Pretschner[1,2]

[1] fortiss, Research Institute of the Free State of Bavaria, Munich, Germany
{munaro,muntean,pretschner}@fortiss.org
[2] Department of Computer Science, Technical University of Munich, Munich, Germany

Abstract. Simulation-based Fault Injection (FI) is highly recommended by functional safety standards in the automotive and aerospace domains, in order to "support the argumentation of completeness and correctness of a system architectural design with respect to faults" (ISO 26262). We argue that a library of failure models facilitates this process. Such a library, firstly, supports completeness claims through, e.g., an extensive and systematic collection process. Secondly, we argue why failure model specifications should be executable—to be implemented as FI operators within a simulation framework—and parametrizable—to be relevant and accurate for different systems. Given the distributed nature of automotive and aerospace development processes, we moreover argue that a data-flow-based definition allows failure models to be applied to black-box components. Yet, existing sources for failure models provide fragmented, ambiguous, incomplete, and redundant information, often meeting neither of the above requirements. We therefore introduce a library of 18 executable and parameterizable failure models collected with a systematic literature survey focusing on automotive and aerospace Cyber-Physical Systems (CPS). To demonstrate the applicability to simulation-based FI, we implement and apply a selection of failure models to a real-world automotive CPS within a state-of-the-art simulation environment, and highlight their impact.

Keywords: Functional Safety · Failure Models · Simulation-based Fault Injection · Cyber-Physical Systems

1 Introduction

Functional safety—the "absence of unreasonable risk due to hazards caused by malfunctioning behaviour of electric/electronic systems," as defined by the ISO 26262 standard—is one of the main concerns when designing and developing safety-critical Cyber-Physical Systems (CPS), especially for automotive, aviation, and spaceflight applications. The increasing complexity of such systems poses novel challenges in validating their availability, reliability, and safety.

In particular, the behaviour of safety mechanisms and their effect at the system level is no longer fully comprehensible by means of analytical approaches, such as Failure Mode and Effect Analysis (FMEA) and System Theoretic Process Analysis (STPA) [58,68]. Thus, it is "highly recommended" by the ISO 26262 and other relevant standards in the automotive and aerospace domains (e.g., NASA-GB-8719.13, IEC 61508, and DO-178C) to complement static safety analysis techniques with simulation-based fault injection (FI) testing to "support the argumentation of completeness and correctness of a system architectural design with respect to faults" as stated in the ISO 26262-4 standard (Clause 6) [13].

To establish the *correctness* of safety mechanisms, simulation-based FI tests apply executable failure models to trigger and observe the error detection and reaction processes. However, adhering to the ISO 26262 vocabulary, simulation-based FI shall also support the argumentation of *completeness* w.r.t. faults. Yet, we are confronted with the lack of failure model collections providing (1) executable behaviour specifications and, crucially, (2) a convincing argument for their comprehensiveness. We address this gap by introducing a library of 18 executable and parametrizable failure models inherently suited for simulation-based FI and whose methodical derivation provides a starting point for the argument as to why a FI testing effort involving *all* of the presented failure models supports the argumentation of completeness w.r.t. faults.

Central Concepts. The effects of various forms of *defects* are mitigated by *safety mechanisms* the adequate functioning of which is assessed by *simulation-based fault injection (FI)*:

As defined by the ISO 26262 standard, a *failure* is the termination of an intended behaviour of a software/hardware component; an *error* is an incorrect internal state. If activated, an error may lead to a failure. The cause of an error is a *fault*. A *failure mode* is the manner in which a component fails to provide the intended behaviour. *Safety mechanisms* increase the dependability of a system by ensuring that a system continues to operate in the presence of faults. E.g., by deploying replicas of a software unit to different processing units, the loss of one or more of these hardware components can be tolerated. A well-established technique for assessing safety mechanisms is *fault injection (FI)* testing: Controlled experiments performed at various levels of abstraction and at different stages of development to observe the behaviour of systems in the presence of faults. While *physical* approaches to FI (e.g., irradiation of control units) are more representative w.r.t. hardware failures, due to their overhead, cost, and duration they are commonly applied only manually and in late development stages. In contrast, *simulation-based* methods allow front-loading of FI testing, are highly automatable, inexpensive, non-destructive, highly reproducible, and offer high degrees of controllability and observability while still considering specific properties of the hardware platform based on appropriate models.

The information on how to inject particular failures during simulation is encoded by means of *executable* failure models: They specify how the system under test shall be altered (e.g., through source code modifications) to reproduce the observable behaviour of a failure. *FI operators* implementing these failure models are responsible for applying these alterations during simulation. As a failure that is critical to one application might be irrelevant to the next [23], it is crucial to identify *relevant* failure modes—i.e., likely to affect the functional safety of the system under test. Moreover, the same failure mode (e.g., a value drift) is likely to manifest itself differently on different systems (e.g., a subtle yet accelerating drift caused by actuator wear [64] vs. a more significant yet constant drift due to a rounding error [33]). Hence, to be relevant and accurate in either case, failure models must be able to accommodate these differences in manifestation—e.g., by means of *parametrization*.

Problem. With available sources (e.g., existing surveys, case studies, or incident reports) providing only fragmented (e.g., formulated in a use-case-specific manner), ambiguous (e.g., in natural language), incomplete (e.g., including only individual examples for an evaluation), and redundant (e.g., most failures mentioned across sources can be traced to the same failure models) information on potentially relevant failure models, **compiling a collection of executable and parametrizable failure models supporting the argument of *completeness* presents itself as a substantial challenge for which no suitable starting point is available.** We moreover argue that a *data-flow-based* definition is key to leveraging failure models in distributed automotive and aerospace development processes as it allows FI to be applied to black-box components: Without the possibility to access a component's implementation due to intellectual property concerns, the observable behaviour of a failure can only be reproduced by manipulating the outgoing data at the component's interface.

Contribution. To address the research gap delineated in Sect. 2, we introduce a **library of 18 executable and parameterizable failure models** for simulation-based FI of automotive and aerospace systems with potentially black-box components. Crucially, the library **covers all failure modes defined in peer-reviewed publications** collected based on the query defined in Sect. 3, thus providing a compelling starting point and blueprint for the "argumentation of completeness" suggested by the ISO 26262 standard.

2 Related Work

As noted in Sect. 1, we could not identify any collections of failure models (1) directly applicable for simulation-based FI through, e.g., executable specifications and (2) supporting "the argumentation of completeness and correctness of a system with respect to faults" as suggested by the ISO 26262 standard. We did, however, identify four types of closely related work:

Simulation-Based Fault Injection Frameworks: Selecting and implementing FI operators is an essential part of developing FI frameworks. However, the underlying failure models are typically either implicit, introduced in natural language only, or specific to components and contexts (cf., e.g., [1,10,32,68]). Moreover, none of the identified frameworks provide any argumentation w.r.t. the relevance or comprehensiveness of their failure models.

Context-Specific Fault Model Definitions: [24,41], and [34] introduce *fault* models used to test for the absence of known defects in the functionality of control systems, object-oriented programs or access control policies. Many other fault models, relevant for different systems and use cases, have been presented in the past. However, using *fault models* (encoding, e.g., a division by zero) for testing safety mechanisms requires elaborate operationalizations to ensure that the injected faults are activated, and the error propagation leads to a failure the safety mechanism is expected to handle.

Surveys on Defects in Software and Systems: In [23], the authors interview practitioners to identify common software development defects. While this results in a list of 14 frequent technical faults, all of them must be operationalized as described above. Moreover, the study does not comprise hardware-related faults and failures. [72] and [26], in contrast, focus specifically on CPS: Both derive defect taxonomies by analysing data on real defects of CPS. However, as these studies also provide informal fault models rather than executable failure models, their applicability for simulation-based FI is just as limited.

Standards: The AUTOSAR E2E Protocol Specification provides a collection of informally defined faults and failures, each associated with the suitable safety mechanism for detecting them. Similarly, the Error Model Annex of the Architecture Analysis and Design Language (AADL) (cf. SAE AS5506/1A) introduces an extensive "error library"—a taxonomy of failure models defined in natural language. As both collections require elaborate operationalization of their fault and failure models, [48] introduces a declarative formal specification of selected AADL failure models. However, as it covers only a fraction of the standard's "error library", this formalization can hardly support any completeness arguments.

3 A Failure Model Library for Simulation-Based FI

The following sections introduce the methodology applied to systematically compile a collection of failure models (Sect. 3.1), and their execution semantics tailored for simulation-based FI. (Sect. 3.2).

3.1 Failure Model Collection

To identify the most pertinent defects affecting cyber-physical systems in the automotive and aerospace domains, we perform a *developmental literature review* [66]. Accounting for synonyms, the following Scopus[1] query was used to determine the initial body of literature:

```
TITLE-ABS-KEY ("cyber-physical"
AND ("automotive" OR "aerospace" OR "avionics" OR "aviation")
AND ("fault" OR "failure" OR "defect" OR "bug" OR "error"))
```

---

[1] https://www.scopus.com.

At the time of writing, the query returns 277 conference papers, articles, and book chapters published between 1990 and 2023. 151 of these publications mention the keywords in contexts not relevant to our goal (e.g., referring to manufacturing defects) and were thus excluded. While malicious attacks are out of scope (e.g., spoofing), their safety-related effects are considered (e.g., increased network latency). To specify the observable behaviour of the collected failure models, we focus on sources *defining* failure modes (formally or informally), thus excluding 69 entries only mentioning failure modes. Finally, all failure mode descriptions were extracted from the remaining 57 documents. As the extracted information is highly redundant yet often profoundly context- and system-specific, the data synthesis started by consolidating the failure models based on their observable behaviour. The resulting collection of distinct failure modes is presented in Table 1.

Table 1. Distinct failure modes identified by means of the systematic literature review described in Sect. 3.1, each with the body of literature defining its observable behaviour.

STUCK-AT/SATURATION		[3–7, 12, 19–21, 25, 27, 28, 36, 39, 40, 45, 50, 55, 57, 63, 72]	
OMISSION/LOSS		[2, 4, 19, 25, 30, 40, 44, 54, 62, 63, 72]	
OFFSET/BIAS		[2, 6, 9, 11, 15, 16, 27–30, 42, 43, 45, 54, 57, 63, 64, 69]	
DRIFT/RAMP		[8, 9, 12, 16, 28, 29, 33, 36, 42, 43, 57, 62, 64]	
BIT-FLIP/SINGLE EVENT UPSET (SEU)		[6, 14, 18–20, 25, 37, 40, 45, 50, 52, 56, 65, 67, 70, 72]	
NOISE		[4, 6, 9, 17, 18, 39, 40, 51, 54, 57, 59, 64, 69]	
COUPLING/BRIDGING		[3, 19, 21, 25, 55]	
PACKET DROP		[6, 19, 44, 53, 60]	
NEG./INV./ABS.	[3, 6, 45]	OVERFLOW	[15, 35]
OSCILLATION	[35, 64]	COMMISSION	[2]
LOSS OF PRECISION	[9, 15, 54]	INCORRECT ROUTING	[19]
DELAY	[5, 6, 35, 40, 44, 46, 54, 60–62, 71, 72]	EARLY/LATE	[2, 38, 49, 54, 61, 63]
JITTER	[9, 31, 49, 53, 60]	FAST/SLOW	[2, 38, 49, 61, 63]

3.2 Failure Model Semantics

In Sect. 1 we established that failure models shall be *executable* to be applicable for simulation-based FI, *parametrizable* to be relevant and accurate for different systems, and *data-flow-based* to be applied to black-box components. To this end, we base the here introduced failure models on the generic representation of *defect models* introduced in [47]. It defines faults and failure models for possibly non-deterministic *behaviour descriptions*: Timed sequences of input/output pairs specifying the behaviour of a system, subsystem, or atomic hardware or software component. In practice, behaviour descriptions are commonly implemented in the form of executable models (e.g., a MATLAB Simulink model of a feedback control mechanism), or program code (e.g., a C++ implementation of a network gateway). A *FI operator* for a failure model K is defined as a mapping α_K from a *correct* implementation of a behaviour specification (w.r.t. a given specification) to an *incorrect* implementation behaviour description (w.r.t. the same specification). Rather than specifying FI operators in terms of modifications to an implementation, we instantiate them as functions modifying the respective input/output sequences w.r.t. both value and timing.

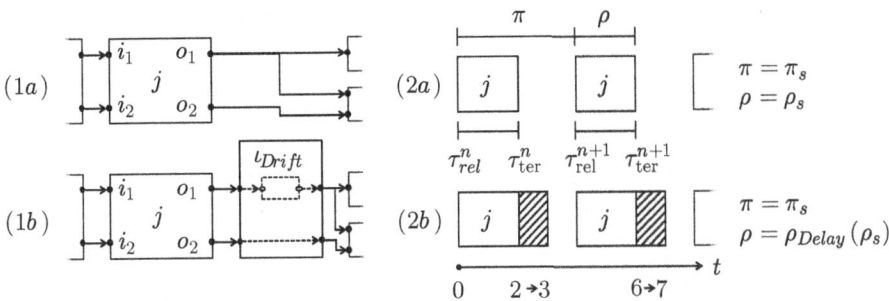

Fig. 1. The data-flow diagrams **(1a)** and **(1b)** depict the application of the FI operator ι_K to the output o_1 of a component j. The Gantt charts **(2a)** and **(2b)** illustrate the periodic execution of j before and after the application of a delay of one time unit.

To replicate a failure mode's effect on the *values* of a component's[2] output sequences, we define a FI operator for value-based failure models applying a function ι_K to a subset of the component's output sequences. The function ι_K is specific to each failure model and modifies the output values to replicate the failure mode. Failure models influencing the *temporal* behaviour of components are defined based on the execution semantics for real-time systems introduced in Giotto [22]: Without loss of generality, a periodically executed component (with period π) reads its input at the time of its release $\tau_{rel} \equiv 0 \pmod{\pi}$ and produces its output after termination at $\tau_{ter} = \tau_{rel} + \rho$, with ρ being the worst-case execution and/or transmission time. We now define a FI operator for time-based failure models using two functions π_K and ρ_K overriding the nominal period π_s and duration ρ_s of a behaviour specification s. Figure 1 illustrates the application of both value- and time-based failure models. By activating and deactivating the injectors, permanent, transient, and intermittent failures can be simulated. To facilitate the operationalization of failure models, we introduce a pseudocode-based representation of ι_K, ρ_K, and π_K rather than a formal one. Here, the `param` keyword indicates parameters of a failure model dictated by the underlying fault (e.g., the drift rate). Thus, they do not change during a FI. `local variables` are used to store time-varying information (e.g., the accumulated error). Tables 2 and 3 present implementations of the failure modes identified in Sect. 3.1 based on the commonalities of the respective descriptions of observable behaviour. Crucially, the failure models are interchangeable, executable, composable, and applicable to black-box components.

---

[2] We refer to behaviour description implementations as *components* from here on.

Table 2. Value-based Failure Models

$K =$ Stuck-at / Saturation	$K =$ Loss of Precision
| ```
param mode;
local lastValue;
ι_K(signal) {
  if mode == ZERO {return 0;}
  if mode == ONE {return 1;}
  if mode == MIN {
    return signal.min;}
  if mode == MAX {
    return signal.max;}
  if mode == LAST {
    return lastValue;}}
``` | ```
param mode; // Rounding mode
param prec; // Remaining precision
ι_K(signal) {
  if mode == ROUND {
    return round(signal, prec);}
  if mode == CEIL {
    return floor(signal, prec);}
  if mode == FLOOR {
    return ceil(signal, prec);}
  if mode == TRUNCATE {
    return trunc(signal, prec);}}
``` |
| $K =$ Offset / Bias | $K =$ Drift / Ramp |
| ```
param mode, offset;
ι_K(signal) {
  if mode == ABSOLUTE {
    return signal + offset;}
  if mode == RELATIVE {
    return signal * offset;}}
``` | ```
param mode; // Progression
param rate; // Drift rate
param init; // Initial error
local error = init; // Accumulator
// E.g: Exp. and lin. progression
ι_K(signal) {
  if mode == EXP {error *= rate;}
  elif mode == LIN {error += rate;}
  return signal + error;}
``` |
| $K =$ Bit-Flip / SEU | $K =$ Coupling / Bridging |
| ```
param bitMask; // Bit mask
param bitOp; // Bit-wise operator
ι_K(signal) {
  return bitOp(signal, bitMask);}
``` | ```
param coupledSignal;
ι_K(signal) {
  // &: Bit-wise AND
  return signal & coupledSignal;}
``` |
| $K =$ Noise | $K =$ Packet Drop |
| ```
// Random distribution
param randomDist;
ι_K(signal) {
  error = randomDist.sample();
  return signal + error;}
``` | ```
param rate; // Range (0,1)
ι_K(signal) {
  rand = uniform(0,1).sample();
  if rand < rate {return empty();}
  return signal;}
``` |
| $K =$ Neg. / Inv. / Abs. | $K =$ Overflow |
| ```
param mode;
ι_K(signal) {
  if mode == NEG {return -signal;}
  if mode == INV {return !signal;}
  if mode == ABS {return |signal|;}}
``` | ```
// Bit representation to simulate
param bits;
ι_K(signal) {
  return signal % (2 ^ bits);}
``` |
| $K =$ Oscillation | $K =$ Commission |
| ```
param amplitude, frequency;
local time; // Current time
// E.g.: Sine-based oscillation
ι_K(signal, amplitude, frequency) {
  deviation = amplitude * sin(
    time * frequency);
  return signal + deviation;}
``` | ```
param probability; // Range (0,1)
param value; // Value to commit
ι_K(signal) {
  if signal = ⟨⟩ {
    rand = uniform(0,1).sample();
    if (rand < probability) {
      return value;}}
  return empty()}
``` |
| $K =$ Omission / Loss | $K =$ Incorrect Routing |
| ```
ι_K(signal) {
  // Do not return any value
  return empty();}
``` | ```
// Incorrectly routed signal
param otherSignal;
ι_K(signal) {
  return otherSignal;}
``` |

Table 3. Time-based Failure Models

$K =$ Delay	$K =$ Jitter

`// Ticks to be added to ρ`
`param delay;`
`ρ_K(nominal) {`
`    return nominal + delay; }`
	`input dist; // Random distribution`
	`ρ_K(nominal) {`
	`    offset = dist.sample();`
	`    return nominal + offset; }`

$K =$ Early / Late	$K =$ Fast / Slow

`param offset; // Ticks to offset ρ`
`ρ_K(nominal) {`
`    return nominal + offset; }`
`// Permanent injection`
`// results in a delay.`
	`param factor; // Fast < 1 < Slow`
	`π_K(nominal) {`
	`    return nominal * factor; }`
	`ρ_K(nominal) {`
	`    return nominal * factor; }`

Listing 1. Excerpt (adapted for readability) of the case study's C++ FI operator implementing the Drift/Ramp failure model (cf. Sect. 4). Here, the error added to the signal values is accumulated for each connection and set to grow exponentially. The `inject` function is applied to the targeted signals at each (fixed) communication step.

```
double initialError, driftRate;
std::map<Connection, double> accumulatedError;
// ...
void inject(Connection connection, Signal &sig) {
    std::map<Connection,double>::iterator match =
        accumulatedError.find(connection);
    if (match == accumulatedError.end()) {
        accumulatedError.emplace(connection, initialError);
    } else {
        match->second = match->second * driftRate;
    }
    double currentError = match->second;
    signal.setValue(signal.getValue() * currentError);
}
```

4   Case Study

To demonstrate the application of the presented failure models for triggering and assessing safety mechanisms using simulation-based FI, we perform an exemplary case study on a real-world automotive CPS: More precisely, we execute exactly the same simulation-based test case four times—once without FI, then three times applying different failure models from Tables 2 and 3 to the same black-box component lacking safety mechanisms. After determining the failures' effects, we augment the System Under Test (SUT) with adequate safety mechanisms and repeat the same FIs.

The use case consists in an industry-oriented autonomous model vehicle[3] equipped with Automated Driving (AD) functions. Its digital twin, our SUT, consists of a black-box co-simulation of its subsystems acting within a state-of-the-art CARLA[4] environment. The executed test case is based on a scenario

---

[3] https://www.fortiss.org/en/research/fortiss-labs/detail/mobility-lab.
[4] http://carla.org/.

aimed at testing the vehicle's Adaptive Cruise Control (ACC) system: The SUT is supposed to follow a vehicle which abruptly reduces its speed at time $\tau = 5s$ while keeping a safe distance at all times. Three failure models (OMISSION, DRIFT, and DELAY) are implemented as C++ FI operators[5], altering the output of targeted components as it is forwarded by the co-simulation framework. Listing 1 illustrates one such FI operator: As envisioned, its implementation does not require modifications to components under test, is neither specific to the use case nor to the components it is applied to and can make full use of the failure models' composability as it operates in-place.

One at a time, these FI operators are applied to the *target speed* calculated by the vehicle's black-box ACC system. This output is used as a setpoint for the motor controller. The results are shown in Fig. 2 (left): As expected, the delay leads to a later and stronger deceleration compared to the nominal execution. While the value drift does not affect the instant the ACC wants to decelerate, by then, the accumulated error neutralizes the reduction in target speed. As intended, the omission failure model simulates the loss of the ACC component by suppressing any target speed signals, effectively causing the vehicle not to decelerate at all. All FIs result in collisions with the leading vehicle (cf., e.g., Fig. 2 (right)).

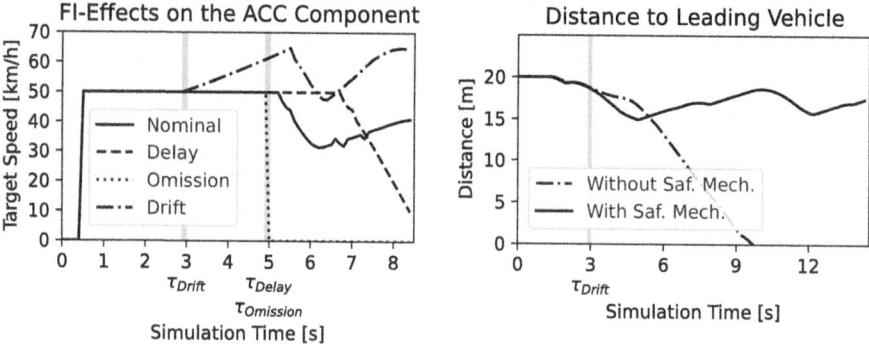

Fig. 2. Effects of different failure models on the target speed calculated by the ACC component lacking any safety mechanisms (left) and distance to the leading vehicle in the presence of a DRIFT failure before and after implementing safety mechanisms (right). The omission and the delay are injected at $\tau = 5\ s$, the drift already at $\tau = 3\ s$.

To address the loss of the ACC component, a replica of the task is introduced as well as a monitor switching to the secondary instance if the primary's output is omitted. The drift is addressed by extending the monitor to perform boundary checks on the target speed. Finally, the monitor is equipped with a redundant clock signal to detect delays as well. Crucially, reapplying the failure models

---

[5] The FI operators' source code is available at https://git.fortiss.org/ffl/rosco.

allows us to trigger and observe these error detection and reaction mechanisms, and establish that the augmented version of the SUT is capable of safely handling the three injected failures (see, e.g., Fig. 2 (right)).

5 Limitations

While we argue that the extensive systematic literature review provides a compelling argument for the relevance of the identified failure models and resembles a first step towards building an argument for completeness, we cannot claim that the presented library itself is complete—in particular due to the query's limitation to *cyber-physical* systems, introduced to ensure the feasibility of the review given the numerous sources referring to *embedded systems*. Moreover, the literature review is based only on the Scopus database which, however, covers all major publishers of peer-reviewed publications.

6 Conclusion

We envision the presented failure model library as a first step towards supporting "the argumentation of *completeness*" of a system architectural design w.r.t. faults (ISO 26262): The systematic literature review provides a compelling argument for the library's coverage of relevant failure modes affecting automotive and aerospace CPS. We demonstrate the application of the failure models for triggering and assessing safety mechanisms by applying C++ implementations thereof to a real-world automotive CPS with black-box components using simulation-based FI. With a library of parametrizable, interchangeable, executable, composable failure models now available, in future research we plan on improving the argumentation of a system architecture's *correctness* w.r.t. faults: Leveraging search-based test case generation techniques, we aim at efficiently identifying particularly challenging test cases (i.e., underlying scenarios and FI parameters).

Disclosure of Interests. The authors have no competing interests to declare that are relevant to the content of this article.

Acknowledgments. Carolin Ganahl provided valuable support in extending the case study's system under test with safety mechanisms.

References

1. Aliabadi, M.R., Pattabiraman, K.: FIDL: a fault injection description language for compiler-based SFI tools. Comput. Saf., Reliab. Secur. **9922**, 12–23 (2016). https://doi.org/10.1007/978-3-319-45477-1_2
2. Amorim, T., et al.: Runtime safety assurance for adaptive cyber-physical systems: conserts M and ontology-based runtime reconfiguration applied to an automotive case study, pp. 137–168. IGI Global (2017). https://doi.org/10.4018/978-1-5225-2845-6.ch006

3. Arlat, J., et al.: Fault injection for dependability validation: a methodology and some applications. IEEE Trans. Softw. Eng. **16**(2), 166–182 (1990). https://doi.org/10.1109/32.44380
4. Bakker, T., Leccadito, M.T., Klenke, R.H.: Flexible FPGA based hardware in the loop simulator for control, fault-tolerant and cyber-physical systems. In: AIAA SciTech Forum - AIAA Aerospace Science Meeting. AIAA (2017). https://doi.org/10.2514/6.2017-0549
5. Banerjee, A., Maity, A., Gupta, S.K., Lamrani, I.: Statistical conformance checking of aviation cyber-physical systems by mining physics guided models. In: IEEE Aerospace Conference Proceedings, vol. 2023-March. IEEE Computer Society (2023). https://doi.org/10.1109/AERO55745.2023.10115613
6. Bartocci, E., Mariani, L., Ničković, D., Yadav, D.: FIM: fault injection and mutation for simulink. In: 30th ACM Joint Euro Software Engineering Conference and Symposium on the Foundations of Software Engineering, ESEC/FSE 2022, pp. 1716–1720. ACM, New York (2022). https://doi.org/10.1145/3540250.3558932
7. Baumeister, J., Dauer, J.C., Finkbeiner, B., Schirmer, S.: Monitoring with verified guarantees. Int. J. Softw. Tools Technol. Transf. **25**(4), 593–616 (2023). https://doi.org/10.1007/s10009-023-00712-3
8. Ben Hamouda, L., Ayadi, M., Langlois, N.: Fault tolerant fuzzy-based model predictive controllers for automotive application. In: International Conference on Control, Decision on Information Technology, CoDIT, pp. 117–122. IEEE (2016). https://doi.org/10.1109/CoDIT.2016.7593546
9. Bhatt, D., Schloegel, K., Madl, G., Oglesby, D.: Quantifying error propagation in data flow models. In: International Symposium and Workshop on Engineering of Computer Based Systems, pp. 2–11 (2013). https://doi.org/10.1109/ECBS.2013.7
10. Carreira, J., Madeira, H., Silva, J.: Xception: a technique for the experimental evaluation of dependability in modern computers. IEEE Trans. Softw. Eng. **24**, 125–136 (1998). https://doi.org/10.1109/32.666826
11. Chen, W.D., Niu, B., Wang, H.Q., Li, H.T., Wang, D.: Adaptive event-triggered control for non-strict feedback nonlinear cpss with time delays against deception attacks and actuator faults. IEEE Trans. Autom. Sci. Eng. 1–11 (2023). https://doi.org/10.1109/TASE.2023.3292367
12. Corradini, M.L., Monteriù, A., Orlando, G., Pettinari, S.: An actuator failure tolerant robust control approach for an underwater remotely operated vehicle. In: IEEE Conference on Decision and Control and European Control Conference, pp. 3934–3939 (2011). https://doi.org/10.1109/CDC.2011.6160578
13. Cotroneo, D., Natella, R.: Fault injection for software certification. IEEE Secur. Priv. **11**(4), 38–45 (2013). https://doi.org/10.1109/MSP.2013.54
14. Dietrich, C., Schmider, A., Pusz, O., Vayá, G.P., Lohmann, D.: Cross-layer fault-space pruning for hardware-assisted fault injection. In: ACM/ESDA/IEEE Design Automation Conference (DAC), pp. 1–6 (2018). https://doi.org/10.1109/DAC.2018.8465787
15. Edwards, J., Kashani, A., Iyer, G.: Evaluation of software vulnerabilities in vehicle electronic control units. In: 2017 IEEE Cybersecurity Development Conference, SecDev, pp. 83–84. IEEE (2017). https://doi.org/10.1109/SecDev.2017.26
16. Freeman, P., Balas, G.J.: Analytical fault detection for a small UAV. In: AIAA Infotech@Aerospace (I@A) Conference. American Institute of Aeronautics and Astronautics (2013). https://doi.org/10.2514/6.2013-5217
17. Ghosh, S.K., Jaffer Sheriff, R., Jain, V., Dey, S.: Reliable and secure design-space-exploration for cyber-physical systems. ACM Trans. Embed. Comput. Syst. **19**(3) (2020). https://doi.org/10.1145/3387927

18. Giri, N.K., Munir, A., Kong, J.: An integrated safe and secure approach for authentication and secret key establishment in automotive cyber-physical systems. In: Arai, K., Kapoor, S., Bhatia, R. (eds.) SAI 2020. AISC, vol. 1230, pp. 545–559. Springer, Cham (2020). https://doi.org/10.1007/978-3-030-52243-8_39
19. Gohringer, D., Meder, L., Oey, O., Becker, J.: Reliable and adaptive network-on-chip architectures for cyber physical systems. Trans. Embed. Comput. Syst. **12** (2013). https://doi.org/10.1145/2435227.2435247
20. Guinebert, I., Barrilado, A., Delmas, K., Galtié, F., Pagetti, C.: Quality of fault injection strategies on hardware accelerator. In: Trapp, M., Saglietti, F., Spisländer, M., Bitsch, F. (eds.) Computer Safety, Reliability, and Security, pp. 222–236. Springer, Cham (2022). https://doi.org/10.1007/978-3-031-14835-4_15
21. Hasan, O., Tahar, S., Abbasi, N.: Formal reliability analysis using theorem proving. IEEE Trans. Comput. **59**(5), 579–592 (2010). https://doi.org/10.1109/TC.2009.165
22. Henzinger, T.A., Horowitz, B., Kirsch, C.M.: Giotto: a time-triggered language for embedded programming. In: Henzinger, T.A., Kirsch, C.M. (eds.) EMSOFT 2001. LNCS, vol. 2211, pp. 166–184. Springer, Heidelberg (2001). https://doi.org/10.1007/3-540-45449-7_12
23. Holling, D., Fernández, D.M., Pretschner, A.: A Field Study on the Elicitation and Classification of Defects for Defect Models. In: Abrahamsson, P., Corral, L., Oivo, M., Russo, B. (eds.) PROFES 2015. LNCS, vol. 9459, pp. 380–396. Springer, Cham (2015). https://doi.org/10.1007/978-3-319-26844-6_28
24. Holling, D., Pretschner, A., Gemmar, M.: 8cage. In: 29th ACM/IEEE International Conference on Automated Software Engineering - ASE 2014. pp. 859–862. ACM Press (2014). 10.1145/2642937.2648622
25. Hsueh, M.C., Tsai, T., Iyer, R.: Fault injection techniques and tools. Computer **30**(4), 75–82 (1997). https://doi.org/10.1109/2.585157
26. Huang, F., Huang, B., Wang, Y., Wang, Y.: A taxonomy of software defect forms for certification tests in aviation industry. In: Guiochet, J., Tonetta, S., Bitsch, F. (eds.) SAFECOMP 2023. LNCS, vol. 14181, pp. 55–63. Springer, Cham (2023). https://doi.org/10.1007/978-3-031-40923-3_5
27. Hwang, I., Kim, S., Kim, Y., Seah, C.E.: A survey of fault detection, isolation, and reconfiguration methods. IEEE Trans. Control Syst. Technol. **18**(3), 636–653 (2010). https://doi.org/10.1109/TCST.2009.2026285
28. Isermann, R.: Fault-Diagnosis Systems: An Introduction From Fault Detection to Fault Tolerance. Springer, Heidelberg (2006). https://doi.org/10.1007/3-540-30368-5
29. Isermann, R.: Supervision, fault-detection and fault-diagnosis methods – a short introduction. In: Combustion Engine Diagnosis. A, pp. 25–47. Springer, Heidelberg (2017). https://doi.org/10.1007/978-3-662-49467-7_2
30. Khan, A.H., Khan, Z.H., Khan, S.H.: Optimized reconfigurable autopilot design for an aerospace CPS. Stud. Comput. Intell. **540**, 381–420 (2014). https://doi.org/10.1007/978-981-4585-36-1_13
31. Kukkala, V., Pasricha, S., Bradley, T.: JAMS-SG: a framework for jitter-aware message scheduling for time-triggered automotive networks. ACM Trans. Des. Autom. Electron. Syst. **24**(6) (2019). https://doi.org/10.1145/3355392
32. Lu, Q., Farahani, M., Wei, J., Thomas, A., Pattabiraman, K.: LLFI: an intermediate code-level fault injection tool for hardware faults. In: 2015 IEEE International Conference on Software Quality, Reliabilty and Security, pp. 11–16. IEEE (2015). https://doi.org/10.1109/QRS.2015.13

33. Marshall, E.: Fatal error: how patriot overlooked a scud. Sci. **255**(5050), 1347–1347 (1992). https://doi.org/10.1126/science.255.5050.1347
34. Martin, E., Xie, T.: A fault model and mutation testing of access control policies. In: 16th International Conference on WWW, pp. 667–676. ACM (2007). https://doi.org/10.1145/1242572.1242663
35. Matinnejad, R., Nejati, S., Briand, L.C., Bruckmann, T.: Test generation and test prioritization for simulink models with dynamic behavior. IEEE Trans. Softw. Eng. **45**(9), 919–944 (2019). https://doi.org/10.1109/TSE.2018.2811489
36. McIntyre, M., Dixon, W., Dawson, D., Walker, I.: Fault identification for robot manipulators. IEEE Trans. Robot. **21**(5), 1028–1034 (2005). https://doi.org/10.1109/TRO.2005.851356
37. Meng, X., Tan, Q., Shao, Z., Zhang, N., Xu, J., Zhang, H.: Optimization methods for the fault injection tool SEInjector. In: International Conference on Information and Computer Technology (ICICT), pp. 31–35 (2018). https://doi.org/10.1109/INFOCT.2018.8356836
38. Mitra, S., Wongpiromsarn, T., Murray, R.M.: Verifying cyber-physical interactions in safety-critical systems. IEEE Secur. Priv. **11**(4), 28–37 (2013). https://doi.org/10.1109/MSP.2013.77
39. Moradi, M., Oakes, B.J., Saraoglu, M., Morozov, A., Janschek, K., Denil, J.: Exploring fault parameter space using reinforcement learning-based fault injection. In: Annual IEEE/IFIP International Conference on Dependable Systems and Networks, DSN-W, pp. 102–109. IEEE (2020). https://doi.org/10.1109/DSN-W50199.2020.00028
40. Moradi, M., Van Acker, B., Vanherpen, K., Denil, J.: Model-implemented hybrid fault injection for simulink (tool demonstrations). In: Chamberlain, R., Taha, W., Törngren, M. (eds.) CyPhy/WESE -2018. LNCS, vol. 11615, pp. 71–90. Springer, Cham (2019). https://doi.org/10.1007/978-3-030-23703-5_4
41. Offutt, J., Alexander, R., Wu, Y., Xiao, Q., Hutchinson, C.: A fault model for subtype inheritance and polymorphism. In: 12th International Symposium on Software Reliabilty Engineering, pp. 84–93. IEEE (2001). https://doi.org/10.1109/ISSRE.2001.989461
42. Oucheikh, R., Fri, M., Fedouaki, F., Hain, M.: Deep real-time anomaly detection for connected autonomous vehicles. Procedia Comp. Sci. **177**, 456–461 (2020). https://doi.org/10.1016/j.procs.2020.10.062
43. Park, H., Easwaran, A., Andalam, S.: TiLA: Twin-in-the-loop architecture for cyber-physical production systems. In: IEEE International Conference on Computer Design, ICCD, pp. 82–90. IEEE (2019). https://doi.org/10.1109/ICCD46524.2019.00019
44. Pethő, Z., Szalay, Z., Török, A.: Safety risk focused analysis of V2V communication especially considering cyberattack sensitive network performance and vehicle dynamics factors. Veh. Commun. **37** (2022). https://doi.org/10.1016/j.vehcom.2022.100514
45. Pill, I., Rubil, I., Wotawa, F., Nica, M.: SIMULTATE: a toolset for fault injection and mutation testing of simulink models. In: IEEE International Conference on Software Testing, V&V Workshops (ICSTW), pp. 168–173 (2016). https://doi.org/10.1109/ICSTW.2016.21
46. Poudel, B., Munir, A.: Design and evaluation of a novel ecu architecture for secure and dependable automotive cps. In: 2017 14th IEEE Annual Consumer Communications and Networking Conference, CCNC 2017, pp. 841–847 (2017). https://doi.org/10.1109/CCNC.2017.7983243

47. Pretschner, A.: Defect-based testing. Dependable Softw. Syst. Eng. 224–245 (2015). https://doi.org/10.3233/978-1-61499-495-4-224
48. Procter, S., Feiler, P.: The AADL error library. ACM SIGAda Ada Lett. **39**, 63–70 (2020). https://doi.org/10.1145/3379106.3379113
49. Procter, S., Hatcliff, J., Weininger, S., Fernando, A.: Error type refinement for assurance of families of platform-based systems. In: Koornneef, F., van Gulijk, C. (eds.) SAFECOMP 2015. LNCS, vol. 9338, pp. 95–106. Springer, Cham (2015). https://doi.org/10.1007/978-3-319-24249-1_9
50. Qutub, S., et al.: Hardware faults that matter: understanding and estimating the safety impact of hardware faults on object detection DNNs. In: Trapp, M., Saglietti, F., Spisländer, M., Bitsch, F. (eds.) Computer Safety, Reliability, and Security, pp. 298–318. Springer, Cham (2022). https://doi.org/10.1007/978-3-031-14835-4_20
51. Rahman, Y., Xie, A., Bernstein, D.S.: Retrospective cost adaptive control: pole placement, frequency response, and connections with LQG control. IEEE Control. Syst. **37**(5), 28–69 (2017). https://doi.org/10.1109/MCS.2017.2718825
52. da Rosa, F.R., Garibotti, R., Ost, L., Reis, R.: Using machine learning techniques to evaluate multicore soft error reliability. IEEE Trans. Circuits and Syst. I: Regular Papers **66**(6), 2151–2164 (2019). https://doi.org/10.1109/TCSI.2019.2906155
53. Roy, D., Zhang, L., Chang, W., Mitter, S.K., Chakraborty, S.: Semantics-preserving cosynthesis of cyber-physical systems. Proc. IEEE **106**(1), 171–200 (2018). https://doi.org/10.1109/JPROC.2017.2779456
54. Sabatini, R., Moore, T., Ramasamy, S.: Global navigation satellite systems performance analysis and augmentation strategies in aviation. Prog. Aerosp. Sci. **95**, 45–98 (2017). https://doi.org/10.1016/j.paerosci.2017.10.002
55. Sahoo, S.K., Li, M.L., Ramachandran, P., Adve, S.V., Adve, V.S., Zhou, Y.: Using likely program invariants to detect hardware errors. In: IEEE International Conference on Dependable Systems and Networks With FTCS and DCC (DSN), pp. 70–79 (2008). https://doi.org/10.1109/DSN.2008.4630072
56. Sauer, M., et al.: An FPGA-based framework for run-time injection and analysis of soft errors in microprocessors. In: 2011 IEEE 17th International On-Line Testing Symposium, IOLTS 2011, pp. 182–185 (2011). https://doi.org/10.1109/IOLTS.2011.5993836
57. Scoggin, J., Selmic, R., Oonk, S., Vosburg, N., Maldonado, F.: Sensor networks faults detection and identification: models and software development. In: AIAA Infotech@Aerospace (I@A) Conference. American Institute of Aeronautics and Astronautics (2013). https://doi.org/10.2514/6.2013-5140
58. Sini, J., Violante, M.: An automatic approach to perform FMEDA safety assessment on hardware designs. In: 2018 IEEE 24th International Symposium on On-Line Testing and Robust System Design (IOLTS), pp. 49–52. IEEE (2018). https://doi.org/10.1109/IOLTS.2018.8474217
59. Son, H., Youn, B.D., Kim, T.: Model improvement with experimental design for identifying error sources in a computational model. Struct. Multidiscip. Optim. **64**(5), 3109–3122 (2021). https://doi.org/10.1007/s00158-021-03002-1
60. Starke, A., Kumar, D., Ford, M., McNair, J., Bell, A.: A test bed study of network determinism for heterogeneous traffic using time-triggered ethernet. In: IEEE Military Communications Conference MILCOM, vol. 2017-October, pp. 611–616. IEEE (2017). https://doi.org/10.1109/MILCOM.2017.8170786
61. Steiner, W., Rushby, J.: TTA and PALS: formally verified design patterns for distributed cyber-physical systems. In: AIAA IEEE Digital Avionics Systems Conference, pp. 7B51–7B515 (2011). https://doi.org/10.1109/DASC.2011.6096120

62. Stott, E.A., Wong, J.S., Sedcole, P., Cheung, P.Y.: Degradation in FPGAs: measurement and modelling. In: 18th Annual ACM/SIGDA International Symposium on FPGAs, FPGA '10, pp. 229–238. ACM, New York (2010). https://doi.org/10.1145/1723112.1723152
63. Struss, P.: Model-based analysis of embedded systems: placing it upon its feet instead of on its head: an outsider's view. In: ICSOFT - International Joint Conference on Software Technologies, pp. 284–291 (2013)
64. Syd Ali, B., Ochieng, W., Majumdar, A., Schuster, W., Kian Chiew, T.: ADS-B system failure modes and models. J. Navig. **67**(6), 995–1017 (2014). https://doi.org/10.1017/S037346331400037X
65. Tabacaru, B.-A., Chaari, M., Ecker, W., Kruse, T., Novello, C.: Gate-level-accurate fault-effect analysis at virtual-prototype speed. In: Skavhaug, A., Guiochet, J., Schoitsch, E., Bitsch, F. (eds.) SAFECOMP 2016. LNCS, vol. 9923, pp. 144–156. Springer, Cham (2016). https://doi.org/10.1007/978-3-319-45480-1_12
66. Templier, M., Paré, G.: A framework for guiding and evaluating literature reviews. Commun. Assoc. Inf. Syst. **37**, 112–137 (2015). https://doi.org/10.17705/1CAIS.03706
67. Thomas, T.M., Dietrich, C., Pusz, O., Lohmann, D.: ACTOR: accelerating fault injection campaigns using timeout detection based on autocorrelation. In: Trapp, M., Saglietti, F., Spisländer, M., Bitsch, F. (eds.) Computer Safety, Reliability, and Security, pp. 252–266. Springer, Cham (2022). https://doi.org/10.1007/978-3-031-14835-4_17
68. Uriagereka, G.J., Lattarulo, R., Rastelli, J.P., Calonge, E.A., Lopez, A.R., Ortiz, H.E.: Fault injection method for safety and controllability evaluation of automated driving. In: 2017 IEEE Intelligent Vehicle Symposium (IV), pp. 1867–1872. IEEE (2017). https://doi.org/10.1109/IVS.2017.7995977
69. Vatanparvar, K., Al Faruque, M.A.: Self-secured control with anomaly detection and recovery in automotive cyber-physical systems. In: Design Automation and Test in Europe Conference and Exhibition, DATE 2019, pp. 788–793 (2019). https://doi.org/10.23919/DATE.2019.8714833
70. Xu, X., Li, M.L.: Understanding soft error propagation using efficient vulnerability-driven fault injection. In: IEEE/IFIP International Conference on Dependable Systems and Networks (DSN 2012), pp. 1–12 (2012). https://doi.org/10.1109/DSN.2012.6263923
71. Yiu, C.Y., et al.: A digital twin-based platform towards intelligent automation with virtual counterparts of flight and air traffic control operations. Appl. Sci. (Switz.) **11**(22) (2021). https://doi.org/10.3390/app112210923
72. Zampetti, F., Kapur, R., Di Penta, M., Panichella, S.: An empirical characterization of software bugs in open-source Cyber-Physical Systems. J. Syst. Softw. **192** (2022). https://doi.org/10.1016/j.jss.2022.111425

Strategic Resilience Evaluation of Neural Networks Within Autonomous Vehicle Software

Anna Schmedding[1]($\boxtimes$), Philip Schowitz[2], Xugui Zhou[3], Yiyang Lu[1], Lishan Yang[4], Homa Alemzadeh[3], and Evgenia Smirni[1]

[1] William and Mary, Williamsburg, USA
{akschmedding,ylu21,exsmir}@wm.edu
[2] University of British Columbia, Vancouver, Canada
philipns@cs.ubc.ca
[3] University of Virginia, Charlottesville, USA
{xz6cz,ha4d}@virginia.edu
[4] George Mason University, Fairfax, USA
lyang28@gmu.edu

Abstract. Self-driving technology has become increasingly advanced over the past decade due to the rapid development of deep neural networks (DNNs). In this paper, we evaluate the effects of transient faults in DNNs and present a methodology to efficiently locate critical fault sites in DNNs deployed within two cases of autonomous vehicle (AV) agents: Learning by Cheating (LBC) and OpenPilot. We locate the DNN fault sites using a modified Taylor criterion and strategically inject faults that can affect the functioning of AVs in different road and weather scenarios. Our fault injection methodology identifies corner cases of DNN vulnerabilities that can cause hazards and accidents and therefore dramatically affect AV safety. Additionally, we evaluate mitigation mechanisms of such vulnerabilities for both AV agents and discuss the insights of this study.

Keywords: Autonomous Vehicles · Fault Tolerance · DNNs

1 Introduction

Autonomous vehicles (AVs) are real-world safety-critical systems of increasing importance. With the growing complexity of software and use of deep neural networks (DNNs) for perception and control in AVs, many factors can threaten their safe operation, such as software bugs [1] and transient faults in hardware [2], leading to mis-classifications by DNNs and potential safety hazards.

Transient faults (i.e., soft errors) originating from cosmic radiation [3] or from operating under low voltage [4] have been shown to threaten the functionality of DNN hardware and software [5]. Transient faults in the main memory (DRAM) can manifest as single- or multi-bit flips, specifically in neurons or weights of DNN models [5] and may cause silent data corruption (SDC) where the output is faulty despite a seemingly "correct" execution. Since DRAM is used in AVs[1], this safety-critical application inherits the reliability challenges of DRAM faults. Transient faults have already contributed to vehicle crashes [7].

Table 1. Fault Space for OpenPilot Supercombo and LBC.

	LBC	Supercombo
# CONV Layers	40	70
# Weights	21,268,928	5,811,616
# Single-Bit Fault Sites	680,605,696	185,971,712
# Double-Bit Fault Sites	21,098,776,576	5,765,123,072
# Triple-Bit Fault Sites	632,963,297,280	172,953,692,160

Locating critical faults that cause safety hazards or accidents is necessary in such safety-critical systems. A major challenge here is the vast fault site space, often in the order of billions (see Table 1) which would require thousands of years to exhaustively analyze. This makes identifying corner cases where faults may affect the functional safety of a self-driving vehicle [8] similar to searching for a needle in a haystack. Within the AV ecosystem, the classical statistical fault injection [9], cannot discover the critical corner cases that could lead to safety violations. Past works focus on specific DNN tasks (e.g., image classification) and examine DNN resilience *without the context of the entire AV system* [10, 11]. Recent AV resilience assessment works have focused on input, models, or outputs [12–14], but not on its DNN components that are at the core of AV operation. *Our aim is to develop a method to efficiently locate these critical fault sites in the DNN components of AVs.* We evaluate these fault sites by injecting transient faults in DNN weights and determining whether their effects propagate to other AV components and eventually result in hazards or accidents.

We present a *strategic fault injection* method, called *Taylor-Guided Fault Injection (TGFI)*, that identifies and targets the DNN weights that are of high importance to reliable inference. This is done using a modified Taylor criterion [15] which ranks all the DNN weights with respect to their relative importance to inference accuracy. We inject faults in those important weights and show that our strategic fault injection method can efficiently discover safety-critical vulnerabilities in AVs. We specifically focus on locating critical fault sites within two AV systems: (i) Learning-by-Cheating (LBC), a fully autonomous self-driving agent [16] that is widely used in academic studies [2], and (ii) Open-Pilot, a popular driver-assistance system that is used in over 250 existing car models on the road [17]. By using two systems with different levels of autonomy, we demonstrate the ability of TGFI to generalize to different AV DNNs.

---

[1] For example, the NVIDIA Jetson AGX Orin Series, which is used by NVIDIA DRIVE, uses 32GB or 64GB of DRAM [6].

We also characterize the effect of mitigation in the cases where these critical faults occur. For LBC, we consider a state-of-the-art fault tolerance method based on neuron value range restriction for CNNs, called Ranger [11]. For OpenPilot, we examine the existing system safety checks which return control to the human driver. Additionally, we examine the effects of considering contextual factors such as the location of faults and environmental conditions that impact the input in order to offer insights into the practical reliability challenges of deploying AVs on the road. Both mitigation methods show improvement in resilience, while TGFI is still able to find critical corner cases.

2 Autonomous Driving Frameworks

The Society of Automotive Engineers (SAE) defines 6 levels of driving automation for AVs, from Level 0 (L0, no driving automation) to Level 5 (L5, full driving automation) [18]. L0 to L2 assume that there is a human in the loop who controls the automotive environment by supervising or taking over the autonomous features. For higher levels, the car autonomously controls the driving environment without human involvement. With DNN models incorporated, an L4 AV may use end-to-end ML models for perception and planning without human intervention, while L0-L2 levels use DNNs for driver assistance.

2.1 L4 System: LBC

Learning by Cheating (LBC) is a pretrained end-to-end agent for L4 autonomous driving and is widely used in research studies [2,19,20]. Figure 1a shows the model structure of LBC, which is built on a ResNet34 [21] backbone to process input images from a front-facing camera on the vehicle. The model takes two additional inputs: vehicle speed and a high-level command vector generated by the planner, instructing the vehicle to follow the lane, turn left/right

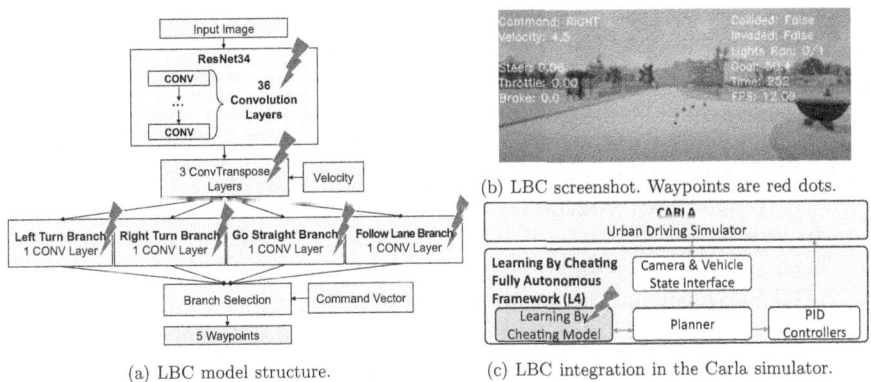

(a) LBC model structure.

(b) LBC screenshot. Waypoints are red dots.

(c) LBC integration in the Carla simulator.

Fig. 1. LBC framework.

or go straight at an intersection. After the ResNet34 backbone, the model splits into four parallel branches that each corresponds to a high-level command. The final output is a set of five waypoints of the AV path (see Fig. 1b), which are passed to a low-level controller that produces the steering, throttle, and braking commands.

2.2 L2 System: OpenPilot

OpenPilot is an L2 Autonomous Driving Assistance System (ADAS) that supports more than 250 popular makes and models of cars [17]. A high-level overview of OpenPilot is shown in Fig. 2c. Car sensor data such as images and vehicle state information are passed into the Supercombo model. The output of Supercombo is sent to the planners. The PID Controller uses the results from the planner to decide actuator actions. The generated commands (e.g., brake) are passed through the Panda CAN interface to the actuators to perform the commands.

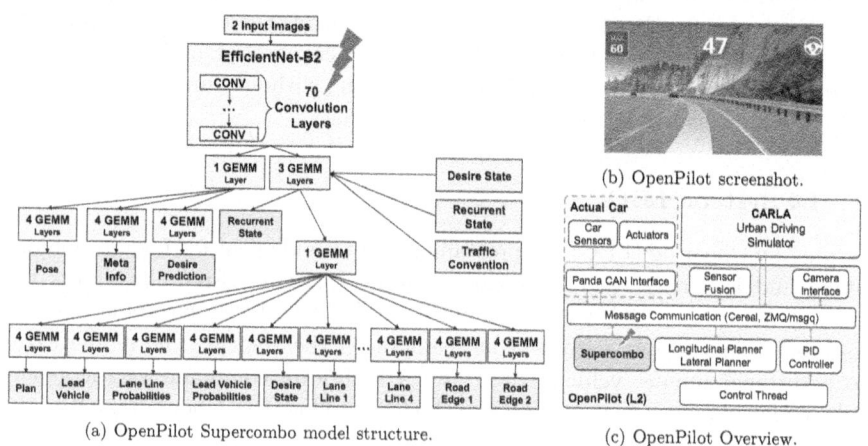

Fig. 2. OpenPilot framework.

The Supercombo model is at the core of the perception module and provides fifteen output fields. As shown in Fig. 2a, Supercombo utilizes an EfficientNet-B2 [22] base CNN for processing the incoming images from the car sensors. Then it uses additional inputs of the traffic convention, desire state, and recurrent state to incorporate the state of the vehicle and environment. Once all inputs have been tied in, Supercombo branches into separate general matrix multiplication (GEMM) computations to generate the plan, lanelines, laneline probabilities, road edges, lead vehicles, lead probabilities, desire state, meta information, vehicle pose, and recurrent state. Figure 2b shows an example screenshot.

2.3 Driving Simulator: CARLA

CARLA is an open-source simulator for autonomous driving research, design, and testing [23]. It provides a realistic urban environment for a vehicle to navigate with features such as variable road and weather conditions. CARLA provides a wealth of data at each simulation timestamp including whether a collision, lane invasion, or red light violation has occurred. CARLA is integrated to Openpilot and LBC, see Fig. 2c and Fig. 1c, respectively.

3 Methodology

Fault Model. We use fault injection in a single 32-bit floating point weight of the DNN to simulate commonly occurring transient faults in DRAM (Dynamic Random Access Memory). DNN weights typically reside in DRAM. A fault site in a neural network weight is defined by the weight id and the bit position(s) of the weight to be flipped. The size of the fault site space for single-, double-, and triple-bit faults for OpenPilot Supercombo and LBC is tremendous, see Table 1 and prevents its exhaustive exploration.

For the majority of the analysis (Sects. 4 and 5) we focus on double-bit flips that are detectable but not correctable by ECC. This is consistent with other reliability studies [11,24] and also consistent with DRAM faults in the wild [25]. For a broader view of the effect of bit flips, we also do experiments with single-bit flip (detectable and correctable) and triple-bit flips (undetectable, their safety implications are similar to double-bit ones, see Sect. 6).

Fault Injection Method. Fault injection is implemented as a two-stage process: *Preparation* and *Injection*. In the *Preparation* stage, before the actual simulation run, we load the (correct) neural network and select an injection site. Portions of the network where faults may be injected are denoted by a lightning bolt in Fig. 1a for LBC and Fig. 2a for OpenPilot Supercombo. We corrupt a weight tensor of the neural network by altering the values of an individual weight (depending on the type of experiment, we induce one single-bit, one double-bit, or one triple-bit fault), and save the corrupted tensor.[2] In the *Injection* stage, the corrupted model generated by the fault injector is used by the AV control software while performing the driving task. The corrupted Pytorch or ONNX files are loaded at the beginning of each LBC or OpenPilot experiment.

---

[2] Once the weight is altered, the modified model is written to a file in the necessary format (ONNX [26] or PyTorch [27], for OpenPilot and LBC, respectively). To corrupt Pytorch models, we load the model, alter the state dictionary associated with it and then save the new faulty model. For ONNX models, we read the model file as bytes, locate the plain text layer ID, and modify the bits corresponding to the target weight in the binary data.

Fault Injection Outcomes. Throughout the simulation, we focus on events that indicate abnormal behavior. A *lane invasion* occurs when the vehicle crosses into a neighboring lane erroneously. Since a lane invasion can occur when the car barely crosses the lane line, a lane invasion alone is not a hazard. We define a *hazard* to be one of the following situations:

H1: The vehicle violates safe following-distance constraints with the lead vehicle: *Relative Distance / Speed* $\leqslant t_{safe}$ and *Speed* > *Lead Speed*.

H2: The vehicle drives out of lane beyond a threshold (e.g., 0.1 m) while speed is higher than β, these are predetermined values given by the driving scenario.

We record all hazards that occur within the experiment as well as their time stamps. These hazards can lead to the following *accidents*:

A1: Collision with the lead vehicle.
A2: Collision with road-side objects or other vehicles in the neighboring lane.

An accident terminates the simulation, at which point the time stamp and the nature of the accident are recorded. The simulation also terminates if the vehicle successfully reaches its goal location. These definitions are based on the STPA [28] hazard analysis method, which has been utilized in other studies [12, 29]. The hazards considered here are indicative of failures of lane keep assist (LKA) and adaptive cruise control (ACC), two main functionalities of L2 AVs.

3.1 Vulnerable Weights: Taylor Guided Fault Injection (TGFI)

In a fault space as vast as the one reported on Table 1, the odds that the standard practice of 1,000 random fault injections [30] capture rare corner cases that result in catastrophic driving scenarios are low. Since our target is the identification of the aforementioned corner cases, we rank the importance of the weights in the neural network using a modified Taylor criterion [15] which estimates the first-order Taylor expansion of the contribution of each weight to the accuracy of the neural network. The more a weight contributes to the accuracy of the network, the more critical it is and the more it can affect accuracy if a fault occurs there. We relatively rank all weights in the two target models using

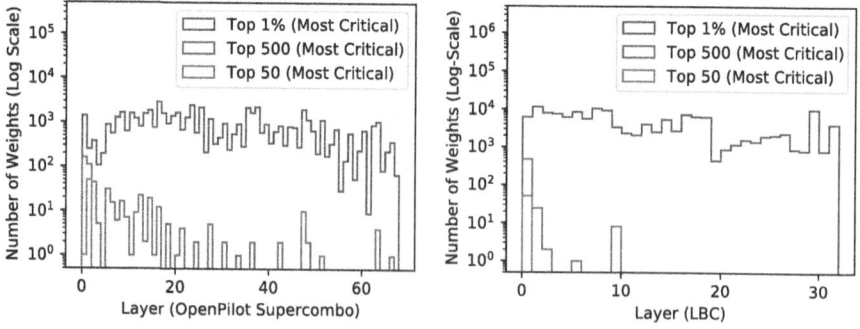

Fig. 3. Locations of top 50, top 500, and top 1% of important weights in OpenPilot Supercombo (left) and LBC (right).

the *importance scores* generated by the Taylor criterion [15]. The importance of a weight using this criterion can be estimated using the following equation:

$$I_m^{(1)}(W) = (g_m w_m)^2, \qquad (1)$$

where I is the importance, W is the set of network parameters, g_m are elements of the gradient, and w_m are the weight values.

We use the Comma2k19 data set [31] of real-world driving footage as input to OpenPilot Supercombo and LBC, and compute the importance of their weights [15]. Figure 3 illustrates the location (layer, x-axis) and number (y-axis logscale) of the most critical weights for OpenPilot Supercombo and LBC. We make the following observations: 1) critical weights may be located in any layer and 2) the few *most* critical weights are concentrated in the earlier layers of both models. We perform fault injection experiments on the most critical weights as guided by Eq. 1, called *Taylor-Guided Fault Injection (TGFI)*.

3.2 Experimental Campaigns

Every distinct fault site uses the same map for the AV to traverse, the same starting location of the car(s) in the simulator, the same goal location, and the same random seed, initial velocity, and weather. The fault sites are selected according to the fault injection (FI) campaign: 1) **Random**: The weight where the double bit flip occurs is chosen using a uniform distribution [30]: we select 1,000 random fault sites to obtain results with 95% confidence intervals and $\pm 3\%$ error margins. 2) **TGFI-top500**: We select the top 500 most critical weights. 3) **TGFI-top50**: We select the top 50 most critical weights.

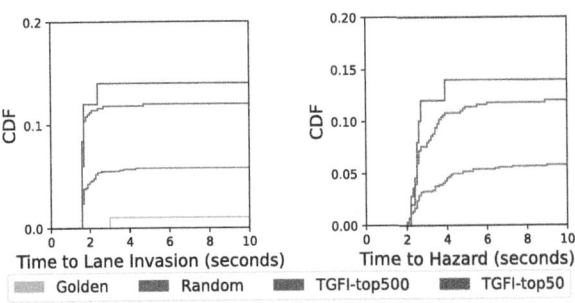

Fig. 4. LBC time to lane invasion and time to hazard. No time to driver intervention because of L4 autonomy. *Note that CDFs do not reach 1.0 since a portion of the experiments do not experience a lane invasion or hazard.*

4 Resilience Evaluation

Following the methodology laid out in Sect. 3, we perform three FI campaigns for LBC and OpenPilot. These FI campaigns highlight different methods for selecting the target weight for fault injection (i.e., Random FI, TGFI-top500, TGFI-top50) using a two-bit fault model. We perform golden runs (i.e., fault-free) to capture normal behavior. The driving scenario is on a curved road in cloudy weather, which represents ideal driving conditions (i.e., good visibility with no glare). Variations of experimental setup are explored in Sect. 6.

4.1 Resilience of L4 LBC

Lane invasions are the least severe violation, since not every lane invasion leads to a hazard. Only 1% of experiments have lane invasions in golden runs, indicating that these are rare in fault-free cases. The time to lane invasion is shown in Fig. 4(a). Across all three fault injection campaigns, the lane invasion tends to occur early in the simulation. Table 2 shows the percentage: TGFI experiments double the percentage of lane invasions comparing to the random FI experiments.

Table 2. Percentage of various events. There is no front vehicle in LBC, hence H1 and A1 cannot happen. Driver intervention: the L2 system returns control to the driver.

ADS	Fault-Site Selection	H1	H2	A1	A2	Lane Inv.	Driver Int.
	Golden run	N/A	0%	N/A	0%	1%	N/A
L4 (LBC)	Random FI	N/A	5.9%	N/A	5.9%	6%	N/A
	TGFI-top500	N/A	12%	N/A	12%	12%	N/A
	TGFI-top50	N/A	14%	N/A	14%	14%	N/A
	Golden run	0%	0%	0%	0%	10%	0%
L2 (OpenPilot	Random FI	2.0%	0.6%	0%	0.1%	23.4%	21.4%
Supercombo)	TGFI-top500	3.3%	8.8%	0.1%	0.2%	29.62%	21.0%
	TGFI-top50	7.9%	16.3%	0.2%	0.3%	62.5%	18.3%

No hazards are detected in the golden fault-free runs, but Random FI and TGFI cause H2 hazards. The time to hazard is shown in Fig. 4. For both random FI and TGFI, hazards are encountered between the one and two second mark. There is a near-complete overlap between the experiments that have lane

Fig. 5. Fault injection in left-turn control command branch in LBC. Four waypoints (circled in green) are correct but one (circled in yellow) is incorrect. (Color figure online)

invasions and the experiments that have hazards, i.e., if the fault causes a lane invasion, then the fault is also severe enough to cause a hazard shortly after. Similar to the lane invasions, TGFI finds more hazards than Random FI.

Every hazard for Random FI and TGFI scenarios results in an accident, indicating that LBC has a very limited ability to function under critical faults. Indeed, LBC does not have alerts or safety checks. TGFI triggers more accidents than Random FI.

We also evaluate the impact of faults in the convolution branches at the end of the network (see Fig. 1a) corresponding to high-level control commands. Figure 5 presents the results of a fault injected in the branch corresponding to the left turn control command: four waypoints are correct, but one (yellow circle) is clearly incorrect. When the car turns the corner, it tries to adhere to the trajectory generated by fitting a curve to the points, but in doing so turns the corner too widely and crashes into the wall on the side of the road.

4.2 Resilience of L2 OpenPilot

In OpenPilot, lane invasions occur often but do not always result in a hazard or accident. 10% of golden runs and 23.4% of Random FI experiments have lane invasions, see Table 2. TGFI-top500 shows slightly more lane invasions and TGFI-top50 nearly triples the number of lane invasions compared to Random FI. The time to lane invasion is shown in Fig. 6(a): most of the lane invasions occur between the 15 and 20 s marks.

The golden runs never result in a hazard, but 2.6% of Random FI experiments result in a hazard that typically occurs between the 5 and 10 s marks, see Fig. 6(b). 10.8% of TGFI-top500 experiments and 22.7% of TGFI-top50 experiments result in a hazard (1.3% and 1.5% of them experience both H1 & H2 hazards for TGFI-top500 and TGFI-top50, respectively). Hazards only occur after second 15 in the simulation and show a steeper increase towards second 50. The H1 hazard that occurs when the vehicle follows another car too closely

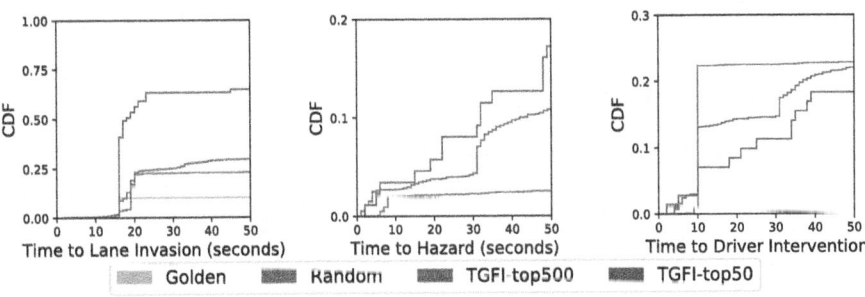

Fig. 6. OpenPilot Supercombo: Time to lane invasion, hazard, and driver intervention. No hazards or driver interventions were observed in golden runs. *CDFs do not reach 1.0 since some experiments do not have a lane invasion, hazard, or driver intervention.*

occurs in 7.9% of simulations for TGFI-top50 and the H2 hazard, which indicates a significant lane invasion, occurs in 16.3% of experiments for TGFI-top50, see Table 2. TGFI results in more hazards than Random FI.

Since OpenPilot requires the driver to resume control of the vehicle in the case of dangerous situations, hazards are often masked and accidents rarely happen, see Table 2. We will discuss this mitigation technique in detail in 5.2.

5 Mitigation

The effects of faults can be mitigated through several approaches, many of which are orthogonal to one another [11,24,32]. In this section, we examine different mitigation techniques for the two AV cases examined here.

5.1 L4 LBC: Ranger

Ranger [11] is a popular, state-of-the-art fault corrector which employs range restriction on neuron activation values to protect ML models from faults with negligible overhead. Here, we present a proof-of-concept mitigation of applying Ranger to LBC. To implement Ranger, we insert range restriction into the model following activation layers at crucial points in the network, such as after convolution layers. Each protection layer has a pair of minimum and maximum activation values to use as bounds, which are set after profiling through golden runs under cloudy (ideal) weather. This step is performed once, before the deployment of the protected model with Ranger. When Ranger is active, any activation values outside the ranges defined by Ranger are clipped to the bound.

We examine the effectiveness of Ranger using the TGFI-top500 experiments, see Fig. 7. We examine three CARLA driving scenarios: curved road, turn at intersection, and straight road. A combination of clear and inclement weather conditions (cloudy, rainy, sunset, wet) is also used to offer insight into how well the fault mitigation functions in unseen conditions. Ranger improves the resiliency of LBC across all three driving scenarios and all weather conditions. Significant improvements are observed under cloudy (ideal) weather conditions.

5.2 L2 OpenPilot: Driver Intervention

As an L2 ADAS, OpenPilot raises driver alerts and returns control to the driver when a problem is detected by its safety checker, i.e., *driver intervention*. We record the *time to driver intervention* as a CDF in Fig. 6(c). The golden runs never require driver intervention and are therefore not plotted. Random FI results in driver intervention in 21.4% of experiments, occurring close to the 10 s mark. The TGFI-top500 and TGFI-top50 experiments show an initial jump in driver intervention early in the simulation. OpenPilot has 3 types of driver alerts which may be displayed when control of the vehicle is returned to the driver, see Table 3. In fault injection experiments, PlannerError/noEntry alerts are the most

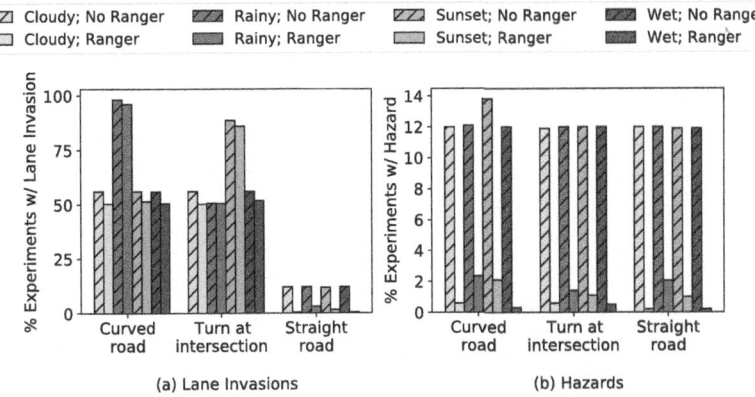

Fig. 7. Comparison of LBC without Ranger vs. LBC with Ranger for the TGFI-500 scenario.

common ones. These alerts indicate that the Model Predictive Control (MPC) cannot find a feasible solution for lateral planning (steering) and longitudinal planning (gas/brake) and releases control to the driver, successfully mitigating the fault before any hazard is triggered. The CanError/ImmediateDisable alerts indicate communication errors and occur the least frequently. The steerSaturatedWarning alerts also rarely occur and they indicate that the car is swerving sharply in the presence of the fault. *This indicates that TGFI finds faults that are more rare and more challenging for OpenPilot safety checks to detect and mitigate.*

Table 3. Experiments where the driver is alerted to a problem in OpenPilot.

Alert	Golden Runs	Random	TGFI-top500	TGFI-top50
plannerError/noEntry	0%	19.8%	10.88%	4.01%
canError/immediateDisable	0%	0.5%	0.1%	0.4%
steerSaturated/warning	0%	1.9%	2.98%	3.85%

6 Case Studies and Discussion

In this section, we use LBC as a case study to evaluate and discuss the impact of different faults, layers of DNN and bit positions on AV resilience.

6.1 Importance of Layer Depth for Resilience

Table 4 shows four distinct experiments with corrupted LBC models that identify the importance of the layer depth where the fault occurs. When Ranger is

Table 4. Case study of four distinct corrupted LBC models.

Exp. ID	Layer ID	Lane Invasion (No Ranger)	Accidents (No Ranger)	Lane Invasion (Ranger)	Accidents (Ranger)
1	10	100%	100%	0%	0%
2	30	100%	100%	0%	0%
3	37	92.5%	100%	91%	100%
4	38	53.3%	84.7%	32.7%	78.4%

not applied and if the fault is injected in the earlier DNN layers (experiments 1 and 2), resilience is severely affected: all faults result to accidents. Faults occurring earlier in the network have more time to propagate horizontally across the neurons, resulting in more severe corruption. This propagation still occurs even if the fault site is relatively deep into the ResNet34 section of the network, as experiment 2 with injection in layer 30 shows. The fault sites of experiments 3 and 4 are in a the final layers in the network, therefore error propagation is minimal and the severity of corruption in the final output is lessened.

With Ranger, results for faults injected in layer 10 and 30 improve from 100% accidents to 0% accidents. For faults in layer 37 and 38, applying Ranger still results in a majority of accidents. Since these two layers are at the end of the network, the clipped Ranger bound used as output, differs significantly from the ideal one.

6.2 Sensitivity to Single and Multi-bit Faults

We analyze the sensitivity of LBC to single- and multi-bit faults [33]. We evaluate how the bit position(s) and the number of bit flips affect hazards in the LBC model under cloudy ("easy") and rainy ("challenging") weather conditions using TGFI-500, as shown in Fig. 8. We compare fault injection outcomes using different numbers and locations of bit flips: single-bit exponent, single-bit mantissa, double-bit (the fault model majorly used in this paper), and triple-bit. These experiments show that single-bit exponent causes the most hazards, followed by triple-bit, double-bit, and finally single-bit mantissa. This experiment also shows that lane invasions and hazards for *non-detectable triple* bit faults are prominent, thus their mitigation for safety is of paramount importance.

6.3 Lessons Learned from L4 LBC and L2 OpenPilot

L4 LBC and the L2 OpenPilot have similarities: both ML models are structured around a backbone CNN which is more vulnerable to faults in earlier layers than later ones. Faults are commonly masked when they occur in weights of low importance. Meanwhile, structural differences of the DNNs affect their resilience: LBC has high-level command branches that activate or deactivate parts of the network depending on the driving scenario, thus possibly masking faults during

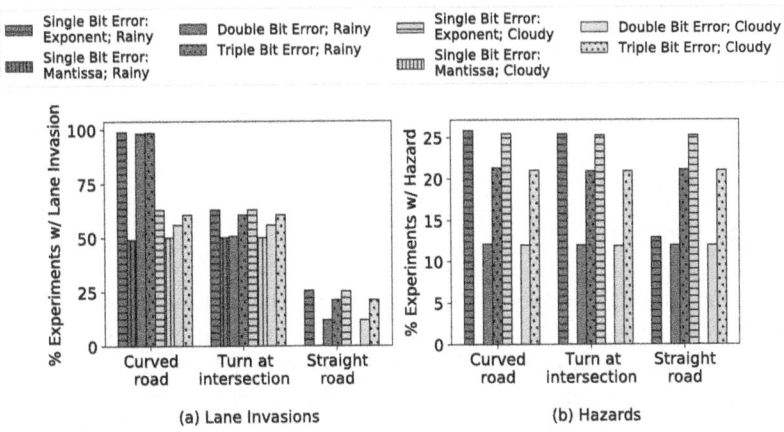

Fig. 8. Bit position analysis for LBC and the TGFI-top500 experimental campaign.

inference. OpenPilot uses all layers in every inference, meaning that faults can only be masked by the network itself, rather than circumstances around its use. Secondly, the format of the DNN output is different. LBC outputs waypoints while OpenPilot's Supercombo outputs information about lanelines, road edges, lead vehicle position, and many other variables. The availability of these different outputs in OpenPilot allows for the implementation of safety checks.

For OpenPilot, we find that many accidents are not mitigated by the system safety checks. This speaks to both the need for the improvement of existing checks, and for the implementation of new ones. Adding DNN-oriented error detection/correction mechanisms would improve AV resilience. High-level checks, like many of those employed in OpenPilot, cannot detect potential incorrect outputs but while imperfect, they provide a blueprint for developing resilience in L4 systems like LBC.

7 Related Work

Past studies have shown that faults [34] can cause hazardous behavior in AVs. DeepTest [1] focuses on testing the reliability of autonomous driving systems and safety engineering techniques for autonomous systems are investigated in [35], but these works do not consider soft errors. Other works examine the fault space for AVs through Bayesian fault injection [34] and rely on large amounts of random fault injection experiments. [36] explores the problem of effective safety checks for an ML-based AV.

[2] uses duplication of computation and temporal data diversity to improve the resilience of AVs with LBC but requires additional hardware. On the OpenPilot side, [12] focuses on hazard coverage and fault injection, but does not study its ML model. [37,38] present attacks on the CAN bus or camera input for a vehicle using OpenPilot. Unlike these works, we focus on the effects of unintended faults rather than attacks.

Most importantly, both for LBC and Openpilot, we identify corner cases that are otherwise hard to find: we identify which portions of their DNNs are vulnerable to faults and result in AV safety violations, i.e., we do not simply examine the accuracy of DNNs for classification but their holistic effect into the AV operation.

8 Conclusions

We perform strategic resilience evaluation of the neural networks of two autonomous vehicles against transient faults. We focus on an L4 system widely used in academia and an L2 ADAS system which is widely deployed on the road. We find that both systems are vulnerable to single- and multi-bit faults which may induce hazards/accidents. We use the Taylor criterion to strategically identify the most important weights for reliability in the DNNs used in the L4 LBC and L2 Openpilot and inject errors on those weights using TGFI. TGFI is efficient in identifying vulnerabilities that result in hazards and accidents. We also examine the effectiveness of mitigation in AVs. For L4 self-driving, mitigation techniques such as Ranger can be effective at minimizing the impact of faults. Driver intervention is a crucial contributing factor to the security of L2 systems.

Acknowledgments. This material is based upon work supported by two Commonwealth Cyber Initiative (CCI) grants (#HC-3Q24-047 and COVA C-Q122-WM-02).

References

1. Tian, Y., Pei, K., Jana, S., Ray, B.: DeepTest: automated testing of deep-neural-network-driven autonomous cars. In: Proceedings of the 40th International Conference on Software Engineering, pp. 303–314 (2018)
2. Jha, S., et al.: Exploiting temporal data diversity for detecting safety-critical faults in AV compute systems. In: 2022 52nd Annual IEEE/IFIP International Conference on Dependable Systems and Networks (DSN), pp. 88–100. IEEE (2022)
3. Fratin, V., Oliveira, D., Lunardi, C., Santos, F., Rodrigues, G., Rech, P.: Code-dependent and architecture-dependent reliability behaviors. In: 2018 48th Annual IEEE/IFIP International Conference on Dependable Systems and Networks (DSN), pp. 13–26. IEEE (2018)
4. Ganapathy, S., Kalamatianos, J., Beckmann, B.M., Raasch, S., Szafaryn, L.G.: Killi: runtime fault classification to deploy low voltage caches without MBIST. In: 2019 IEEE International Symposium on High Performance Computer Architecture (HPCA), pp. 304–316. IEEE (2019)
5. Li, G., et al.: Understanding error propagation in deep learning neural network (DNN) accelerators and applications. In: Proceedings of Supercomputing, pp. 1–12 (2017)
6. Karumbunathan, L.S.: NVIDIA Jetson AGX Orin Series: A Giant Leap Forward for Robotics and Edge AI Applications (2022). https://www.nvidia.com/content/dam/en-zz/Solutions/gtcf21/jetson-orin/nvidia-jetson-agx-orin-technical-brief.pdf

7. Yoshida, J.: Toyota Case: Single Bit Flip That Killed (2013). https://www.eetimes.com/toyota-case-single-bit-flip-that-killed
8. Road vehicles - Functional safety. Standard, International Organization for Standardization, Geneva, CH, December (2018)
9. Leveugle, R., Calvez, A., Maistri, P., Vanhauwaert, P.: Statistical fault injection: quantified error and confidence. In: 2009 Design, Automation & Test in Europe Conference & Exhibition, pp. 502–506. IEEE (2009)
10. dos Santos, F.F., et al.: Analyzing and increasing the reliability of convolutional neural networks on GPUs. IEEE Trans. Reliab. **68**(2), 663–677 (2018)
11. Chen, Z., Li, G., Pattabiraman, K.: A low-cost fault corrector for deep neural networks through range restriction. In: 2021 51st Annual IEEE/IFIP International Conference on Dependable Systems and Networks (DSN), pp. 1–13. IEEE (2021)
12. Rubaiyat, A.H.M., Qin, Y., Alemzadeh, H.: Experimental resilience assessment of an open-source driving agent. In: 2018 IEEE 23rd Pacific Rim International Symposium on Dependable Computing (PRDC), pp. 54–63. IEEE (2018)
13. Jha, S., et al.: Kayotee: a fault injection-based system to assess the safety and reliability of autonomous vehicles to faults and errors (2019). arXiv preprint arXiv:1907.01024
14. Jha, S., et al.: ML-driven malware that targets AV safety. In: 2020 50th Annual IEEE/IFIP International Conference on Dependable Systems and Networks (DSN), pp. 113–124. IEEE (2020)
15. Molchanov, P., Mallya, A., Tyree, S., Frosio, I., Kautz, J.: Importance estimation for neural network pruning. In: Proceedings of the IEEE/CVF Conference on Computer Vision and Pattern Recognition, pp. 11264–11272 (2019)
16. Chen, D., Zhou, B., Koltun, V., Krähenbühl, P.: Learning by cheating. In: Conference on Robot Learning, pp. 66–75. PMLR (2020)
17. Comma.ai. Supported Cars by OpenPilot. https://github.com/commaai/openpilot/blob/master/docs/CARS.md
18. SAE International. SAE Levels of Driving Automation^{TM} Refined for Clarity and International Audience (2021). https://www.sae.org/blog/sae-j3016-update
19. Filos, A., Tigkas, P., McAllister, R., Rhinehart, N., Levine, S., Gal, Y.: Can autonomous vehicles identify, recover from, and adapt to distribution shifts? In: International Conference on Machine Learning, pp. 3145–3153 (2020)
20. Toromanoff, M., Wirbel, E., Moutarde, F.: End-to-end model-free reinforcement learning for urban driving using implicit affordances. In: Proceedings of the IEEE/CVF Conference on Computer Vision and Pattern Recognition, pp. 7153–7162 (2020)
21. He, K., Zhang, X., Ren, S., Sun, J.: Deep residual learning for image recognition. In: Proceedings of the IEEE Conference on Computer Vision and Pattern Recognition, pp. 770–778 (2016)
22. Tan, M., Le, Q.: EfficientNet: rethinking model scaling for convolutional neural networks. In: International Conference on Machine Learning, pp. 6105–6114 (2019)
23. Dosovitskiy, A., Ros, G., Codevilla, F., Lopez, A., Koltun, V.: CARLA: an open urban driving simulator. In: Conference on Robot Learning, pp. 1–16 (2017)
24. Kadam, G., Smirni, E., Jog, A.: Data-centric reliability management in GPUs. In: 2021 51st Annual IEEE/IFIP International Conference on Dependable Systems and Networks (DSN), pp. 271–283. IEEE (2021)
25. Beigi, M.V., Cao, Y., Gurumurthi, S., Recchia, C., Walton, A., Sridharan, V.: A systematic study of DDR4 dram faults in the field. In: 2023 IEEE International Symposium on High-Performance Computer Architecture (HPCA), pp. 991–1002. IEEE (2023)

26. The Linux Foundation. Open neural network exchange: The open standard for machine learning interoperability (2019). https://onnx.ai/
27. Paszke, A., et al.: PyTorch: an imperative style, high-performance deep learning library. In: Advances in Neural Information Processing Systems, vol. 32 (2019)
28. P Leveson, N., Thomas, J.: An STPA Primer. Cambridge, MA (2013)
29. Zhou, X., Ahmed, B., Aylor, J.H., Asare, P., Alemzadeh, H.: Data-driven design of context-aware monitors for hazard prediction in artificial pancreas systems. In: 2021 51st Annual IEEE/IFIP International Conference on Dependable Systems and Networks (DSN), pp. 484–496. IEEE (2021)
30.) Nie, B., Yang, L., Jog, A., Smirni, E.: Fault site pruning for practical reliability analysis of GPGPU applications. In: 2018 51st Annual IEEE/ACM International Symposium on Microarchitecture (MICRO), pp. 749–761 (2018)
31. Schafer, H., Santana, E., Haden, A., Biasini, R.: A commute in data: The comma2k19 dataset (2018). arXiv preprint arXiv:1812.05752
32. Yang, L., Nie, B., Jog, A., Smirni, E.: Enabling software resilience in GPGPU applications via partial thread protection. In: 43rd IEEE/ACM International Conference on Software Engineering, ICSE 2021, Madrid, Spain, 22-30 May 2021, pp. 1248–1259 (2021)
33. Yang, L., Nie, B., Jog, A., Smirni, E.: Practical resilience analysis of GPGPU applications in the presence of single- and multi-bit faults. IEEE Trans. Comput. **70**(1), 30–44 (2021)
34. Jha, S., et al.: ML-based fault injection for autonomous vehicles: a case for Bayesian fault injection. In: 49th Annual IEEE/IFIP International Conference on Dependable Systems and Networks, DSN 2019, Portland, OR, USA, June 24-27, 2019, pp. 112–124 (2019)
35. Osborne, M., Hawkins, R., McDermid, J.: Analysing the safety of decision-making in autonomous systems. In: Trapp, M., Saglietti, F., Spisländer, M., Bitsch, F. (eds.) Computer Safety, Reliability, and Security. SAFECOMP 2022. LNCS, vol. 13414. Springer, Cham (2022). https://doi.org/10.1007/978-3-031-14835-4_1
36. Terrosi, F., Strigini, L., Bondavalli, A.: Impact of machine learning on safety monitors. In: Trapp, M., Saglietti, F., Spisländer, M., Bitsch, F. (eds.) Computer Safety, Reliability, and Security. SAFECOMP 2022. LNCS, vol. 13414. Springer, Cham (2022). https://doi.org/10.1007/978-3-031-14835-4_9
37. Zhou, X., et al.: Strategic safety-critical attacks against an advanced driver assistance system. In: 2022 52nd Annual IEEE/IFIP International Conference on Dependable Systems and Networks (DSN), pp. 79–87. IEEE (2022)
38. Zhou, X., et al.: Runtime stealthy perception attacks against DNN-based adaptive cruise control systems (2024). arXiv preprint arXiv: 2307.08939

System and Software Safety Assurance

Reconciling Safety Measurement and Dynamic Assurance

Ewen Denney and Ganesh Pai[✉][iD]

KBR/NASA Ames Research Center, Moffett Field, CA 94035, USA
{ewen.denney,ganesh.pai}@nasa.gov

Abstract. We propose a new framework to facilitate dynamic assurance within a safety case approach by associating safety performance measurement with the core assurance artifacts of a safety case. The focus is mainly on the *safety architecture*, whose underlying risk assessment model gives the concrete link from safety measurement to operational risk. Using an aviation domain example of autonomous taxiing, we describe our approach to derive safety indicators and revise the risk assessment based on safety measurement. We then outline a notion of *consistency* between a collection of safety indicators and the safety case, as a formal basis for implementing the proposed framework in our tool, AdvoCATE.

Keywords: Dynamic assurance · Safety cases · Safety measurement · Safety metrics · Safety performance · Safety risk assessment

1 Introduction

Software-based self-adaptation and machine learning (ML) technologies for enabling autonomy in complex systems—such as those in civil aviation—may induce new and unforeseen ways for operational safety performance to deviate from an approved baseline of acceptable risk. This phenomenon, known as *practical drift* [13], emerges from the inevitable variabilities in real-life operations to meet service expectations in an operating environment that is inherently dynamic. Conceptually, it can be understood as progressively imperceptible reductions in the safety margins built into a system in part due to initially benign operational tradeoffs between safety and performance. A system therefore appears to be operating safely but, in fact, is operating at a higher level of safety risk than what was originally considered acceptable, or approved for service. Left unchecked, practical drift may suddenly manifest as a serious *incident* or *accident*. Assessing the change in operational safety risk is thus key to identifying practical drift, its impact, and the mitigations needed.

Related Work. The conventional approach to operational safety assurance in aviation largely relies upon hazard tracking and safety performance monitoring and measurement, as part of a larger *safety management system* (SMS) [10].

The contemporary safety case approach to assurance has similarly employed safety monitoring and measurement: for example, our earlier work on *dynamic safety cases* [5] first suggested connecting safety monitoring to assurance argument modification actions. Subsequently, an approach to defining performance metrics and monitors by identifying the defeaters and counterarguments to a safety case has been developed in [12]. The concept has since also been applied to safety assurance of self-adaptive software [4], and to detect operational exposure to previously unknown hazardous conditions [18]. More recently, the use of *safety performance indicators* (SPIs)—a concept with a well-established history of use in aviation safety [13]—has been proposed for evaluating safety cases for autonomous vehicles [15]. These approaches all share a common motivation: using measurement based assessment to confirm at deployment, and maintain in operation, the validity of the assurance arguments of a safety case.

Although such an approach suggests which parts of an argument may have been invalidated, and thus require changing, the nature and extent of the change to operational safety risk levels is left implicit. Such analyses can also meaningfully inform what modifications may be needed to the system and its safety case, especially when—due to practical drift—improved system performance is observed without detrimental safety effects, even though parts of the safety argument have become invalid. Current safety case approaches that use safety performance measurement to validate assurance arguments give limited guidance on how to facilitate what this paper considers as *dynamic* assurance (see Fig. 3): *continued, justified confidence that a system is operating at a safety risk level consistent with an approved risk baseline.*

There are other variations of the dynamic assurance concept [17,21] that aim to optimize operational system performance, and thus opt for situation-specific runtime tradeoffs between safety and functional performance, instead of designing for the worst case. However, such tradeoffs may result in the initiating conditions for practical drift. Our proposed framework rather aims to identify and contain practical drift, whilst considering that a safety case for a system is always for a design that accounts for the worst credible safety effects. In [18], dynamic assurance refers to the automated aspects of so-called *continuous assurance*: a concept that, in effect, extends our prior work [5], by using monitors for different kinds of uncertainty that then trigger modifications to the system and its assurance case. The relationship of testing and operational metrics to safety assurance has been explored in [19], similar to our work in this paper (see Sect. 4), though there the focus is on providing confidence that a system meets its safety target, given evidence of mishap-free operation. In contrast, our focus here is on determining how safety risk has changed given similar measurement evidence.

Contributions and Paper Organization. To facilitate a framework for dynamic assurance within a safety case approach, the focus of this paper is on associating safety performance measurement with the *safety architecture* of a system, in addition to assurance arguments (Sect. 2). Using an aviation domain system (Sect. 3) as motivation, we present our approach to define safety metrics and indicators, through a concept of *safety measurement basis* (SMB), then revise the operational safety risk assessment based on safety measurement, and characterize the change to safety risk levels (Sect. 4). Additionally, we give illustrative numerical examples. Then (Sect. 5) we formalize a notion of *consistency* between the SMB for a system and the arguments of its safety case. We conclude (Sect. 6) by describing a preliminary implementation in AdvoCATE, and with a discussion of our future plans to further advance this work. The contributions above differentiate our work from prior related research.

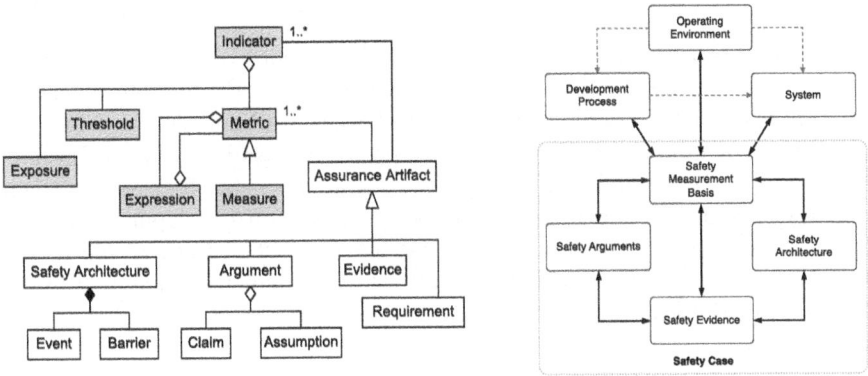

Fig. 1. Fragment of AdvoCATE safety case metamodel (on the left) extended with measurement concepts that are part of a safety measurement basis (shown on the right), which is the interface between quantities in the system, its environment, its development process, and the assurance artifacts comprising a safety case (solid arrows denote consistency relations).

2 Conceptual Background

Safety Case Metamodel. Our safety case concept [1] communicates confidence in safety through multiple viewpoints via a collection of core, interlinked *assurance artifacts*, namely: hazard, requirement, and evidence logs, a safety architecture, and an assurance rationale. Of those, the last two are particularly relevant for this paper. Assurance rationale captured as structured arguments expresses the reasoning why safety claims ought to be accepted on the basis of the evidence supplied. A safety architecture [6,8] models the mitigations (and their interrelations) to the events characterizing the operational *risk scenarios* for a system, thereby offering a system-level viewpoint on how safety risk is reduced.

Figure 1 shows a fragment of the metamodel associated with our safety case concept (as unshaded class nodes), for which we have a model-based implementation in our tool, AdvoCATE [7]. We use the *goal structuring notation* (GSN) [20] to represent structured arguments, and *bow tie diagrams* (BTDs) to represent views of a safety architecture. Those views capture a causal chain (e.g., see Fig. 2) of *threats* (initiating events) causing a *top event* (a hazard) that can lead to *consequence events* (undesired safety effects), along with the *barriers* (mitigations) necessary to reduce the safety risk posed. Each such event chain requires a combination of *hazardous activity, environmental condition,* and *system state* (together representing the operating context[1]), and can admit an arbitrary number of intermediate events between the initiating threat and terminating consequence events. Each barrier is itself a system comprising underlying *controls*; thus, it can have its own associated safety architecture, giving the overall model a layered structure that can mirror the system hierarchy.

A risk assessment model underlying a safety architecture gives the formal basis to: (i) characterize the extent of risk reduction, and (ii) link safety metrics and indicators to operational safety risk (see Sect. 4). In brief, this model relates the risk of consequence events, i.e., their probability and severity, to that of the precursor events, and to the *integrity*[2] of the applicable barriers and their constituent controls. Depending on the stage of system development, we can interpret each of an event probability and barrier/control integrity both as a design target and verification goal. For the rest of this paper, we mainly consider the risk reduction contribution of barriers.

Safety Measurement. We extend the safety case metamodel in AdvoCATE-with concepts for safety performance measurement (shown by the shaded class nodes in Fig. 1) as follows: we link the *indicators* to the core assurance artifacts—in particular, the event and barrier elements of a safety architecture, the claims and assumptions in arguments, to requirements, and to evidence artifacts. An *indicator* consists of a *metric* along with a *threshold*, representing the target that a metric should (or should not) reach, over a specified *exposure*, expressed either as a duration of continuous time or a specified number of occurrences of a discrete event. Indicators that have a bearing on safety can be called *safety indicators* (SIs) or *safety performance indicators* (SPIs). *Metrics* are computed values based on *measures*—directly observable parameters of the system, its environment, and its development process—and other metrics, which we represent using an expression language. Thus, they are arithmetic expressions over measured variables drawn from the most recent *mission*—which we term as a *data run*—or the missions conducted over the lifetime of the system. They can also refer to values referenced in assurance artifacts.

As shown in Fig. 1, a safety case can be seen as comprising a *dynamic* portion (indicators, metrics, and measures) and a static portion (safety arguments and safety architecture), with links associating the two. We refer to the set of interconnected indicators, metrics, and measures, along with their traceability links to the assurance artifacts of a safety case as a *safety measurement basis* (SMB). Roughly speaking, the connection between the dynamic and static por-

---

[1] Also known as an *operational design domain* (ODD) for systems integrating ML [14].
[2] Integrity is the probability that a barrier or control is not breached, i.e., it delivers its intended function for reducing risk in the specified operating context and scenario [8].

tions is that the indicators represent the objectively quantifiable content of the arguments and the safety architecture which, in turn, give the justification for how those indicators collectively provide safety substantiation. Put another way, we want the SMB to be *consistent* with the static portions of the safety case, especially the arguments and the safety architecture (see Sect. 5).

3 Motivating Example

We motivate this work using an aviation domain use case of autonomous aircraft taxiing [1]. This system uses a *runway centerline tracking* function comprising a classical controller coupled to a deep convolutional neural network that estimates aircraft position from optical sensor data. The functional objective is to maintain both the *cross-track error* (CTE) and the *heading error* (HE) within pre-defined bounds. CTE is the horizontal distance between the runway centerline and the aircraft body (or *roll*) axis; HE is the angle between the respective headings of the runway centerline and the roll axis. The safety objective is to avoid a lateral *runway overrun* (also known as a *runway excursion*), i.e., departing the sides of the runway.

Figure 2 shows a BTD fragment for this example (annotated to show its graphical elements and their identifiers) as a view of its wider safety architecture (not shown), which composes [8] similar such BTDs, albeit for different operating contexts, threats, top events, and consequences. Here, the operating context involves a relatively low speed (25 kn), low visibility taxi operation on a wet runway, at dusk, under no crosswind conditions. The hazard to be controlled (E_3) is a violation of the allowed lateral offset from the runway centerline, failing which a lateral runway overrun (E_4) could occur. Two (out of many) initiating causes for this hazard have been shown: a controller malfunction that steers the aircraft away from the centerline when not required (E_1); and runway centerline markings that are not visible, or are obscured (E_2).

3.1 Baseline Safety

To characterize the safety risk level of an operating scenario, we use the risk assessment model associated with the safety architecture to establish a *baseline level* of operational safety risk for the identified safety effects.

For the scenario in Fig. 2, the *initial risk level* (IRL) of the consequence event F_4 is labeled 4A(Medium). That is, E_4 has a *medium* level of unmitigated risk, and is assigned the *risk classification category* 4A. That refers to a region of the overall risk space that has been discretized using a classical 5 × 5 risk matrix of categories of consequence event probability, ranging from *Frequent* (A) to *Extremely Improbable* (E), and consequence event severity, ranging from *Minimal* (5) to *Catastrophic* (1). For a definition of those categories, see [10]. A similar interpretation applies to *residual risk level* (RRL) which, for E_4, is shown as 4D(Low), representing the risk remaining after mitigation using the indicated barriers and the associated controls. Specifically, B_1: *Runtime Monitoring*, B_2:

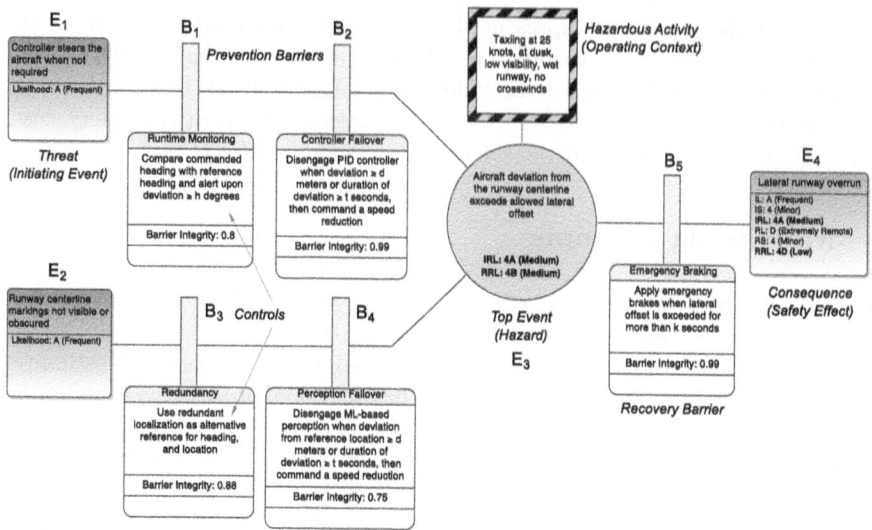

Fig. 2. Annotated BTD fragment for an autonomous taxiing capability, showing how a lateral runway overrun is mitigated under specific initiating events leading to centerline tracking violation.

Controller Failover, B_3: Redundancy, and B_4: Perception Failover, serve as prevention barriers for exceeding the allowed CTE, while B_5: Emergency Braking is a recovery barrier invoked after the top event occurs.

For aeronautical applications, civil aviation regulations and the associated certification or approval processes generally establish what constitutes *acceptable* and *approved* baseline risk levels respectively. The two can be the same (though they need not be) and, typically, are given in terms of a so-called *target level of safety* (TLOS), which specifies the (maximum acceptable) probability of the undesired safety effect per unit of operational exposure, e.g., 10^{-6} lateral runway excursions per taxi operation. How TLOS is established and approved is out of scope for this paper[3]; as such, in Fig. 2, either of the values of the IRL, 4A(Medium), or the RRL, 4D(Low), may plausibly meet the TLOS, and therefore could be an approved baseline level of safety risk. For the purposes of this example, we assume that the RRL shown is the approved baseline that meets the TLOS. Once a system is deployed, note that the RRL for an event is, in fact, *dynamic*, i.e., as a sequence of values starting from the approved baseline, it represents how the risk of that event evolves over the system lifetime (also see Fig. 3).

---

[3] Interested readers may refer to [3].

3.2 Practical Drift

Some barriers or controls in the safety architecture of a system may be relaxed in operation to improve the performance of system services and/or to make local optimizations that address the operating context. In our running example, for instance, to increase runway throughput whilst operating on large runways in better environmental conditions (e.g., clear weather, and dry runway surface), the time an aircraft spends on a runway could be reduced. For that purpose, suppose that disengagement of the perception function or the controller is delayed (see Fig. 2), or that more permissive CTE bounds are admitted. In those cases, the system may enter certain states that would have been prohibited otherwise. In particular, such states represent violations of the barrier/control requirements that were stated as claims in the pre-deployment safety case.

However, when there is improved system performance without observed safety consequences or mishaps, those states are not perceived as violations that increase residual risk. This can lead to misplaced assurance in operational safety when the system as operated deviates from its safety case. Practical drift can then emerge when multiple safety mitigations may be progressively loosened, and continued, incident-free system operations under such changes obscure the increase in operational safety risk. It is important to emphasize that relaxing mitigations to improve performance represents an operational tradeoff rather than a deliberate attempt to subvert safety. An analogy, for example, is highway driving at the speed of traffic that exceeds the posted speed limits—a practice that is not always unsafe, but poses higher risk in general.

4 Framework

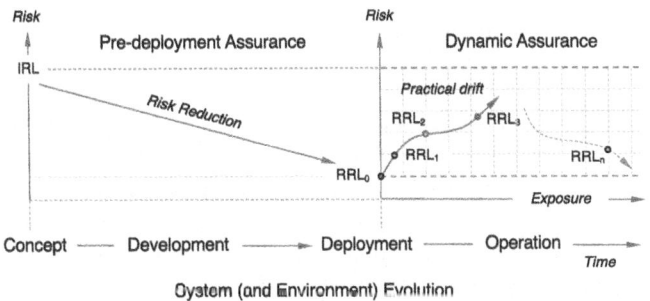

Fig. 3. Pre-deployment assurance gives justified confidence in the reduction of an initial risk level (IRL) of the safety effects in a system concept, through system development, to a baseline residual risk level (RRL_0) that meets the TLOS at deployment. Dynamic assurance provides confidence in operation that the approved baseline is maintained, by identifying and managing practical drift.

Dynamic assurance within a safety case approach gives a proactive means to assess and contain practical drift through continued assurance that the operational safety risk level for the system is aligned with its approved baseline (see Fig. 3). A framework that enables this must at least: (i) characterize how operational safety risk levels have changed; (ii) determine which mitigations, if any, may be legitimately relaxed without safety deteriorating; and (iii) identify the necessary modifications to both the system and its safety case, so that the two are mutually consistent during system operation. Next, we discuss how relating safety performance measurement to the safety architecture in a safety case, in addition to its arguments, gives the necessary elements and technical foundations for the first of the preceding three requirements—the main focus in this paper. The examples presented next are meant to be illustrative and not comprehensive.

4.1 Defining Safety Metrics and Indicators

A safety architecture and its associated risk assessment model [8] give a basis to allocate safety targets to the safety functions, and subsequently confirm them (analytically and empirically). TLOS is a system-level safety target always assigned to consequence event probability. Decomposing and allocating the TLOS across the elements of the safety architecture gives the safety integrity targets for barriers and controls, along with precursor event probabilities that we interpret as *scenario-specific safety targets*. Relating safety targets to safety performance measurement in general, and safety indicators (SIs) in particular, facilitates tracking and confirming that the mitigations are performing in operation as intended. One way to embed TLOS into an SI is by simply converting the corresponding probability value into an event frequency threshold applied to an appropriate safety metric used during development or in operation. In this section we focus on the operational safety metrics, addressing the metrics used during development in Sect. 4.2.

TLOS and the corresponding SIs can be *generic*, i.e., apply to all relevant operating contexts of a safety architecture, or *scenario-specific*, i.e., applicable to a particular operating context. For instance, let the TLOS for lateral runway overrun under all relevant operating conditions of the example system be 10^{-6} per taxi operation. We can then define a corresponding generic SI, $\mathcal{I}_{\text{LRE}}$: opLatRwyEx $\leq$ 1 in 10^6 taxi operations, where opLatRwyEx is an operational safety metric[4] for the *number of lateral runway overrun events in operation*, whose threshold value is 1, measured over an exposure of 10^6 taxi operations. Another commonly used unit of exposure is *flight hours* [11], and the SI can be given equivalently as $\mathcal{I}_{\text{LRE}}$: opLatRwyEx $\leq$ 1 in $10^6 \times t$ flight hours, where t is the average time in flight hours of a taxi operation.

Scenario-specific SI definition proceeds in the same way, but is applied to specific operating contexts after first decomposing and allocating the TLOS of a consequence event to its scenario-specific instance. For example, if 10% of all

---

[4] Henceforth, identifiers with the prefix 'dev' refer to metrics used during system development, and the prefix 'op' indicates an operational safety metric.

taxi operations occur under the operating context of Fig. 2, then we can modify the exposure of $\mathcal{I}_{\text{LRE}}$ to 10^5 taxi operations to get the scenario-specific SI for the consequence event E_4.

Similarly, we can define generic and scenario-specific SIs for the remaining safety architecture elements by converting the associated event probability and barrier integrity values as applicable. Moreover, recalling that a barrier can have its own safety architecture (Sect. 2), we can iteratively define SIs for the lower layers of a system hierarchy. Thus, in Fig. 2, we can define the scenario-specific SI for the barrier B_4: *Perception Failover* as $\mathcal{I}_{\text{PFO}}$: opPcpDisEngF $\leq y$ in n taxi operations. In Sect. 4.4, we illustrate one approach to instantiate y and n.

Here, opPcpDisEngF is a metric related to the integrity of B_4 (itself a metric) that counts the *number of failed disengagements of ML-based perception in operation*; its threshold value is y, to be measured over an exposure of n taxi operations conducted in the stated operating context for the specified scenario. This metric relies upon a precise definition of a *failed disengagement* (not given here), which may itself be given in terms of other metrics, e.g., those associated with its *functional deviations* (i.e., violation of the requirements for the barrier, its constituent controls, or their verification), and its *failure modes* (of the physical systems to which the barrier function is allocated). Additional operational safety metrics related to barrier integrity include opTxLowVisW, counting the *number of taxi operations conducted at dusk under low visibility, no crosswind, and wet runway conditions* (i.e., the operating context of Fig. 2), from which we may infer the *number of successful disengagements of ML-based perception* as the metric opPcpDisEngS = opTxLowVisW − opPcpDisEngF.

4.2 Updating and Revising the Operational Risk Assessment

A pre-deployment safety case represents what (we believe) a system design achieves at deployment, and will continue to achieve in operation. Some of the metrics and SIs applicable during system development constitute measurement evidence verifying safety performance, e.g., during pre-deployment system testing or flight testing. Thus, by associating those metrics and SIs with the safety architecture, we get the *prior* values of event probability and barrier integrity. For the scenario and operating context of Fig. 2, some of the metrics used during system development for the barrier B_4 are: devTxLowVisW: the *number of tests for* $\mathsf{B}_4 = t$ (say); devPcpDisEngS: the *number of successful disengagements of ML-based perception* $= s$; and devPcpDisEngF: the *number of failed disengagements of ML-based perception* $= (t - s) = f$.

If the test campaign during system development is designed as a Bernoulli process [16] then we can model the sequence of test results as a binomial distribution, $\text{Binom}(\chi : \eta, \theta)$, whose parameters are χ: the number of successes, η: the number of independent trials, and θ: the probability of success in each trial. Hence, we can assign the values of the metrics devPcpDisEngS and devTxLowVisW, respectively, to the first two parameters as $\chi := s$, and $\eta := t$. Let $\theta := p$, the unknown (fixed) probability that each test produces a successful disengagement. We can model p as the conjugate prior beta distribution,

$\pi(p) \sim \text{Beta}(\alpha, \beta)$. The *hyperparameters* (i.e., the parameters of the prior distribution) represent our prior knowledge of the number of successful and failed tests during development. Hence we assign to them the values of the metrics devPcpDisEngS and devPcpDisEngF, respectively, as $\alpha := s$, and $\beta := f$. The beta distribution mean, $\mu_p = s/t$, gives a point estimate of the prior barrier integrity, and its variance, $\sigma_p^2 = sf/t^2(t+1)$, gives the uncertainty in that estimate.

In operation, safety performance measurement yields a sequence of observations of the state of the safety system. We can transform this data into a *likelihood function*, i.e., a joint probability of the observations given as a function of the parameters of a model of the underlying data generation process. In our example, a binomial probability density function (PDF) is a reasonable *initial* model (i) assuming that the pre-deployment safety case provides the argument and evidence that testing is representative of actual operations (as would likely be necessary), and (ii) since a binomial distribution models the sequence of test results. Thus, supposing that over n taxi operations conducted in the operating context of Fig. 2, there were x failures to disengage ML-based perception on demand. We now have the operational safety metrics opTxLowVisW $= n$, and opPcpDisEngF $= x$, so that opPcpDisEngS $= (n-x) = y$, and the likelihood function is $\mathcal{L}(p|n, y) = \binom{n}{y} p^y (1-p)^x$. As before, p is the unknown probability of a successful disengagement of ML-based perception on a random demand, representing a surrogate measure of barrier integrity.

Bayesian inference gives the formal procedure to update the priors into *posterior* values of barrier integrity (and event probability), which represent what the operational system *currently* achieves. Thus, for our running example, the posterior integrity for B_4 is given by (the proportional form of) Bayes' theorem as $\pi(p|y) \propto \mathcal{L}(p|n,y) \times \pi(p)$. Since the beta prior and the binomial likelihood are a conjugate pair, the posterior has a closed form solution, $\pi(p|y) \sim \text{Beta}(s + y, (t - s) + x)$. The distribution mean, $\mu_{p|y} = (s+y)/(t+n)$, is the updated point estimate of barrier integrity. To get a revised assessment of the operational safety risk level for the system, we propagate the posterior barrier integrity through the risk assessment model underlying the safety architecture.

4.3 Characterizing the Change to Safety Risk

We use *risk ratio* (RR), a metric of relative risk, to quantify the change in operational safety risk. In operation, the RR for a consequence event is the ratio of its current estimated probability of occurrence and the approved baseline. More generally, we will (re)compute the RR for any event of interest in the safety architecture, typically after the operational risk assessment has been revised (as in Sect. 4.2) as the ratio of its updated (i.e., prior or posterior, as appropriate) probability to its (scenario-specific) safety target. Denoting the RR for event E_i by RR(E_i), RR(E_i) > 1 indicates an increase in the safety risk of E_i. Similarly, RR(E_i) < 1 indicates a decrease, while RR(E_i) $= 1$ indicates no change. By itself, RR reflects how effective the safety architecture is in reducing the risk of the identified safety effects.[5] By considering the trend of RR over time, we can construct a powerful SI of practical drift, e.g., by fitting a linear trend line to a temporally ordered sequence of RR values computed over some pre-determined exposure, the sign and magnitude of the slope indicate, respectively, the direction and the rate of the change in safety risk.

4.4 Numerical Examples

We now give some numerical examples to concretize the preceding discussion.

Example 1 (Prior Barrier Integrity). During the development of our running example system and its pre-deployment safety case, assume we have a total of devTxLowVisW = 32 flight tests in which there are devPcpDisEngF = 8 failing tests for the *Perception Failover* barrier. Thus a prior distribution for its integrity is $\pi(p) \sim \text{Beta}(24, 8)$, whose mean is $\mu_p = 0.75$, and variance is $\sigma_p^2 = 0.0057$. The mean gives a point prior value of barrier integrity which we show in the corresponding node in the BTD of Fig. 2.

Example 2 (Scenario-specific Barrier Safety Indicator). Recall that a scenario-specific SI for the *Perception Failover* barrier is $\mathcal{I}_{\text{PF0}}$: opPcpDisEngF $\leq y$ in n taxi operations (Sect. 4.1). As before, opPcpDisEngF measures the number of failed disengagements of ML-based perception in operation. To determine a suitable exposure n and threshold y, consider that a conservative range of values for p that would provide the same, or better, risk reduction performance as its prior mean is the closed interval $[\mu_p + \sigma_p, 1] = [0.8254, 1]$. In other words, observing 8 or more successful disengagements or, equivalently, 2 or fewer failed disengagements on demand of ML-based perception over at least 10 taxi operations conducted in the specified operating context would validate the safety performance of the barrier. Thus, here, $n = 10$ and $y = 2$.

Example 3 (Likelihood of Data and Posterior Integrity). After system deployment, suppose that to improve runway utilization, the control in B_4 (see Fig. 2) is relaxed such that ML-based perception is disengaged after a larger distance (or duration) of position deviation than what was specified in the safety architecture. The metric that records the number of failed disengagements in operation, opPcpDisEngF (Sect. 4.2), includes violations of the barrier requirement as initially specified, which itself includes violations of the barrier requirement after operational modification. That is, the operational safety metric *should not be modified* even though the barrier function has been operationally changed. Supposing opPcpDisEngF = 4 violations have been observed over opTxLowVisW = 10 taxi operations. Given this data, the likelihood function is $\mathcal{L}(p|6) = \binom{10}{4} p^6 (1-p)^4$, and the posterior distribution is $\pi(p|6) \sim \text{Beta}(30, 12)$, whose mean is $\mu_{p|6} = 0.7143$ and variance is $\sigma_{p|6}^2 = 0.0047$.

---

[5] RR has also been used as a development safety metric, e.g., in designing aircraft collision avoidance systems [9].

Example 4 (Operational Safety Risk Update). We assume prior data is available (from characterizing the ODD [14] for the autonomous taxiing function) on how often runway markings are obscured during taxiing due to runway surface and weather conditions. Hence we can give a prior distribution, say $\pi(E_2) \sim \text{Beta}(10, 190)$, whose mean is the prior point estimate $\text{Pr}(E_2) = 0.05$. Similarly, let $\text{Pr}(E_1) = 0.05$. Given these priors and the barrier integrity values as in Fig. 2, the prior probability of the consequence event is $\text{Pr}(E_4) = 1.5998 \times 10^{-5}$ corresponding to an RRL of 4D(Low). We recall from Example 3 that 4 barrier violations were observed in 10 taxi operations. Hence E_2 must have occurred on $z = 4$ occasions for B_4 to have been invoked and have failed on demand. Thus, we may reasonably model this event as a Bernoulli process with a binomial PDF as the likelihood function for the observed data. Thus, the posterior distribution over E_2 is $\pi(E_2|z) \sim \text{Beta}(14, 196)$ so that $\mu_{E_2|z} = 0.0667$ is the point posterior for $\text{Pr}(E_2)$. Propagating both the posteriors for E_2 and B_4 through the risk assessment model of the safety architecture [8], we get the updated prior $\text{Pr}(E_4) = 2.386 \times 10^{-5}$ for the consequence event. The corresponding RRL remains unchanged suggesting that the operational modification to B_4 may be acceptably safe.

Example 5 (Safety Risk Level Change and Practical Drift). The risk ratio for the consequence event E_4 given the change to barrier B_4 (as in Example 3) is $\text{RR}(E_4) = {}^{2.386}/_{1.5998} \approx 1.49$. Thus, despite an unchanged risk level (see Example 4), the RR indicates increasing safety risk. Now, further suppose that to improve runway utilization, a greater deviation in CTE from the stated bounds is operationally admitted (see the top event E_3 in Fig. 2). Consequently barrier B_5 needs to be relaxed to be invoked after a longer duration than specified (see Fig. 2). Suppose that the posterior integrity computed from operational safety metrics (omitted here due to space constraints) is $\text{Pr}(B_5) \approx 0.96$. In this case, the revised prior for E_4 is $\text{Pr}(E_4) \approx 2.4 \times 10^{-4}$, the revised RRL is 4C(Medium), and $\text{RR}(E_4) \approx 10$. The updated RRL now violates the TLOS even if no safety effects may have been observed. Moreover, the modifications to the barriers B_4 and B_5 are at least *an order of magnitude more likely* to result in a lateral runway overrun, indicating an appreciable increase in safety risk relative to the approved baseline, and suggests practical drift.

5 Towards Formal Foundations

As mentioned earlier (Sect. 2), we want to formalize a notion of *consistency* between the static portion of the safety case (i.e., its assurance artifacts, see Fig. 1) and the collection of indicators that constitute the SMB. The safety metrics and indicators represent the objectively quantifiable content referenced in the arguments and the safety architecture, which in turn provide the justification for how the metrics and indicators collectively provide safety substantiation. Although operational safety measurement entails updating and revising the risk assessment (Sect. 4), changing the SMB may not be necessary. However, in situations where replacing, modifying, or adding metrics and indicators is required—e.g., to reflect new observable phenomena in the environment—the SMB will change and so would the associated assurance artifacts to retain consistency. Note that currently we are not considering changes that would entail modification of the safety architecture (e.g., replacing a barrier). Hence we exclude that from our notion of consistency for now and focus on consistency with arguments.

We can achieve this consistency if the argument structure reflects the risk reduction rationale implicit in the safety architecture. That is, the form of the argument structure proceeds from all terminating consequence events in the safety architecture, working recursively backwards (i.e., leftwards) to all initiating (leftmost) threat events. Thus, each level of the argument has the following form: all consequence events are acceptably mitigated (i.e., the residual risk level meets the allocated TLOS), *which is supported by the argument that*: all their identified precursor events (causes) are acceptably mitigated, *which is supported by the argument that*: (a) all applicable barriers are operational and effective, and (b) all causes have the stated probability of occurrence. In a GSN representation of this argument, the leaves are *solution nodes* [20] that have the following *evidence assertion*: the initiating threat has the stated (assumed) probability.

Thus, the overall argument states that if the barriers are effective and operational, and the events have the assumed probabilities, then the consequences have acceptable risk levels. Indicators map into the corresponding claims of barrier effectiveness and event probabilities, serving to monitor that those values are within the required limits.

Now we briefly outline how to place this consistency on a more rigorous basis. Let **Arg** and **SMB** represent the sets of well-formed arguments and SMBs, respectively, and define mappings $F : \mathbf{Arg} \rightarrow \mathbf{SMB}$ and $G : \mathbf{SMB} \rightarrow \mathbf{Arg}$, such that F extracts the associated indicators from an argument, and G embeds an SMB into a skeleton argument of the form outlined above. Then we require that $F;G \leq I$ and $G;F = I$ (where I is the identity mapping), where arguments are ordered by *refinement*. The first inequality ensures that the argument contains the necessary rationale for the SMB, with the refinement allowing that the argument can contain additional reasoning; the second ensures that all quantifiable components of the argument are represented in the SMB.

6 Concluding Remarks

We have a preliminary implementation of the SMB in AdvoCATE that currently supports the following functionality: real-time import of data (i.e., measures) from multiple data sources (simulations or feeds from external sensors); computation of derived metrics and indicators over multiple data runs; and tracing to assurance artifacts (events and barriers in the safety architecture, and goals and assumptions in the safety arguments). We display indicators and the associated assurance artifacts in a dynamically updated table (Fig. 4 shows an example) that highlights when the conditions on the indicator thresholds have been met (in green) or have not been met (in red). A dashboard (not shown) allows selection between the various metrics of the SMB with charts displaying real-time updates of their values as well as other dynamically updated risk status, such as hazards ordered by risk level, and barriers ordered by integrity.

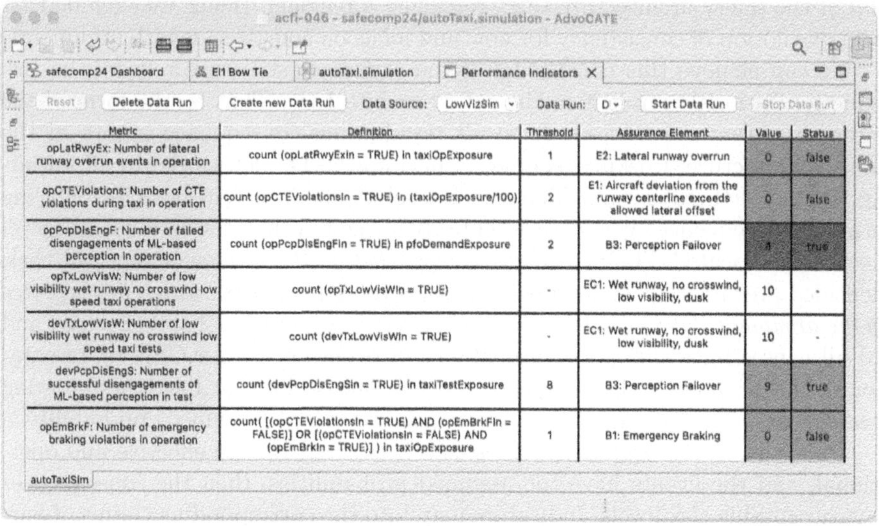

Metric	Definition	Threshold	Assurance Element	Value	Status
opLatRwyEx: Number of lateral runway overrun events in operation	count (opLatRwyExIn = TRUE) in taxiOpExposure	1	E2: Lateral runway overrun	0	false
opCTEViolations: Number of CTE violations during taxi in operation	count (opCTEViolationsIn = TRUE) in (taxiOpExposure/100)	2	E1: Aircraft deviation from the runway centerline exceeds allowed lateral offset	0	false
opPcpDisEngF: Number of failed disengagements of ML-based perception in operation	count (opPcpDisEngFIn = TRUE) in pfoDemandExposure	2	B3: Perception Failover	3	true
opTxLowVisW: Number of low visibility wet runway no crosswind low speed taxi operations	count (opTxLowVisWIn = TRUE)	-	EC1: Wet runway, no crosswind, low visibility, dusk	10	-
devTxLowVisW: Number of low visibility wet runway no crosswind low speed taxi tests	count (devTxLowVisWIn = TRUE)	-	EC1: Wet runway, no crosswind, low visibility, dusk	10	-
devPcpDisEngS: Number of successful disengagements of ML-based perception in test	count (devPcpDisEngSIn = TRUE) in taxiTestExposure	8	B3: Perception Failover	9	true
opEmBrkF: Number of emergency braking violations in operation	count([(opCTEViolationsIn = TRUE) AND (opEmBrkFIn = FALSE)] OR [(opCTEViolationsIn = FALSE) AND (opEmBrkIn = TRUE)]) in taxiOpExposure	1	B1: Emergency Braking	0	false

Fig. 4. AdvoCATE screenshot: Table of safety indicators for the example system in Sect. 3.

The goal of managing practical drift has mainly informed our choice of safety metrics and indicators. We plan to leverage the *Goal Question Metrics* (GQM) approach [2] to define additional metrics suitable for other dynamic assurance goals, e.g., improving functional performance whilst maintaining safety.

A binomial likelihood may be only initially appropriate for certain kinds of measurement data. Indeed, as more data is gathered, the preconditions for using a binomial PDF need to be reconfirmed. As such, it may be necessary to use other PDFs for the likelihood of the data, along with numerical methods for Bayesian inference. Our choice of beta priors is motivated, in part, by computational convenience, its flexibility to approximate a variety of distributions, and the domain-specific interpretation of the distribution parameters in different safety metrics. Although we represent the uncertainty in barrier integrity and event probability by specifying their distributions in the theoretical framework, our prototype implementation currently represents and propagates their point values (i.e., the distribution means) for both the pre-deployment risk assessment, and the revisions of the operational risk assessment. We plan to refine this approach by also propagating the uncertainties through the risk assessment model so as to quantify the corresponding uncertainty in the residual risk of the safety effects of interest. By so doing, we aim to ground the quantification of assurance in safety measurement.

Since TLOS is typically assigned to rare events, legitimate concerns can arise about the credibility of using quantitative methods as in this paper. Though we have yet to explore how *conservative Bayesian inference* [19] could be used in our approach, relative risk metrics such as risk ratio (RR) are a step towards circumventing those concerns.

Practical drift is distinct from *operating environment drift* in that the former results from changes within the system boundary, whereas the latter occurs outside that boundary. We reflect the assumptions about the operating environment in the pre-deployment safety case, for example, as the prior probabilities (conditional on the operating context) associated with the threat events. We can reflect environment drift via the posterior distributions of the corresponding event probabilities updated by operational safety metrics associated with the respective events (see Sect. 4.2, and Example 4). We additionally distinguish *runtime risk assessment* [1], from the update and revision of operational risk as described in this paper: the former occurs during the shorter time span of a mission (e.g., during a taxi operation), whereas the latter occurs over longer time intervals, between missions, and through the lifecycle of the system (e.g., over multiple taxiing operations, possibly involving an aircraft fleet).

The numerical examples (Sect. 4.4) have described a scenario-specific application of our approach, where the event probabilities and barrier integrities are *conditional* on the operating context. For a system-level characterization of how operational risk changes, we must consider the *marginal* probabilities and integrities in the overall safety architecture that composes different risk scenarios. However, we have not considered it in this paper, and it is one avenue for future work.

To further develop our proposed dynamic assurance framework we aim to explore how by thresholding, ranking, and comparing RR under changes made to individual mitigations or their combinations, we may infer: (i) which mitigations may be optimized for system performance whilst maintaining safety (possibly necessitating a change to the safety architecture itself); and, in turn, (ii) which system and safety case changes may be necessary. Some changes may be automated while will induce *tasks* requiring manual attention [5]. Additionally, we aim to define a tool-supported methodology on top of the main components of the framework. This will involve defining and formalizing the methods and procedures to decompose and allocate safety targets, derive safety indicators, and close the safety assurance loop, i.e., maintain consistency of the arguments with the SMB) through targeted changes to the system and its safety case.

Observations of system operations constitute one specific form of evidence that we can use to reason about system safety. We seek to systematize this through a notion of *evidence requirement* that will also cover *static* data. We are also extending the metrics expression language to express trends, although work remains to integrate it into our methodology and to relate it to the concept of *safety objective*. A need to update the SMB, e.g., modify indicators and possibly their thresholds, accompanies operational safety measurement. We aim to better understand the principles that underlie those modifications and, subsequently, implement the corresponding tool features. However, practically deploying this framework will necessitate harmonizing with existing safety management system (SMS) [10] infrastructure, whilst carefully considering the roles of different stakeholders in safety performance monitoring, measurement, and assurance.

Acknowledgments. This work was performed under Contract No. 80ARC020D0010 with the National Aeronautics and Space Administration (NASA), with support from the System-wide Safety project, under the Airspace Operations and Safety Program of the NASA Aeronautics Research Mission Directorate. The United States Government retains and the publisher, by accepting the article for publication, acknowledges that the United States Government retains a non-exclusive, paid-up, irrevocable, worldwide license to reproduce, prepare derivative works, distribute copies to the public, and perform publicly and display publicly, or allow others to do so, for United States Government purposes. All other rights are reserved by the copyright owner.

References

1. Asaadi, E., Denney, E., Menzies, J., Pai, G., Petroff, D.: Dynamic assurance cases: a pathway to trusted autonomy. IEEE Comput. **53**(12), 35–46 (2020)
2. Basili, V., Caldiera, G., Rombach, D.: Goal Question Metric Paradigm, pp. 528–532. Encyclopedia of Software Engineering, John Wiley & Sons, Inc., 2nd edn. (1994)
3. Busch, A.C.: Methodology for Establishing a Target Level of Safety. Technical Report DOT/FAA/CT-TN85/36, US DOT, FAA Technical Center (1985)
4. Calinescu, R., Weyns, D., Gerasimou, S., Iftikhar, M.U., Habli, I., Kelly, T.: Engineering trustworthy self-adaptive software with dynamic assurance cases. IEEE Trans. Softw. Eng. **44**(11), 1039–1069 (2018)
5. Denney, E., Habli, I., Pai, G.: Dynamic safety cases for through-life safety assurance. In: 37th International Conference on Software Engineering - Vol. 2, pp. 587–590. (2015)
6. Denney, E., Johnson, M., Pai, G.: Towards a rigorous basis for specific operations risk assessment of UAS. In: 37th IEEE/AIAA Digital Avionics Systems Conference (2018)
7. Denney, E., Pai, G.: Tool Support for Assurance Case Development. J. Autom. Softw. Eng. **25**(3), 435–499 (2018)
8. Denney, E., Pai, G., Whiteside, I.: The role of safety architectures in aviation safety cases. Reliab. Eng. Syst. Saf. **191**, 106502 (2019)
9. Edwards, M., Mackay, J.: Determining required surveillance performance for unmanned aircraft sense and avoid. In: 17th AIAA Aviation Technology, Integration, and Operations (ATIO) Conference. AIAA 2017-4385 (2017)
10. FAA Air Traffic Organization: Safety Management System Manual (2022)
11. US Department of Transportation, FAA: Safety Risk Management Policy. Order 8040.4C (2023)
12. Hawkins, R., Ryan Conmy, P.: Identifying run-time monitoring requirements for autonomous systems through the analysis of safety arguments. In: Guiochet, J., Tonetta, S., Bitsch, F. (eds.) Computer Safety, Reliability, and Security. SAFECOMP 2023. LNCS, vol. 14181. Springer, Cham (2023). https://doi.org/10.1007/978-3-031-40923-3_2

13. International Civil Aviation Organization (ICAO): Safety Management Manual (Doc 9859), 4 edn. (2018)
14. Kaakai, F., Adibhatla, S., Pai, G., Escorihuela, E.: Data-centric operational design domain characterization for machine learning-based aeronautical products. In: Guiochet, J., Tonetta, S., Bitsch, F. (eds.) Computer Safety, Reliability, and Security. SAFECOMP 2023. LNCS, vol. 14181. Springer, Cham (2023). https://doi.org/10.1007/978-3-031-40923-3_17
15. Koopman, P.: How Safe is Safe Enough? Measuring and Predicting Autonomous Vehicle Safety. 1st edn. (2022)
16. Ladkin, P.: Evaluating software execution as a Bernoulli process. Saf. Crit. Syst. eJournal **1**(2) (2022)
17. Reich, J., Trapp, M.: SINADRA: towards a framework for Assurable situation-aware dynamic risk assessment of autonomous vehicles. In: 16th European Dependable Computing Conference (EDCC), pp. 47–50 (2020)
18. Schleiss, P., Carella, F., Kurzidem, I.: Towards continuous safety assurance for autonomous systems. In: 6th International Conference on System Reliability and Safety (ICSRS 2022), pp. 457–462 (2022)
19. Strigini, L.: Trustworthy quantitative arguments for the safety of AVs: challenges and some modest proposals. In: 1st IFIP Workshop on Intelligent Vehicle Dependability and Security (IVDS) (2021)
20. The Assurance Case Working Group (ACWG): Goal Structuring Notation Community Standard Version 3. SCSC-141C (2021)
21. Trapp, M., Weiss, G.: Towards dynamic safety management for autonomous systems. In: 27th Safety-Critical Systems Symposium (SSS), pp. 193–204 (2019)

Safety Invariant Engineering for Interlocking Verification

Alexei Iliasov[1], Dominic Taylor[2], Linas Laibinis[3(✉)], and Alexander Romanovsky[1,4]

[1] The Formal Route Ltd., London, UK
[2] Consilium Aquis Sulis Ltd., Bath, UK
[3] Institute of Computer Science, Vilnius University, Vilnius, Lithuania
linas.laibinis@mif.vu.lt
[4] School of Computing, Newcastle University, Newcastle upon Tyne, UK

Abstract. The paper discusses our work on formal static verification of interlocking functional safety via inductive safety invariants. The comprehensiveness and fidelity of verification are determined by the scope and adequacy of the safety invariants in question. This becomes a central issue when verification is done in industrial settings, as engineers need to know exactly how the invariants are related to the safety standards, which invariants are verified and, in case of violations, in what specific ways they fail. In our work, formal verification relies on the SafeCap toolset which supports fully automated verification by mathematical proof. The development of safety invariants is a critical part of its design. The main contribution of the paper is the definition of a systematic engineering method for this development and its core stages: invariant elicitation, false positive reduction, reporting all possible violations, and regression testing. We explain how these stages are carried out and which, if any, changes in the toolset they require. The method has been continuously and successfully used in the recent improvements and the extensions of SafeCap while the technology has been applied in numerous live signalling projects.

1 Introduction

Railway Interlocking. Effective signalling is essential to the safe and efficient operation of any railway network. Signalling allows trains to move only when it is safe for them to do so and locks moveable infrastructure, such as the points that form railway junctions, before trains travel over it.

At the heart of any signalling system there are one or more interlockings. These devices constrain authorisation of train movements as well as movements of the infrastructure to prevent unsafe situations arising. One of the earliest forms of computer-based interlocking was the Solid State Interlocking (SSI) [1], developed in the UK in the 1980s. SSI is the predominant interlocking technology used on UK mainline railways. It also has applications overseas, including in Australia, New Zealand, France, Egypt, India and Belgium. Running on bespoke

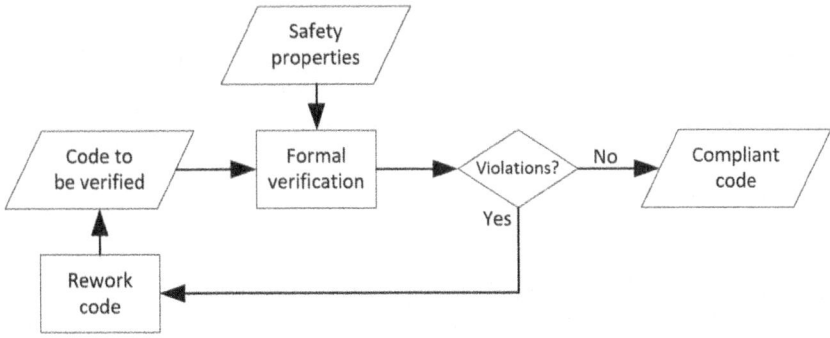

Fig. 1. SafeCap verification process

hardware, SSI software consists of a generic application (common to all signalling schemes) and site specific geographic data. The latter configures a signalling area by defining site specific rules, concerning the signalling and other equipment that the interlocking must obey. Despite being referred to as data, a configuration resembles a program in a procedural programming language and is iteratively executed in a typical loop controlling the signalling equipment.

SafeCap Verification. This paper discusses our work on SSI safety verification in the context of industrial verification of railway signalling interlocking systems. We have been developing and using the SafeCap technology [2] in the verification of live mainline railway projects in the UK for several years. SafeCap provides extra checks of interlocking safety and can detect problems earlier in a development cycle to avoid re-work. It has been approved by Network Rail for use to ensure that automated interlocking verification requirements as set out in Network Rail Standard NR/L2/SIG/11201 Module B11 [3] are met.

SafeCap is an industrial-strength tool that delivers a fully automated, scalable verification of modern SSI systems by mathematical proof. Our earlier paper [2] discusses the motivations for its development, the main technical decisions behind, the research advances made in order to support them, and the advantages that this technology delivers to signalling companies. The tool inputs the digital layout and the SSI interlocking data (code) for a given area and outputs the diagnostics reports which either confirm that the input data possess all safety properties or provide a comprehensive information about all detected property violations in terms of the input data, see Fig. 1.

The safety properties (safety invariants) are formal conditions that the input data have to meet to guarantee the interlocking safety. They play a crucial role in the operation of the SafeCap toolset. These properties are developed by our engineers in close cooperation between railway signalling and formal verification experts. During the initial deployment of SafeCap in industry we came to realise the importance and the intricate complexity of this development.

The main contribution of this paper is in presenting a practical approach to engineering a set of required safety properties in a systematic and continuous manner. The novelty is in the identification of several core steps of this method and in proposing sound and practical solutions to supporting them. Our recent papers [4] and [5] discuss the ongoing work on SafeCap improvement and certification, and provide the latest overview of its industrial applications.

2 Safety Invariants

2.1 Requirements

Idealized Case and Real-World Challenges. For the SafeCap tool to work successfully (Fig. 1) the safety properties need to be given in a formal notation (based on the first order logic and the set theory) and presented as an input to the formal verification process alongside the code to be verified. Any violations found are addressed by modifying the code and repeating the verification process until it is free from violations.

However, industrial application of formal verification presents challenges that are not encountered in this idealized case:

- safety properties are not available in a formal notation and rarely given in isolated, concise fashion; they need to be elicited from various textual sources;
- coding efficiencies can lead to inter-dependencies between safety properties, which need to be accounted for to avoid false positives;
- in exceptional cases, generic safety properties can be intentionally violated where complying with them is not be reasonably practicable;
- delivery time and cost pressures require rework cycles to be minimized.

Real-World Requirements. In our work we identified the following core requirements for developing properties:

1. they must correctly represent required interlocking behaviour;
2. they must be useful for identifying errors;
3. they must enable comprehensive unambiguous reporting/diagnostics of found violations;
4. any modifications to properties must ensure that they remain correct and consistent.

We routinely verify projects with tens of thousands of lines of code often resulting in hundreds of thousands of proof obligations – conjectures that must be discharged to establish satisfaction of all safety invariants. At this scale it is unfeasible to employ any form of interactive proof or manual inspection of failed proofs. This in turn requires a high degree of confidence in the prover to do its work correctly and leave failed proof obligations in a state suitable for reporting.

One of the crucial motivations for the development of SafeCap has been to deploy it with the minimal disruption to the existing industrial practices [2]. This in particular meant that we could not insist on changing existing practices or ask signalling engineers to provide (i.e., formalise) safety properties for us. Instead, we had to develop these properties from textual descriptions in the client's standards rather than receive them directly as logical statements in formal notation.

2.2 Current Solutions

Our analysis shows that the existing approaches to developing safety invariants are not applicable in our context.

Many approaches to formal verification focus on verification of only few (often one!) properties without either relying on any systematic methodologies for engineering them or providing any comprehensive justification for selecting them.

Very often the existing approaches assume that there are complete unambiguous up to date system requirements, including fully defined safety conditions, from which a complete set of safety properties could be easily deduced. Unfortunately this is not often the case in practice. In particular, the mainline railway systems we are dealing with are not closed systems: they always interface with other systems and include a large number of legacy components and subsystems. For this type of systems (e.g. the signalling of a large railway station or junction) it will not be practical to develop and fix the full set of requirements. This is why requirements always refer to compliance with standards.

There are no ready solutions to support continuous work on developing and improving the set of safety properties. The prevalent understanding in both research and application is that the properties are always fully stated and checked in advance, so the assumption is that there is no need to extend or modify them. Unfortunately this is not how the real systems are developed. Our experience in designing, extending and deploying SafeCap shows that developing a full set of formal safety conditions for verification of real signalling systems is a daunting task that needs a long term effort from the verification team. This stems from the complexity of the systems, the legacy of their components, the large number of stakeholders, and the fact that the full technology upgrade in this industry could be up to 30–40 years.

There are only few papers that propose systematic solutions for developing a safety invariant. For example, paper [6] discussed an approach to generating safety conditions from signalling principles using a topological model of a railway yard. This is an interesting approach but we need a solution that links the safety conditions to the generally accepted norms used by all industrial stakeholders.

3 Property Engineering

The proposed engineering method, addressing the most challenging and pressing issues we identified during the commercial use of SafeCap, consists of the following four elements:

1. Elicitation of formal safety properties to be verified (i.e. safety invariants) using the industrial standards;
2. Substantial reduction of the false positives in the verification outcomes;
3. Reporting all possible violations based on precise and rigorous definition of what each property violation is;
4. Regression testing of the developed safety invariants.

They cover the most critical steps of property engineering and support the stakeholders (including, the verification tool developers) in systematic and continuous engineering of the safety properties to be used in large scale industrial development of safety-critical systems. Each constituent part of the method consists of the methodological steps to be conducted by the property designer and, where necessary, the new functionalities of the verification toolset.

3.1 Eliciting Safety Properties from Standards

In the context of our railway signalling interlockings, opportunities to specify a complete set of requirements from scratch are rare and typically only occur on new-build, stand-alone metro lines. Most railway signalling interlockings are constrained by legacy considerations:

- interfacing to a wider railway network;
- supporting operational services whilst a new signalling system is introduced;
- demonstrating the safety of the new system, which is easier if it resembles an old one;
- integrating with established processes, procedures and staff competency.

As a result, safety properties tend to be contained in textual standards that have evolved over many decades through design, risk assessment, operational experience and lessons learnt from accidents. Updating or withdrawing a standard is a time consuming process, because many different stakeholders could be affected by the change and so to be consulted. There is therefore a tendency for new requirements to be embodied in a plethora of overlapping standards. By way of example, the 156 safety properties currently defined for SafeCap are derived from textual clauses in 22 different standards. This is estimated to represent about a third of all properties contained in these standards to which formal verification could be applied, targeted at those properties associated with the most commonly used functionality.

Standards are published by multiple organisations. In the case of GB mainline interlocking, the organisations are Network Rail and Rail Safety and Standards Board (RSSB). Standards cover a range of topics, some of which pertain to interlocking safety properties, others of which do not. For example, Network Rail standard NR/L2/SIG/11201 'Interlockings - Electronic Interlocking Guidelines' [3] contains a few safety properties that must be verified by automated tools alongside guidance about different types of electronic interlocking.

Identification of safety properties pertaining to signalling interlockings requires a thorough review of standards catalogues to identify standards pertaining to signalling interlocking functionality [7,8]. This is followed by a review of those standards to elicit the specific safety properties applicable to interlocking functionality that falls within the scope of SafeCap verification.

SafeCap does not attempt to verify all interlocking functionality, but rather functionality that is most frequently used and hence for which verification benefits most from an automated approach. The scope of verification, and hence

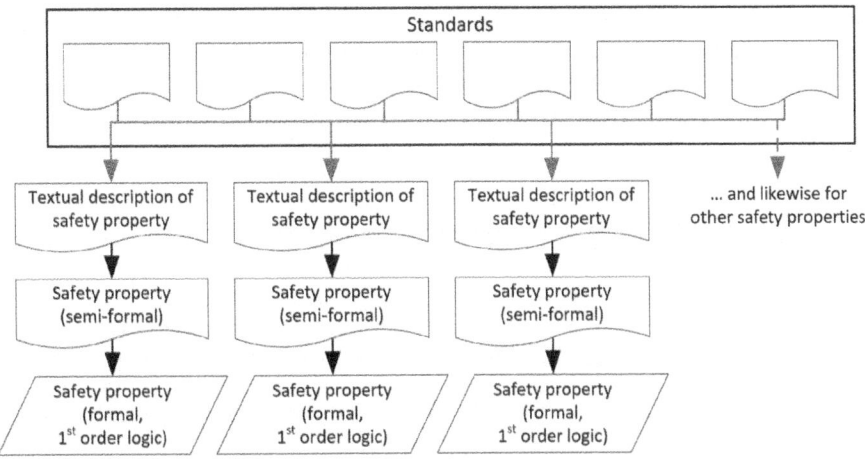

Fig. 2. Traceability

the standards clauses from which safety properties are derived, is continually expanding: initially the scope was constrained to only functionality for which automated verification was explicitly required by standard; it has since expanded to cover the majority 'locking level' functionality related purely to changes in internal states within signalling interlocking; further extensions are covering processing of inputs and generation of outputs by the interlocking. Nonetheless, many clauses in standards are likely to remain permanently outside the scope of SafeCap verification, for example:

- safety integrity of interlocking hardware and generic software;
- accuracy of signalling plans used as inputs into SafeCap;
- correct wiring of physical equipment to the interlocking;
- seldom-used and site-specific functionality for which manual verification is more cost effective than automated verification.

The correctness and completeness (in terms of the scope of SafeCap verification) of safety properties is demonstrated through traceability to the applicable clauses in standards, see Fig. 2. This in turn is subject to independent review by internal and external signalling subject matter experts.

Traceability does not demonstrate the correctness of the standards themselves, which is taken as axiomatic, rather that each safety property correctly embodies the standards clauses that it traces to. As the SafeCap safety properties only embody a subset of interlocking requirements contained within standards, which themselves are only a subset of all the requirements contained in the respective standards, all SafeCap reports are issued with a clear description of exactly which safety properties have been verified. No claims can be made that SafeCap will find violations of safety properties that it is not configured to verify.

An example of traceability of safety properties to standards is given in Table 1. The free text of the standard, generally representing a safety requirement in a technology-agnostic manner, is first written in a semi-formal manner expressing the safety property to be verified in terms of the interlocking data (code) language (currently, the SSI data language). This in turn is expressed in formal, first-order logic as a safety invariant for automatic verification by SafeCap.

Table 1. Example safety properties

Standard			
NR/L2/SIG/11201/ Mod B11	Issue 5	clause 4.5.1 a)	The tools shall be capable of establishing that points cannot be called to move in a direction contrary to any sub route locking that is applied;
First traced safety property: SR/point-simple/normal			
Title			
Whenever individually controlled points are commanded normal then every sub route over them in the reverse position is free.			
Semi-formal			
for every individually controlled points normal command Pxx cn			
it holds that every sub route over the points in the reverse position is free Uxx f			
Formal			
forall p : $Node$ $\quad p/ : dom(\text{``}point : merged\text{''})$ and $\quad point\_c(p)! = \text{``}pointc'p\text{''}(p)$ and $\quad point\_c(p) == NORMAL$ => forall sr : $SubRoute$ $\quad sr : \text{``}subroute : pointreverse\text{''} [\text{``}Node.base\text{''} [p]]/\backslash \text{``}SubRoute.ixl\text{''}$ => $sr/ : subroute\_l$			
Second traced safety property: SR/point-simple/reverse			
As for SR/point-simple/normal with 'normal' replaced by 'reverse' and vice versa			

As a result of the overlapping nature of standards, a given safety property may trace to multiple clauses in standards. A single clause in a standard may also trace to multiple safety properties that all need to be proven to demonstrate compliance with the clause. In the example in Table 1, separate safety properties are needed for commanding points normal and for commanding points reverse in order to demonstrate compliance with the standard clause.

The manner in which clauses from standards are implemented in practical data can also influence the number of safety properties needed to demonstrate compliance. For example, coding efficiencies realised for two sets of points configured to always move simultaneously ('combined points') mean that it is not possible to verify the standard clause in Table 1 using only the safety properites in Table 1. In practical SSI data, rather than explicitly test route locking over both sets of points in combined points, only the last section of route locking ('sub route') over the points is tested in each direction of travel. The other sub route is inferred to be free on the basis that both sub routes are always locked together and only release sequentially as illustrated in Fig. 3.

To prove that combined points can only move when free of route locking, without false positives, it is necessary to prove four safety properties:

- that the interlocking cannot commanded combined points normal when the last sub route over them is locking them reverse;
- that the interlocking cannot commanded combined points reverse when the last sub route over them is locking them normal;
- that sub routes over the points are always locked together;
- that sub routes always release sequentially in the direction of travel.

In some cases, standards allow for safety properties to be overridden subject to site-specific risk assessment. For example, the principle that two trains should not be signalled in opposite directions into the same track section can be overridden to allow one train to couple to the front of another, but only at low speed in specific locations.

3.2 Reducing the Number of False Positives

Three types of false positives are encountered in SafeCap analysis:

1. false positives where SafeCap correctly reports violations of formal safety properties, but those violations do not represent actual violations of the standards from which those safe properties are derived;
2. false positives where SafeCap reports violations of formal safety properties for scenarios that cannot occur in practice;
3. false positives where SafeCap reports violation of formal safety properties where none exist.

The first type of false positives arises from formal safety properties being a conservative simplification of safety properties expressed in standards: actual violations of the original property in standards are a subset of violations of the formal safety property. This occurs, because railway signalling principles are often relaxed in specific circumstances where safety risk is controlled through other means. For example, most of the time it is a requirement that two signalling routes cannot simultaneously be locked in opposite directions over the same section of track. However, this principle can be relaxed in the specific case of low speed shunting moves on non-passenger lines in order to allow a vehicle to shunt backwards and forwards without the signaller needing to set a route each time it changes direction. Refinement of safety properties can reduce or even eliminate this type of false positive at the expense of the properties themselves becoming more complex and hence harder to demonstrate correct traceability.

The second type of false positive arises from the effect of safety over approximation in the safety invariant. The method of static verification based on preservation of safety invariant is not able to inherently differentiate between violations

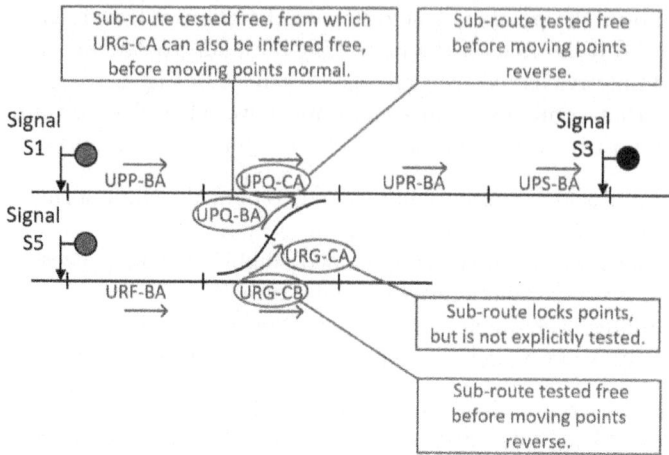

Fig. 3. Coding efficiencies for combined points

that are reachable in realistic setting and those that are not. Thus the tool might report a safety violation corresponding to a scenario that can be shown to be outright unreachable by considering bigger part of interlocking state and possibly other parts of interlocking logic. We pursue two directions to address safety over approximation: more elaborate safety properties that take into an account more details of reachable interlocking cases; emphasising the need for a *defensive style* of interlocking design so that argument necessary to satisfy pertinent safety principles can be localised in its scope to a single request in a single execution cycle. The former approach typically requires complex interdependencies between safety properties in order to demonstrate, through one safety property, that combinations of pre-conditions assumed by another safety property are genuinely unreachable. Such reasoning is inherently challenging even with the support of automated proving tools like SafeCap and likely to be even more so for manual authors and reviewers of the code. The second approach is therefore preferable as a long term solution to the issue.

The third type of false positives arises where theorem proving for the first order logic is undecidable in the sense that there cannot be a terminating procedure that returns whether a conjecture is true or false. We must plan for the possibility that some proof obligations are not discharged because the prover is not powerful enough. However, although the technical limitation underlying this type of false positive is insurmountable, its practical effect has diminished much over time and the prover is now so well tuned that the number of correct but unproved conjectures is vanishingly small. For example, we did not encounter any false positives of this type in the six commercial projects delivered in the last 6 months of 2023. This has been achieved by regularly refining the strategies used by the prover: reported safety property violations are manually analysed and, where it can be logically demonstrated that the violation is a false positive, the reasoning is incorporated within the proof strategies used by the prover.

Dealing with false positives is critical for the industrial use of SafeCap. The development team analysis all false positives found, categorises them and, if necessary, makes the required adjustments (e.g. in the properties or in the prover) to eliminate them in the future. To conclude this subsection, we should mention that false negatives can happen during the SafeCap verification in the rare cases when the detected violation shields another violation.

3.3 Redefining What the Property Violation Is

The efficient definition of what constitutes a property violation is critical for engineering the safety invariants and, as a result, for providing engineers with comprehensive and useful diagnostic reports.

The mainstream approach to defining what the property violation is to simply demonstrate that the given property does not hold for at least one example. Our experience clearly shows that this is not sufficient for industrial scale verification. The first reason is that the engineers need to have all contexts in which the property does not hold reported and corrected. The second one is that in large and complex systems, like railways, there are often a few exceptional cases where the engineers intentionally violate a general property for operational reasons; for example allowing a train to shunt backwards and forwards at low speed on a non-passenger line without the signaller setting routes for each move.

At the core of our solution, presented earlier [9], is the definitions of individual violations in terms of a unique combination of signalling assets with an atomic section of code by writing the safety properties in a way that clearly distinguishes between different violations of the same property. Following this approach, a unique proof obligation is derived for each combination. The reporting algorithms are configured to group all found and reported violations of the same proof obligation. As a result all individual violations are reported.

Since we started using this approach in live projects we have found Safe-Cap to be particularly beneficial in minimising rework cycles and the manual checks in the industrial projects. Reworking of data to address errors is time-consuming and expensive. Where it requires multiple rework cycles to refine data until all safety properties (i.e., the corresponding safety invariants) are fully proven, formal verification can be prohibitively expensive. By contrast, where it can identify a large proportion of errors in a single verification pass, formal verification becomes valuable as a method for reducing rework as well as for the robust safety assurance that it provides. When the violations are grouped to report each unique violation, all can be addressed in a single rework cycle enabling project delivery efficiencies. Referring to the process shown in Fig. 1, after rework to address the violations found by first pass formal verification, a second pass of formal verification can confirm compliant code without the need for further rework.

3.4 Regression Testing of Developed Safety Invariants

Adding a new safety invariant can be a challenging proposition in an industrial setting as any its deficiency might result in missing an error in checked code. The worst possible scenario is when a safety invariant predicate is an incidental tautology (i.e., holds in all practical scenarios). This would appear as if a check of certain principle is done successfully without any further material to assess and validate. Whereas in model-checking setting one can manipulate the specification being checked to preclude such situation and, if necessary, to deliver a positive indication that the specification is not trivially correct (as one example, see [10]), there is no natural equivalent in safety invariant verification.

The other extreme is a property that reports a large number of false positives rendering the whole verification process unmanageable. Between these two extremes lie more subtle cases of missed safety checks and false positives.

For safety property revision our solution is regression testing checking that the previously found and manually confirmed violations are still present and no new violations appear. This in itself is an involved matter as one must have a persistent way to name violations while still be able to change the corresponding safety invariants and proof rules. Unfortunately, many properties are rarely violated so regression testing alone is not sufficient even for property modification.

Errors could be "seeded" by modifying interlocking source code in a way that should evidently (to a qualified engineer) result in an error. We cannot afford to do this manually and hence there must be a procedure for automated seeding of errors. A well known problem with error seeding is the difficulty of systematically and automatically enacting a change that does trigger a safety error. This section details our approach to this problem together with a novel technique for efficient error seeding.

It is not hard to see that a safety invariant and a procedure to seed errors that always trigger some idealised form of a safety property are two complementary ways to characterise the same safety property. This redundancy can be exploited to increase trust in a safety property by employing diverse notations for these characterisations. In other words, we do not attempt to derive error seeding from a safety invariant but rather have a separate formulation in a conceptually different form using an informal statement of the property as a starting point.

Initial trials with source code error seeding demonstrated that in a naive form its computational cost is prohibitive. In SafeCap, interlocking source code is parsed into the intermediate notation and then symbolically executed to derive a symbolic transition system. The proof obligation generator (for any pair of a safety invariant and a symbolic transition) produces proof conjectures that must be discharged by a theorem prover to demonstrate satisfaction of a global safety property. Symbolic execution has a high computational cost (exponential in limit w.r.t. the number of statements) and it is simply not viable to iteratively execute the described process for thousands of seeded errors.

Since SafeCap relies on the formally defined proof semantics, any source code modification must result in an altered set of proof obligations. Hence, we shall try to reproduce the effect of source code error seeding by directly manipulating proof obligations derived from interlocking code without any seeded errors and the safety invariant under test (see Fig. 4). Focus on such proof obligations solves the issue of modifying parts of logic not pertinent to the considered safety predicate – something that we would have to address if we were to enact changes at the level of a transition system.

A source set of proof obligations (proven with no seeded errors) must not contain any contradiction in hypotheses, e.g., due to an unreachable transition derived from interlocking code, as this could later mask the effect of a seeded error. After such filtering, we should have a sufficient number of proof obligations – typically in tens of thousands (as indicated by a dozen of signalling projects). If filtering delivers no proof obligations to work with, it is in itself a sign of an over-constrained property.

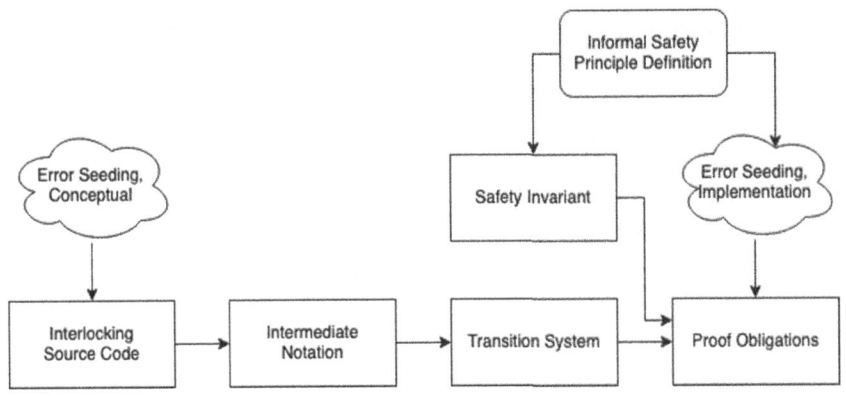

Fig. 4. Source code transformation and error seeding

Manipulating a Proof Obligation. A proof obligation is an term of form $H \vdash G$ where H is the hypothesis and G is the goal, both expressed in the first order logic. Simplistically, H is a symbolic state transition that is being checked and G is derived from a safety predicate. It is convenient to see H as a list of hypotheses $H_1, \ldots, H_n$ with an implicit conjunction operator. In SSI interlockings, most objects have two states and we model such object states as set membership tests. For instance, the fact that some sub route $UAA\text{-}AB$ is locked (not locked) is expressed as predicate $UAA\text{-}AB \in \mathsf{subroute\_l}$ ($UAA\text{-}AB \notin \mathsf{subroute\_l}$). This covers the vast majority of transition pre-conditions for SSI.

Since SSI is a fully deterministic language, transition postconditions are limited to conjunction of equalities of the form $v' = v \cup E_1 \setminus E_2$, where E_1 and E_2 are constant sets of the values to be added or removed. Overall, we expect the following clauses for $H_1, \ldots, H_n$ to appear in a proof obligation:

1. membership clauses $v \in S$, $v \notin S$ or $v \subseteq S$;

2. equality (inequality) clauses $f(v) = c$, $v = c$, $v \neq c$, ...;
3. disjunctive clauses;
4. quantifiers, implications and other forms arising from the SSI axiomatisation and other safety predicates.

By looking at the clause forms 1 and 2, it is possible to deduce, via pattern matching, the sets of objects that are being tested positively or negatively (case 1), added or removed (case 2). Since these forms are prevalent, we will focus on them while describing our approach.

For generality, model variables that we can deduce from the clause forms 1 and 2 shall be referred by indexed set v_i. For instance, v_0 could correspond to concrete model variable subroute_l, v_1 – track_o, and so on.

For the clause form 2, the added and removed values of a model variable v_i are referred to as Z_i^+ and Z_i^-. There is a simple relationship between Z_i^+, Z_i^- and their variable counterpart v_i:

$$v_i = \overline{v}_i \cup Z_i^+ \setminus Z_i^-. \tag{1}$$

Identifier $\overline{v}_i$ stands for the value of v_i in a prior state. For the moment, we only know that it must satisfy the safety invariant.

Clauses of the form 1 typically check variables values in a prior state (i.e., $\overline{v}_i$). Analysing these clauses one can build two sets: values tested to be in a model variable (denoted as $\overline{Z}_i^+$) and values tested to be not in it (denoted as $\overline{Z}_i^-$). The following relates these two sets to an actual prior state of a model variable v_i:

$$\overline{v}_i = x_i \cup \overline{Z}_i^+ \setminus \overline{Z}_i^-, \tag{2}$$

where x_i is some previous (unknown) variable v_i state. Putting (1) and (2) together, we have the following:

$$v_i = x_i \cup (\overline{Z}_i^+ \cup Z_i^+ \setminus Z_i^-) \setminus (\overline{Z}_i^- \cup Z_i^- \setminus Z_i^+).$$

Intuitively, $\overline{Z}_i^+ \cup Z_i^+ \setminus Z_i^-$ and $\overline{Z}_i^- \cup Z_i^- \setminus Z_i^+$ are sets of objects that are known to be added (locked or set) or removed (unlocked, freed) respectively.

To implement error seeding, we shall manipulate the following sets: $\overline{Z}_i^+$ to add or remove positive checks, e.g., remove a check that a track section is clear and add a check that a sub route is locked; $\overline{Z}_i^-$ to add or remove negative checks; Z_i^+ to add commands for locking/setting objects; Z_i^- to add commands for freeing/unsetting objects.

The manipulated sets are inserted at the place of their originals in a proof obligation to produce a mutated proof obligation that, as we hope, now contains a seeded error. Such changes are not simply done at random – we manually define error seeding rules involving model variables and the sets defined above.

Error Seeding Rules Language. As explained, error seeding can be expressed by changing four different sets for each variable. In fact, we can lift the abstraction level and combine manipulation of $\overline{Z}_i^+$ and $\overline{Z}_i^-$, and, separately, of Z_i^+ and Z_i^-, as there is no practical reason for these pairs of sets to have common elements. We introduce the following syntax sugar:

$$\begin{aligned}
\text{pos}(v_i) &\equiv \overline{Z}_i^+ \setminus \overline{Z}_i^- & \text{set of positively checked elements} \\
\text{neg}(v_i) &\equiv \overline{Z}_i^- \setminus \overline{Z}_i^+ & \text{set of negatively checked elements} \\
\text{posp}(v_i) &\equiv Z_i^+ \setminus Z_i^- & \text{set of positively commanded elements} \\
\text{negp}(v_i) &\equiv Z_i^- \setminus Z_i^+ & \text{set of negatively commanded elements}
\end{aligned}$$

where v_i is a model variable. We shall use same notation but now on the left of an assignment operator to denote altering various sets of v_i:

$$\begin{aligned}
\text{pos}(v_i) := S &\equiv (S, \overline{Z}_i^- \setminus S, Z_i^+, Z_i^-) \\
\text{neg}(v_i) := S &\equiv (\overline{Z}_i^+ \setminus S, S, Z_i^+, Z_i^-) \\
\text{posp}(v_i) := S &\equiv (\overline{Z}_i^+, \overline{Z}_i^-, S, Z_i^- \setminus S) \\
\text{negp}(v_i) := S &\equiv (\overline{Z}_i^+, \overline{Z}_i^-, Z_i^+ \setminus S, S)
\end{aligned}$$

Using this notation we can express, for instance, removal of sub route locking tests for some sub routes $UAA\text{-}AB$ and $UAB\text{-}AB$ as pos(subroute_l) := pos(subroute_l) \ $\{UAA\text{-}AB, UAB\text{-}AB\}$).

We call a seeding rule *negative*, if it is expected to cause affected proof obligations to fail, and *positive* otherwise.

Next, consider the following safety property: *for every set route it holds that all sub routes on the route path are locked.* When a set of set routes is empty, the property must obviously hold. We can encode this with a *positive* error seeding rule that must leave any altered proof obligation provable:

$$\text{posp}(\text{route\_s}) := \varnothing$$

This rule captures intuition that any violation could be present only if there is a set route. We can induce a violation by seeding an error that removes all set sub routes:

$$\text{posp}(\text{subroute\_l}) := \varnothing$$

Note that proof obligations would not be generated if route_s or subroute_l are not modified to avoid non-effective error seeding. We can presume that removing some sub route locking should cause the property violation:

$$\text{negp}(\text{subroute\_l}) := \text{negp}(\text{subroute\_l}) \cup \{\text{some}(\text{posp}(\text{subroute\_l}))\}$$

In reality this rule should be refined to only remove sub routes on the paths of a locked route. On the other hand, locking an extra sub route should never result in a failed proof obligation. This justifies the following rule:

$$\text{posp}(\text{subroute\_l}) := \text{posp}(\text{subroute\_l}) \cup \{\text{some}(\text{SubRoute})\}$$

The feedback from error seeding helps us to check intuition behind a property and uncover technical or conceptual shortcomings early on. Systematically writing useful and meaningful rules is not at all easy and we hope to come up with a methodological guidance once we have accumulated enough experience. In practice, the described approach has been already successfully used for revising and adding the SafeCap properties while the tool has been used in several life industrial projects.

4 Discussion and Conclusions

This paper describes practical challenges encountered when engineering safety invariants for verification of industrial code and how these challenges have been overcome in the context of railway signalling interlocking systems.

The first challenge is eliciting formal safety invariants from multiple, textual standards. The correctness and completeness, within a specified scope of verification, of safety invariants can be established through their recorded traceability to clauses in standards with an independent review by subject matter experts. This allows us to define the priorities in our work on extending and improving the verification invariant set. Further evidence of correctness and completeness can be achieved through testing using seeded errors.

The second challenge is avoiding false positives by tailoring safety invariants and the prover to the structure of code being verified. Whilst a very efficient and scalable method of formal verification, static verification based on preservation of safety invariants is not able to inherently differentiate between the violations that are reachable and those that are not. The experience of applying safety invariants lead to refinement of the described verification process to exclude unreachable pre-conditions, to allow exceptional cases in order to avoid false positives, as well as to add proving strategies to avoid failures to verify a safety invariant that is in fact upheld.

The third challenge is providing comprehensive and unambiguous reporting of safety invariant violations. Industrial time and cost pressures require formal verification to find all violations of a safety invariant, so that they can be corrected in a single re-work cycle, rather than simply to report that the invariant has been violated. Reporting all the violations is also important to avoid the risk of a false positive masking a genuine issue. To achieve this, it is necessary to precisely define what constitutes a unique violation and structure safety invariants to uniquely identify and report each of these unique violations (see [9]).

The final challenge is developing an efficient technique for regression testing of the property set when these properties are revised. This is a key to dealing with the properties used in complex industrial systems with a large number of safety conditions, for which complete and rigorous requirements cannot be finalised earlier in the development cycle. This type of technique is critical for successful deployment of verification tools in industry. To address this challenge, we develop a novel technique that helps confirm that, after a revision, all the previously found violations are still present and no new violations appear.

In the past two and a half years the SafeCap safety invariants have been substantially improved and reworked using the engineering method presented in this paper. The number of invariants has grown from 68 to 156. The engineering method presented in the paper proven to be both, efficient and effective. This has allowed us to substantially improve the coverage of the SafeCap verification and the quality of the diagnostics reports delivered to our customers in the eighteen commercial projects completed in 2023 alone.

References

1. Cribbens, A.H.: Solid State Interlocking (SSI): an integrated electronic signalling system for mainline railways. Proc. IEE. **134**(3), 148–158 (1987)
2. Iliasov, A., Taylor, D., Laibinis, L., Romanovsky, A.: Practical verification of railway signalling programs. IEEE TDCS **20**, 695–707 (2023)
3. Network Rail, NR/L2/SIG/11201, Module B11, Interlockings - Electronic Interlocking Guidelines, issue 5, UK (2018)
4. Iliasov, A., Taylor, D., Laibinis, L., Romanovsky, A.B.: Formal verification of railway interlocking and its safety case. In: Proceedings of Safety-Critical Systems Symposium (SSS 2022). Safety-Critical Systems Club, UK (2022)
5. Iliasov, A., Laibinis, L., Taylor, D., Lopatkin, I., Romanovsky, A.: The SafeCap trajectory: industry-driven improvement of an interlocking verification tool. In: Milius, B., Dutilleul, S.C., Lecomte, T. (eds.) RSSRail 2023. LNCS, vol. 14198, pp. 117–127. Springer, Cham (2023). https://doi.org/10.1007/978-3-031-43366-5_7
6. Kanso, K., Moller, F., Setzer, A.: Automated verification of signalling principles in railway interlocking systems. Electron. Notes Theor. Comput. Sci. **250**(2), 19–31 (2009)
7. Rail Safety and Standards Board. Standards Catalogue, UK (2022). www.rssb.co.uk/standards-catalogue
8. Network Rail. NR/CAT/STP/001, Catalogue of Network Rail Standards, Issue 125, 03 September - 02 December, 2022. UK (2022). www.spglobal.com/engineering/en/products/uk-network-rail-standards.html
9. Iliasov, A., Laibinis, L., Taylor, D., Lopatkin, I., Romanovsky, A.: Safety invariant verification that meets engineers' expectations. In: Dutilleul, S.C., Haxthausen, A.E., Lecomte, T. (eds.) RSSRail 2022. LNCS, vol. 13294, pp. 20–31. Springer, Cham (2022). https://doi.org/10.1007/978-3-031-05814-1_2
10. Kupferman, O., Vardi, M.Y.: Vacuity detection in temporal model checking. Int. J. Softw. Tools Technol. Transf. **4**(2), 224–233 (2003)

Assurance Case Synthesis from a Curated Semantic Triplestore

Saswata Paul(✉), Baoluo Meng, Kit Siu, Abha Moitra, and Michael Durling

General Electric Aerospace Research, Niskayuna, NY, USA
{saswata.paul,baoluo.meng,siu,abha.moitra,durling}@ge.com

Abstract. We present a new pipeline for the automatic synthesis of assurance cases in the Goal Structuring Notation (GSN) from a curated semantic triplestore called RACK (Rapid Assurance Curation Kit). RACK allows users to create arbitrary project-specific ontologies to organize and curate certification evidence collected from a heterogeneous set of sources. The arbitrariness of the ontologies makes it challenging to reason about assurance in an automated manner. Our pipeline provides users with the ability to align the arbitrary ontologies to the terminologies GSN standard by creating custom GSN patterns. These patterns are encoded using a special GSN ontology that has the necessary elements to allow such mapping and to also create formal instances of GSN fragments. The pipeline uses these project-specific patterns to generate valid queries to fetch data from RACK's triplestore and instantiate complex GSN trees using the data, where possible, all automatically. Our pipeline is integrated into an open-source tool with an interactive GUI for generating the GSN fragments and analyzing them in a modular fashion.

Keywords: Assurance Case Synthesis · Ontology · Certification · Data Curation · Eclipse IDE · Goal Structuring Notation

1 Introduction

Ensuring the safety, security, and reliability of critical systems is paramount. Assurance cases have emerged in defense, transportation, aerospace, medical device, and energy industries as a powerful framework, and as a supplement to process-based methods, for systematically demonstrating that a system meets its safety, security, and reliability properties. They play a crucial role in building trust among stakeholders by providing a clear, evidence-based, and defensible rationale for the confidence placed in a system. The *Goal Structuring Notation* (GSN) [11] and the *Structured Assurance Case Metamodel* (SACM) [28] are

---

Distribution Statement A - Approved for Public Release, Distribution Unlimited. This research was developed with funding from the Defense Advanced Research Projects Agency (DARPA). The views, opinions and/or findings expressed are those of the authors and should not be interpreted as representing the official views or policies of the Department of Defense or the U.S. Government.

two popular notations for constructing and visualizing assurance cases. In this paper, we focus on the GSN notation which offers a graphical format that facilitates the systematic representation of claims, their supporting evidence, and the relationships among them. The structured nature of GSN provides a visual language for facilitating the systematic representation of goals, evidence, and arguments to effectively convey the dependability of systems.

We present a new pipeline for automatic synthesis of GSN assurance cases from RACK (Rapid Assurance Curation Kit), an evidence curation platform that consists of a semantic triplestore backed by an ontology [4]. RACK is being developed under the DARPA (Defense Advanced Research Projects Agency) Automated Rapid Certification of Software (ARCOS) program for curating software certification evidence. It allows for arbitrary project-specific ontologies that are each specialized to express the verbiage of a project or domain, thereby making it possible to store evidence for any arbitrary project. In most software development projects, there are different types of low-level evidence that trace back to higher-level design and safety goals. Assurance cases can be instantiated using such data that can be traced to one another via the project ontology. In the past, assurance cases would have been constructed manually, which was time-consuming and prone to inconsistencies introduced by human-error. We then attempted to assemble assurance cases by querying RACK with project-specific tools and processing the results [19]. This improved the process, but the capability to automatically synthesize assurance cases in a project-agnostic manner would be more scalable and would reduce the time, effort, and cost involved.

There are several challenges to automating the synthesis of assurance cases from RACK. First, the arbitrary nature of the project ontologies makes it necessary to align the project-specific terminologies with the GSN verbiage so that it can be clearly specified how the ontology classes can be used to create goals, strategies, and solutions. Second, RACK queries are ontology-dependent, so the same set of queries cannot be used across arbitrary ontologies, making it difficult to automate the process of fetching project-specific data from RACK. Finally, a generic pipeline for synthesizing GSN fragments from RACK data that can work with any arbitrary RACK project must be sufficiently configurable to allow engineers to use it across projects. Our pipeline is based on a generic GSN ontology that can be used to formalize GSN fragments and create abstract project-specific *GSN patterns* that can align any arbitrary project ontology to the generic GSN verbiage. Using these project-specific patterns, the pipeline can automatically synthesize queries to fetch data from RACK and construct goals, strategies, and evidence nodes to form GSN trees. The pipeline is integrated into an open-source Eclipse-based toolchain called RITE (RACK Integrated Certification Environment) [21] that provides an interactive GUI for analyzing the synthesized GSN fragments in a modular fashion. To the best of our knowledge, this is the only open-source tool in the literature that can allow fully automated GSN synthesis from a semantic triplestore like RACK which supports arbitrary ontologies.

The rest of the paper is structured as follows: Sect. 2 gives background on RACK; Sect. 3 describes our pipeline; Sect. 4 presents related work; and finally Sect. 5 presents a summary and a discussion on future work.

2 The Rapid Assurance Curation Kit (RACK)

RACK is a data curation system that organizes and stores assurance case evidence. It has an underpinning data model that describes concepts relevant to software development of high assurance systems (*e.g.*, aerospace software). The foundation of the RACK ontology is a provenance model for describing how a piece of evidence came to be—what activities generated the evidence; who was associated with the activity; and what was used as inputs to generate the evidence[1]. We leverage the W3C PROV provenance model [20] shown in Fig. 1 with the fundamental classes and relationships between those classes. We implemented the ontology in the Semantic Application Design Language (SADL) [3,8], a controlled English language with formal semantics that automatically translates to the Web Ontology Language (OWL), another W3C standard [25].

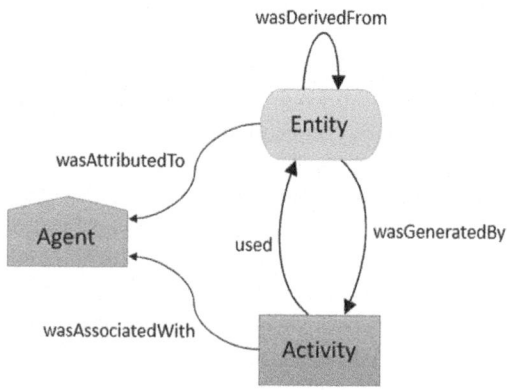

Fig. 1. The W3C provenance model.

The RACK ontology can be extended by custom project-specific ontologies that allow engineers to extend the ontology with project or domain-specific verbiage. These project-specific ontologies are called "overlays" and they allow RACK to handle domain-specific content on top of the underlying provenance focused layer [17]. To create an overlay, the RACK ontology classes are sub-typed to express custom terminologies that refer to software lifecycle components and the artifacts produced during their development, with the appropriate properties. As RACK is designed to be used for arbitrary software development projects,

---

[1] Readers can refer to [2,18] for more background on semantic web techniques.

the overlays too can be arbitrary in nature, thereby allowing more flexibility to engineers in terms of what can be expressed.

In addition to providing access to the underlying triplestore [4], RACK supplies all the services to ingest data and the verification checks to make sure that all incoming data is consistent with respect to the ontology (such as type checking and cardinality) [19]. Queries in RACK are expressed using *nodegroups* [12] that are built by linking classes from the ontology via selected properties. For example, a nodegroup might select REQUIREMENTs that have *verify*ing TESTs with *Passed* TEST_RESULTs. Nodegroups are automatically translated into SPARQL queries, can be saved to a nodegroup store, and can be invoked by their unique IDs. The results returned by query nodegroups can be saved to various formats.

One of the primary benefits of generating assurance cases from RACK is the underlying triplestore. A number of approaches, including the one we propose in this paper, involve the use of defining and using patterns or templates in order to programmatically generate assurance cases. These patterns are composed of concepts with well-defined types that are combined in a specified way to infer some meta-level properties of the system under consideration. When these patterns are instantiated with project-specific instance data, it is critical that the data used satisfies the relevant concept types. This is ensured in our approach by the triplestore provided by RACK since the queries generated for the patterns use the ontology to fetch the correct type of data.

Another benefit of RACK is that its queries are specified in SPARQL, which allows for a very natural way of specifying recursive queries [4]. And recursive queries are required as, for example, a high level requirement is decomposed into lower level requirements which may then be decomposed further. For a large system, this decomposition can involve an arbitrary number of levels corresponding to the system structure. For assurance cases, we need to roll-up properties like requirement satisfaction recursively. Recursive queries are very difficult to specify in traditional data stores, such as SQL-based databases, while they can be handled directly and elegantly using SPARQL.

RITE [7] is an open-source Eclipsed-based IDE for system assurance and certification developed by GE Aerospace Research. It assists different users involved in certifying or developing assurance cases for safety critical systems. RITE seamlessly connects with RACK, offering a user-friendly GUI for easy ingestion of data into RACK, SADL ontology development, test data development, ingestion package creation, ontology analysis, executing nodegroups, etc. Our assurance case synthesis pipeline is integrated into RITE. It leverages the utility functions and SADL environment provided by RITE to interact with RACK.

3 Automated GSN Synthesis from RACK

RACK can be used by engineers to store certification evidence from different tools during the software development lifecycle. The evidence is interconnected using ontologies that help establish traceability between the data using appropriate classes and relationships between them. Our goal was to develop a generic

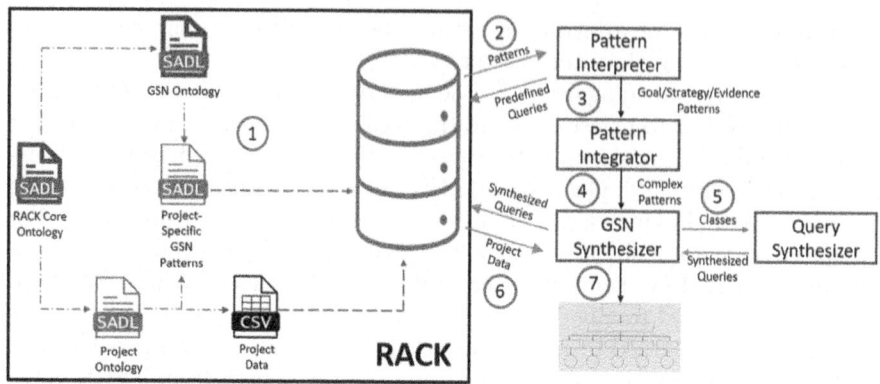

Fig. 2. Workflow of the assurance case generation pipeline from RACK.

pipeline for the automatic generation of GSN fragments for RACK projects. However, to generate GSN fragments using the ontology-backed evidence inside RACK, it is necessary to meaningfully align the domain and project-specific classes and relationships to the terminologies in the GSN notation, which are more generic. The pipeline also needed to be sufficiently configurable so that it could be used for any RACK project without being hindered by the potential arbitrariness of the project ontologies that RACK implicitly allows for.

The fundamental component of our pipeline (Fig. 2) is a generic GSN ontology designed for RACK that can be used to formalize GSN fragments in SADL[2]. This ontology is further extended to allow for the creation of abstract project-specific *GSN Patterns* that can align any arbitrary project ontology to the generic GSN verbiage. The patterns provide guidance on how the project-specific classes and relationships must be used to construct goals, strategies, and evidence nodes in a GSN. A GSN synthesis algorithm uses the project-specific GSN patterns to create GSN fragments from the evidence in RACK.

3.1 The GSN Ontology

The GSN ontology for RACK (Fig. 3) is an extension of a simple GSN ontology presented in [15] that has the classes Goal, Strategy, Solution, Context, Assumption, and Justification to sufficiently annotate all the different GSN terminologies. Relationships like supportedBy and inContextOf allow linkages between the different classes as per the GSN rules. A GSN is defined at the highest-level as a THING, which is one of the fundamental classes in the RACK core ontology. Additionally, there are relationships such as developed, to capture the completion status of a node; supportive, to signify if a node can support its parent; context, to add additional context; and solution, to connect a node with a tangible evidence from RACK (Fig. 4). These constructs allow formal instantiations of concrete GSN artifacts in SADL using evidence data.

---

[2] Complete SADL code presented in this paper is available at [7].

```
//******************* Connecting GSN terminologies to RACK core ontology classes
{
  Goal
                        (note "a requirements statement; a claim forming part of an argument"),
  Strategy
                        (note "mediates between a goal and its subgoals, explaining basis of
inference"),
  Solution
                        (note "presents a reference to an evidence item"),
  Context
                        (note "applies to argument substructure without repetition"),
  Assumption
                        (note "a goal or strategy in the context of an assumption is only applicable
when the assumption is true")
                        (note "applies to argument substructure without repetition")
                        (note "complete sentences in the form of a noun-phrase + a verb-phrase"),
  Justification
                        (note "applies only to goal or strategy and not to argument substructure")
                        (note "complete sentences in the form of a noun-phrase + a verb-phrase")
} are types of THING.

GSN (note "a graphical argument notation, capturing a hierarchy of claims") is a type of THING,
      described by rootGoal with a single value of type Goal.

//******************* Internal relationships between GSN elements
  supportedBy (note "indicated by solid arrow head")
        describes {Goal or Strategy} with values of type {Goal or Strategy or Solution}.
  inContextOf (note "indicated by hollow arrow head")
        describes {Goal or Strategy} with values of type {Context or Assumption or
Justification}.
```

Fig. 3. The SADL GSN ontology adapted for RACK.

Apart from allowing instantiation of concrete GSN artifacts, the extended GSN ontology also has some classes and relationships to encode project-specific GSN patterns. Patterns are abstract GSN fragments that connect arbitrary project-specific ontologies to the GSN verbiage. The `isPat` relationship is used to specify that a GSN node instance is a pattern and not a tangible GSN artifact. The `pGoal` and `pSubGoal` relationships allow linking two project ontology classes using GSN's goal/sub-goal concept. The relationship between these classes that can be used to synthesize a nodegroup to pull relevant data is annotated using the `pGoalSubGoalConnector` relationship. Finally, it is necessary to indicate the ontology classes that can be used as solutions. This can be done using the `GsnEvidence` class which has the `statusProperty` field to indicate which property of that class must be checked to see if the solution is supportive and the `passValue` field to indicate the valid supportive value.

```
//******************* Some properties of GSN elements needed to instantiate GSN fragments
developed (note "to specify if a goal or strategy is completely supported by its children")
          describes {Goal or Strategy} with values of type boolean.

supportive (note "to specify if a solution is supportive of its parent")
           describes Solution with values of type boolean.

context (note "to connect a goal element to a context")
        describes Context with values of type THING.

solution (note "This is used to connect a GSN solution to an evidence element")
         describes Solution with values of type THING.

isPat (note "Is a pattern") describes {Goal or Strategy} with a single value of type boolean.

pGoal (note "Goal class for the pattern's scope") describes {Goal or Strategy} with a single value
of type string.

pSubGoal (note "Subgoal class for the pattern's scope") describes Strategy with a single value of
type string.

pGoalSubGoalConnector (note "Property for the pattern's scope") describes Strategy with a single
value of type string.


GsnEvidence (note "To store info about the evidence classes in RACK") is a class
            described by classId (note "The id of the evidence class") with values of type string
            described by statusProperty (note "The property of the evidence class that decides if it
supports the strategy") with values of type string
            described by passValue (note "The value of statusProperty required for passing") with
```

Fig. 4. Extensions to the GSN ontology to support GSN fragments and patterns.

3.2 Project-Specific GSN Pattern Example

We will use an example OEM (original equipment manufacturer) project ontology (Fig. 5) to show how patterns are created. The OEM ontology has the following classes—SRS_Req for Software System Specification Requirements (SRS); SubDD_Req for and Subsystem Design Document Requirements (SubDD); and SBVT_Test for Software Baseline Verification Tests (SBVT).

```
SRS_Req (note "A Requirement identified in the Software Requirements Specification")
     is a type of REQUIREMENT.

SubDD_Req (note "A Requirement identified in the Subsystem Design Document")
     is a type of REQUIREMENT.
          Rq:satisfies of SubDD_Req only has values of type SRS_Req.

SBVT_Test (note "A test identified in the Software Baseline Verification Tests")
     is a type of TEST.
          verifies of SBVT_Test has at least 1 value.
```

Fig. 5. Snippet from the OEM project ontology.

Let us assume a hypothetical OEM-specific assurance claim that an SRS has passed baseline verification because all the SubDDs that trace to it have passed

SBVT tests. In the GSN notation, this claim can be represented using a goal-strategy pair where the goal states that an SRS has passed baseline verification and the strategy states that all relevant SubDDs have passed baseline verification. This goal-strategy pair can be encoded as an OEM-specific GSN pattern as shown in Fig. 6. Here, the Goal class of the extended GSN ontology is used to create a goal pattern, the isPat field is used to indicate that it is a pattern and not an actual goal instance, the description field is used to indicate the goal statement, and the pGoal field is used to indicate the class from the OEM ontology whose instances must be used to instantiate this pattern. Similarly, in the strategy pattern, the description field indicates the argument[3] to use, the pGoal field indicates the class whose instances can form parent goals of the strategy instances, the pSubGoal field indicates the class whose instances can form subgoals for the strategy instances[4], and pGoalSubGoalConnector indicates the actual relation in the ontology that connects the goal and subgoal classes[5].

```
srs-GP is a Goal
        with isPat true
        with identifier "srs-GP"
        with description "Claim: SRS Requirement has passed baseline verification"
        with pGoal "SRS_Req".
srs-tested-SP is a Strategy
        with isPat true
        with identifier "srs-tested-SP"
        with description "Argument: All relevant SubDDs have passed baseline verification"
        with pGoal "SRS_Req"
        with pSubGoal "SubDD_Req"
        with pGoalSubGoalConnector "satisfies".
evidence-sbvttr is a GsnEvidence
        with classId "TEST_STATUS"
        with statusProperty "identifier"
        with passValue "Passed".
```

Fig. 6. Snippet from the OEM-specific GSN patterns.

The goal-strategy pair patterns can be used to create more complex GSN patterns. *E.g.*, if we have a second goal-strategy pair pattern where the goal states that a particular SubDD has passed baseline verification and the strategy says that there is a passed SBVT test for that SubDD, then this pattern, in combination with the first goal-strategy pair pattern, can form the GSN pattern shown in Fig. 7 where there are multiple levels of goals and strategies.

Finally, solutions appear as leaf nodes in a GSN tree and can be used to support a parent node. Fig. 6 shows an evidence pattern that states that the TEST_STATUS class, which is a part of the RACK core ontology, can be used as a supportive evidence for the OEM GSN patterns when it has a value "Passed".

---

[3] Arguments of the form *all/forall, at least one,* & *exactly one* are supported.
[4] This simply indicates the goal-subgoal GSN relationship between ontology classes.
[5] Metadata needed to auto-generate appropriate RACK queries to fetch evidence.

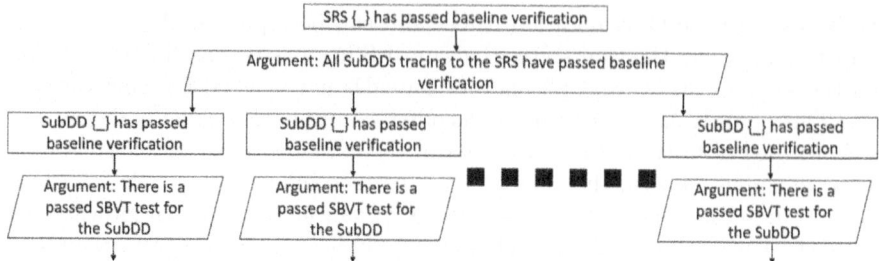

Fig. 7. Integration of goal-strategy patterns to form complex GSN patterns.

Using the extended GSN ontology, users can create project-specific GSN patterns for arbitrary project ontologies, as shown above. This allows the pipeline to be a generic RACK tool, in contrast to a project-specific tool that can only work with a fixed project ontology. In a development workflow, the patterns would be manually designed by engineers while designing the project ontology in the beginning of a project. This would be a one-time effort and the patterns can be reused across other projects as long as the context and related ontology terminologies remain unchanged. The patterns allow the project-specific evidence to be automatically analyzed by the tool for automated synthesis of GSN fragments.

3.3 Automated Synthesis of GSN Fragments Using Patterns

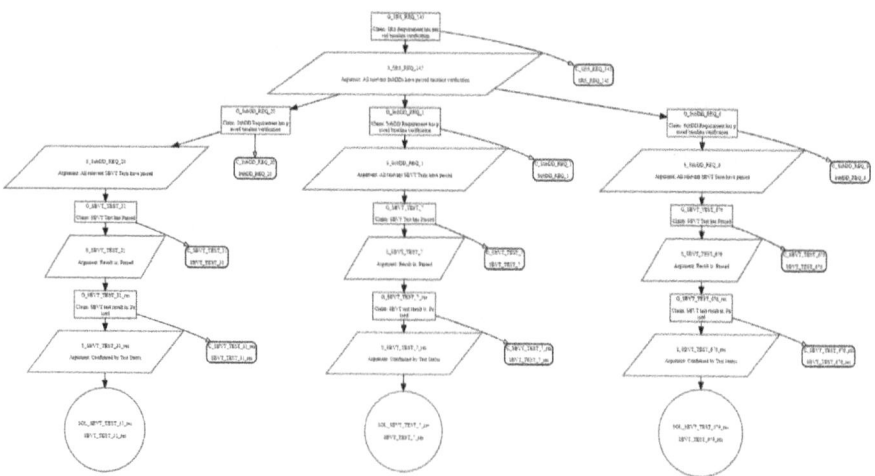

Fig. 8. An graphical GSN fragment created from RACK data.

The project-specific GSN patterns can be instantiated with appropriate data inside RACK to synthesize complete GSN fragments. The workflow[6] of this pipeline (Fig. 2) involves RACK, a *pattern interpreter*, a *pattern integrator*, a *query synthesizer*, and a *GSN synthesizer*. First, the project-specific GSN patterns and the evidence are automatically ingested into RACK using *evidence ingestion pipelines* [4]. The rest of the workflow is activated when users launch the GSN generator GUI (presented in Sect. 3.4) and click on a button. As a response to the click, the pattern interpreter executes a set of preset queries on RACK to fetch the patterns. This creates a set of pattern artifacts, which are then sent to the pattern integrator. The pattern integrator uses hard-coded algorithms to combine the patterns into more complex GSN patterns artifacts as shown in the example in Fig. 7. These complex patterns are then sent to the GSN synthesizer, which sends the project-specific class information in the pattern artifacts to the query synthesizer. The query synthesizer then synthesizes a set of queries[7] to pull the evidence from RACK. The GSN synthesizer then executes the queries to fetch the data as a set of evidence artifacts. A simple backtracking-based algorithm exhaustively evaluates the evidence artifacts to instantiate the complex patterns into actual GSN fragments. It should be noted that after the ontologies, patterns, and evidence have been ingested into RACK, the workflow is initiated by a single button click and executes completely automatically in the backend without any user interaction[8].

The pipeline generates graphical GSN fragments by using Graphviz [6] and also creates SADL instances of the fragments using the GSN ontology. The SADL fragments can be ingested into RACK for archival and further analysis. The graphical fragments use the colored GSN notation presented in [16] where a node that is completely supported by its sub-nodes is colored with green and a node that is not completely supported by its sub-nodes is colored in red[9]. Such a GSN fragment is shown in Fig. 8 and a snippet from its SADL encoding is shown in Fig. 9. One advantage of the colored GSN notation is that if a goal is not satisfied, users can easily find the root cause by visual inspection of the colored branches. This is useful for complex non-trivial assurance arguments [16] where purely textual representations of failed assurance arguments would not provide such visual-aid for convenient fault-tracing. Furthermore, the structured GSN

---

[6] Creation of the project ontology and the project-specific GSN patterns, and the generation of evidence are not part of the workflow. At the beginning of a project, engineers manually create the project ontology and the project-specific GSN patterns and appropriate tools are used to generate evidence for the system under review.

[7] Synthesized queries use the ontology class information from the patterns, allowing them to fetch evidence curated with arbitrary ontologies. This ensures that the correct evidence is fetched for the project-specific patterns. Conversely, preset queries use the RACK core ontology and cannot fetch data curated with arbitrary project ontologies.

[8] A concrete instantiated example using the OEM use case that shows all the steps of the workflow with sample artifacts can be found in [7]

[9] The colored GSN fragments in [16] were synthesized by the VERDICT toolchain, which does not support ontology-based tool-agnostic evidence curation like RACK.

```
G-SRS_REQ_145 is a Goal
        with identifier "G-SRS_REQ_145"
        with description "Claim: SRS Requirement has passed baseline verification"
        with developed true.
S-SRS_REQ_145 is a Strategy
        with identifier "S-SRS_REQ_145"
        with description "Argument: All relevant SubDDs have passed baseline verification"
        with developed true.
G-SubDD_REQ_20 is a Goal
        with identifier "G-SubDD_REQ_20"
        with description "Claim: SubDD Requirement has passed baseline verification"
        with developed true.
S-SubDD_REQ_20 is a Strategy
        with identifier "S-SubDD_REQ_20"
        with description "Argument: All relevant SBVT Tests have passed"
        with developed true.
G-SBVT_TEST_31 is a Goal
        with identifier "G-SBVT_TEST_31"
        with description "Claim: SBVT Test has Passed"
        with developed true.
S-SBVT_TEST_31 is a Strategy
        with identifier "S-SBVT_TEST_31"
```

Fig. 9. Snippet from the SADL encoding of the GSN shown in Fig. 8.

arguments use well-defined project-specific strategies to justify evidence-based claim satisfiability, which makes them more insightful than tools that perform simple traceability analysis between elements (such as [21]).

3.4 GUI Support for GSN Generation and Analysis

The pipeline is integrated into the RITE tool for RACK.[10] In the presence of a large number of levels and subgoals, GSN fragments can oftentimes become too complicated and intractable when viewed as a single SVG graphic. Therefore, the tool provides an interactive interface for generating GSN fragments from the data inside RACK and viewing them in a modular fashion. There are two primary types of views—the GSN Generator view and the GSN Navigator view.

The GSN Generator view (Fig. 10) provides a chart and a list along with buttons to fetch all possible root-goal nodes from RACK. Once populated, the chart shows the number of root-goals of each ontology class possible from the patterns and the evidence stored inside RACK while the list shows the identifier of each root-goal. There is also a filter that allows users to select which classes they wish to view in the list. Users can select any item from the list of root-goals and click the button to generate SADL and SVG GSN artifacts that start at that root-goal. Once the artifacts for a root-goal are generated, users can click on the GSN Navigator button to open the navigator view.

---

[10] Interested users can download the tool and the examples in this paper at [7].

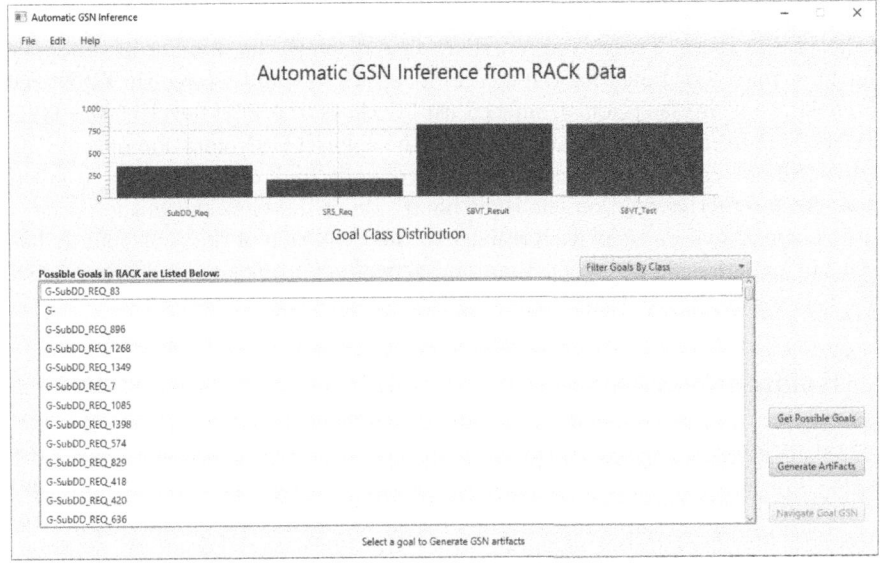

Fig. 10. The GSN Generator GUI in RITE.

The GSN Navigator view (Fig. 11) allows users to *"traverse"* or *"navigate"* the GSN tree at a level-by-level basis, starting at the root-goal. It shows the

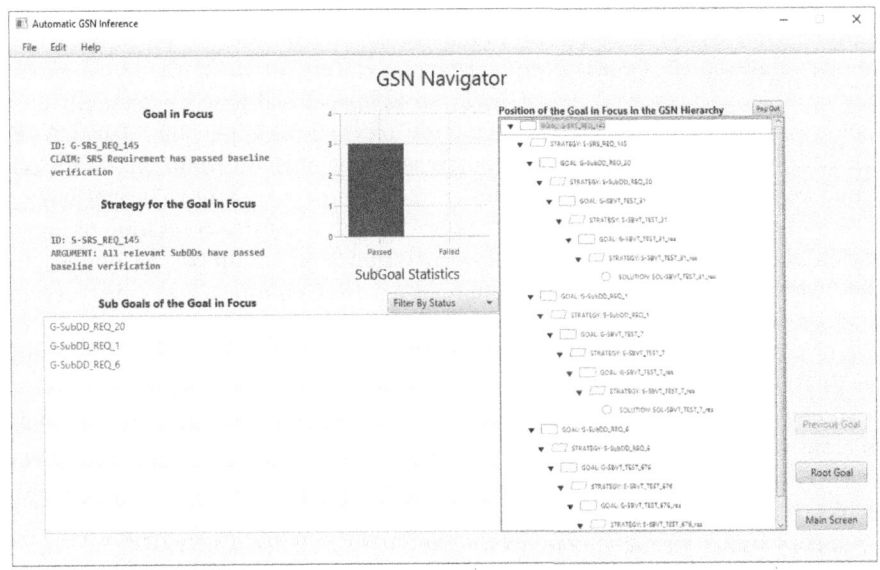

Fig. 11. The GSN Navigator GUI (the GSN in Fig. 8 is shown).

current goal in focus, the strategy for that goal, and a list of all the sub-goals of the goal in focus. There is also a chart that displays the number of sub-goals that have passed or failed for the goal in focus. Users can traverse the GSN tree by clicking on any sub-goal from the list to check the next level in the GSN tree starting from that sub-goal. It follows the colored GSN notation of the SVG figures, where green represents a completed node and red represents a failed node. An interactive field on the right hand side of the view shows the position of the current goal in focus with respect to the GSN tree of the root-goal. It has been our experience that when there are several thousands of nodes in a GSN tree, this interactive view makes it easier to analyze the GSN tree.

4 Related Work

Several techniques for the construction and synthesis of assurance cases exist in the literature. Lin et al. [13] presented a framework for the creation and management of assurance cases using safety-case pattern catalogues that are automatically instantiated using an Atlas Transformation Language program. Zeroual et al. [29] proposed a method to automatically instantiate security case patterns using requirements specified in the Alloy language [9] and verification evidence generated by Alloy. Kapinski et al. [10] developed domain-specific patterns and arguments to generate assurance cases for time-sensitive networks. Beyene et al. [1] presented a formal language for specifying safety case patterns and safety cases to allow automated generation of safety cases. Wang et al. [26,27] proposed a framework for the automatic generation of assurance cases by formalizing argument patterns as hierarchical contract networks using the Resolute language and using an SMT solver to instantiate the patterns. They also use Bayesian Networks to quantify uncertainty in their assurance cases. Denney et al. [5] presented AdvoCATE, an Eclipse-based tool for manually creating assurance case patterns and instantiating them as applicable. AdvoCATE can also automatically instantiate patterns using outputs from the AutoCert formal verification tool. Ramakrishna et al. [22] created a tool that synthesizes safety cases from a "curated evidence store", which is a table consisting of information about a system and all the evidence that support its correct operation. The AMASS [24] initiative supports ontology-driven assurance case construction using the *Common Assurance and Certification Metamodel* (CACM). Meng et al. [16] presented an approach for the auto-generation of assurance cases from the VERDICT toolchain [14,23] using the colored GSN notation (see Sect. 3).

In contrast, we explored the synthesis of GSN assurance cases from a tool-agnostic semantic triplestore where heterogeneous evidence is curated using arbitrary project and domain-specific ontologies. The extended GSN ontology enables this by allowing arbitrary project and domain-specific ontologies and arguments to be formally mapped to the GSN verbiage and reasoning style. Our pipeline is integrated into an Eclipse-based open-source tool that provides an interactive GUI for managing and analyzing the GSN fragments. Moreover, as RACK may be integrated into continuous integration/continuous delivery

(CI/CD) workflows [21], our pipeline can be harnessed to generate assurance cases that can automatically evolve during the software development lifecycle.

5 Conclusion

We presented a pipeline for automatically synthesizing GSN assurance cases from the RACK semantic triplestore using a special GSN ontology. The GSN ontology can map arbitrary project-specific verbiage and reasoning to the GSN standard using project-specific patterns, thereby allowing the pipeline to work with arbitrary project ontologies. The capability is presented as an open-source Eclipse-based toolkit that provides a GUI for easy management and analysis of the GSN fragments. To the best of our knowledge, this is the only open-source tool in the literature that can automatically synthesize GSN fragments from evidence curated in a semantic triplestore using arbitrary ontologies.

Currently, the pipeline only supports strategies where the goal and subgoals are concerned with a single ontology class each. Complex strategies where a goal may be dependent on the status of multiple sub-classes are not currently handled. Therefore, a potential direction of future work would be to support such strategies. Replacing the backtracking-based algorithm for GSN synthesis with an SMT solver will be another interesting problem to investigate. It is also important to include the concepts of uncertainty and defeaters in the reasoning capabilities in the future. Finally, it will be interesting to explore more robust GUI formats to enable easy comprehension and analysis of GSN fragments.

References

1. Beyene, T.A., Carlan, C.: CyberGSN: a semi-formal language for specifying safety cases. In: 51st Annual International Conference on Dependable Systems and Networks Workshops (DSN-W), pp. 63–66. IEEE (2021)
2. Crapo, A.W., Moitra, A.: Using OWL ontologies as a domain-specific language for capturing requirements for formal analysis and test case generation. In: 13th International Conference on Semantic Computing (ICSC), pp. 361–366. IEEE (2019)
3. Crapo, A., Moitra, A.: Toward a unified English-like representation of semantic models, data, and graph patterns for subject matter experts. Int. J. Semant. Comput. **7**(03), 215–236 (2013)
4. Cuddihy, P., Russell, D., Mertens, E., Siu, K., Archer, D., Williams, J.: Aviation certification powered by the semantic web stack. In: Payne, T.R., et al. (eds.) ISWC 2023. LNCS, vol. 14266, pp. 345–361. Springer, Cham (2023). https://doi.org/10.1007/978-3-031-47243-5_19
5. Denney, E., Pai, G., Pohl, J.: AdvoCATE: an assurance case automation toolset. In: Ortmeier, F., Daniel, P. (eds.) SAFECOMP 2012. LNCS, vol. 7613, pp. 8–21. Springer, Heidelberg (2012). https://doi.org/10.1007/978-3-642-33675-1_2
6. Ellson, J., Gansner, E., Koutsofios, L., North, S.C., Woodhull, G.: Graphviz—open source graph drawing tools. In: Mutzel, P., Jünger, M., Leipert, S. (eds.) GD 2001. LNCS, vol. 2265, pp. 483–484. Springer, Heidelberg (2002). https://doi.org/10.1007/3-540-45848-4_57

7. GE Research: RITE SAFECOMP 2024 Release. https://github.com/ge-high-assurance/RITE/releases/tag/safecomp24
8. GE Research: SADL: Semantic Application Design Language. https://github.com/SemanticApplicationDesignLanguage/sadl. Accessed 28 Apr 2023
9. Jackson, D.: Alloy: a language and tool for exploring software designs. Commun. ACM **62**(9), 66–76 (2019)
10. Kapinski, R., Pantelic, V., Bandur, V., Wassyng, A., Lawford, M.: Assurance cases for timing properties of automotive TSN networks. In: Guiochet, J., Tonetta, S., Schoitsch, E., Roy, M., Bitsch, F. (eds.) SAFECOMP 2023. LNCS, vol. 14182, pp. 26–31. Springer, Cham (2023). https://doi.org/10.1007/978-3-031-40953-0_3
11. Kelly, T., Weaver, R.: The goal structuring notation–a safety argument notation. In: Proceedings of the Dependable Systems and Networks 2004 Workshop on Assurance Cases, vol. 6. Citeseer (2004)
12. Kumar, V., Cuddihy, P., Aggour, K.: NodeGroup: a knowledge-driven data management abstraction for industrial machine learning. In: Proceedings of the 3rd International Workshop on Data Management for End-to-End Machine Learning, pp. 1–4 (2019)
13. Lin, C.-L., Shen, W., Yue, T., Li, G.: Automatic support of the generation and maintenance of assurance cases. In: Feng, X., Müller-Olm, M., Yang, Z. (eds.) SETTA 2018. LNCS, vol. 10998, pp. 11–28. Springer, Cham (2018). https://doi.org/10.1007/978-3-319-99933-3_2
14. Meng, B., et al.: Verdict: a language and framework for engineering cyber resilient and safe system. Systems **9**(1), 18 (2021)
15. Meng, B., et al.: Towards developing formalized assurance cases. In: 2020 AIAA/IEEE 39th Digital Avionics Systems Conference (DASC), pp. 1–9. IEEE (2020)
16. Meng, B., Paul, S., Moitra, A., Siu, K., Durling, M.: Automating the assembly of security assurance case fragments. In: Habli, I., Sujan, M., Bitsch, F. (eds.) SAFECOMP 2021. LNCS, vol. 12852, pp. 101–114. Springer, Cham (2021). https://doi.org/10.1007/978-3-030-83903-1_7
17. Moitra, A., et al.: A semantic reference model for capturing system development and evaluation. In: 16th International Conference on Semantic Computing (ICSC), pp. 173–174. IEEE (2022)
18. Moitra, A., et al.: Enabling development of an extensible, multi-perspective ontology. In: 18th International Conference on Semantic Computing (ICSC), pp. 77–80. IEEE (2024)
19. Moitra, A., et al.: RACK: a semantic model and triplestore for curation of assurance case evidence. In: Guiochet, J., Tonetta, S., Schoitsch, E., Roy, M., Bitsch, F. (eds.) SAFECOMP 2023. LNCS, vol. 14182, pp. 149–160. Springer, Cham (2023). https://doi.org/10.1007/978-3-031-40953-0_13
20. Moreau, L., Groth, P., Cheney, J., Lebo, T., Miles, S.: The rationale of PROV. J. Web Semant. **35**, 235–257 (2015)
21. Paul, S., et al.: Automated DO-178C compliance summary through evidence curation. In: 42nd Digital Avionics Systems Conference (DASC), pp. 1–10. IEEE (2023)
22. Ramakrishna, S., Hartsell, C., Dubey, A., Pal, P., Karsai, G.: A methodology for automating assurance case generation. In: Thirteenth International Tools and Methods of Competitive Engineering Symposium (TMCE 2020) (2020)
23. Siu, K., et al.: Architectural and behavioral analysis for cyber security. In: 2019 IEEE/AIAA 38th Digital Avionics Systems Conference (DASC), pp. 1–10. IEEE (2019)

24. de la Vara, J.L., Parra, E., Ruiz, A., Gallina, B.: AMASS: a large-scale European project to improve the assurance and certification of cyber-physical systems. In: Franch, X., Männistö, T., Martínez-Fernández, S. (eds.) PROFES 2019. LNCS, vol. 11915, pp. 626–632. Springer, Cham (2019). https://doi.org/10.1007/978-3-030-35333-9_49
25. W3C: OWL - Semantic Web Standards. https://www.w3.org/OWL
26. Wang, T.E., Daw, Z., Nuzzo, P., Pinto, A.: Hierarchical contract-based synthesis for assurance cases. In: Deshmukh, J.V., Havelund, K., Perez, I. (eds.) NFM 2022. LNCS, vol. 13260, pp. 175–192. Springer, Cham (2022). https://doi.org/10.1007/978-3-031-06773-0_9
27. Wang, T.E., et al.: Computer-aided generation of assurance cases. In: Guiochet, J., Tonetta, S., Schoitsch, E., Roy, M., Bitsch, F. (eds.) SAFECOMP 2023. LNCS, vol. 14182, pp. 135–148. Springer, Cham (2023). https://doi.org/10.1007/978-3-031-40953-0_12
28. Wei, R., Kelly, T.P., Dai, X., Zhao, S., Hawkins, R.: Model based system assurance using the structured assurance case metamodel. J. Syst. Softw. **154**, 211–233 (2019)
29. Zeroual, M., Hamid, B., Adedjouma, M., Jaskolka, J.: Constructing security cases based on formal verification of security requirements in alloy. In: Guiochet, J., Tonetta, S., Schoitsch, E., Roy, M., Bitsch, F. (eds.) SAFECOMP 2023. LNCS, vol. 14182, pp. 15–25. Springer, Cham (2023). https://doi.org/10.1007/978-3-031-40953-0_2

CyberDS: Auditable Monitoring in the Cloud

Lev Sorokin[(✉)] and Ulrich Schoepp

Fortiss, Research Institute of the Free State of Bavaria, Guerickestraße 25,
80805 Munich, Germany
{sorokin,schoepp}@fortiss.org

Abstract. When deploying safety-critical systems in the cloud, where deviations may have severe consequences, the assurance of critical decisions becomes essential. Typical cloud systems are operated by third parties and are built on complex software stacks consisting of e.g., Kubernetes, Istio, or Kafka, which due to their size are difficult to be verified. Nevertheless, one needs to make sure that safety-critical choices are made correctly. In this paper, we propose CyberDS, a flexible runtime monitoring approach designed to transparently monitor safety and data-related properties in the Cloud. CyberDS is based on combining distributed Datalog-based programs with tamper-proof storage based on Trillian to verify the premises of critical actions. We demonstrate our monitoring approach on an industrial use case that uses a cloud infrastructure for the orchestration of unmanned air vehicles.

1 Introduction

Cloud systems are used in various domains, e.g., in healthcare, entertainment, in the financial or the automotive sector. Following the HashiCorp cloud report[1] over 94% of all enterprises world-wide use cloud services. Deployment of systems in the cloud promises cost savings, scalability – by automatic provisioning of resources depending on the request load – and high availability.

While cloud technologies have many advantages, using them for safety-critical applications presents many challenges. We need to address several risks, such as hardware, software faults or attacks that can have an impact on the integrity of processed information or the availability of a service. However, cloud software stacks based on components like Kubernetes or Kafka are complex [17] and are updated frequently, which makes assurance tasks difficult.

If it is not feasible to verify the implementation, then one approach to guaranteeing the safety of critical decisions is to track the events and actions of the implementation in an independent runtime monitoring layer and to allow only safe decisions based on the collected data.

Runtime monitoring is already standard in cloud systems in the form of Security Information and Event Management (SIEM) [3,14], where log data is aggregated and analyzed continuously to identify potential security issues. However,

---
[1] https://www.hashicorp.com/state-of-the-cloud.

information flows only from the components to the SIEM, so one can identify problems only after the fact. For safety-critical decisions it is important to check their correctness before they are made. One approach is to perform the analysis during system operation and provide the results to the components so that they can make their decisions safely.

An example instance of this idea is the Certificate Transparency[2] (CT) project, where HTTPS certificates are stored in a public ledger that allows their verification, and browsers can make the decision on whether to accept a certificate based on information in such a ledger.

The problem of analysing the behavior of a system with *runtime monitoring* has been the topic of extensive work, see, e.g., [7]. For example, monitoring approaches exist based on generating monitors from automata or using event stream processing [6,9–11]. However, to assure the safety of safety-critical decisions, we are interested in a particular monitoring problem that does not seem to be accounted for very well by such approaches: we want to record the relevant history of data leading up to a safety-critical decision and then verify that this data supports the decision before executing it. For instance, in a cloud service that optimizes the operation of a world-wide fleet of unmanned air vehicles, one wants to verify all preparation steps before starting any drone. To do this, one typically needs data from the event history, e.g., the JSON body of API requests, that finite state automata abstract.

In addition, auditability and documentation of process steps are becoming more and more important, as seen in regulatory requirements for the privacy protection (GDPR) or the protection of health data (HIPAA), among others. Traceability of events is in the interest of a service provider to have a convincing argument in case of disputes. Existing approaches that consider auditability [4, 20,23], are based on distributed ledger technology and are only applicable to low-volume processes, such as logistics or manufacturing, because of high transaction latencies and low transaction rates (in the order of 1000 transactions/s).

We extend the approach from the CT project to monitor properties in a distributed setup. Similar to our approach, CT relies on a tamper-proof storage of data but the existing framework is not directly applicable for monitoring cloud system events. Our framework **CyberDS** makes the following contributions:

Cyberlog. A Datalog-based language simplifying property specifications in the cloud. Cyberlog enables temporal property, parallel process, and data invariant specification, by using attestations.

Transparent Monitoring. We deploy distributed Cyberlog monitors configured with Cyberlog specifications. These monitors capture and continuously and transparently analyze messages exchanged between components at runtime. We integrate the tamper-proof storage Trillian into the monitoring architecture for the logging of claims to allow an auditable storage of system events and messages.

---

[2] https://certificate.transparency.dev/.

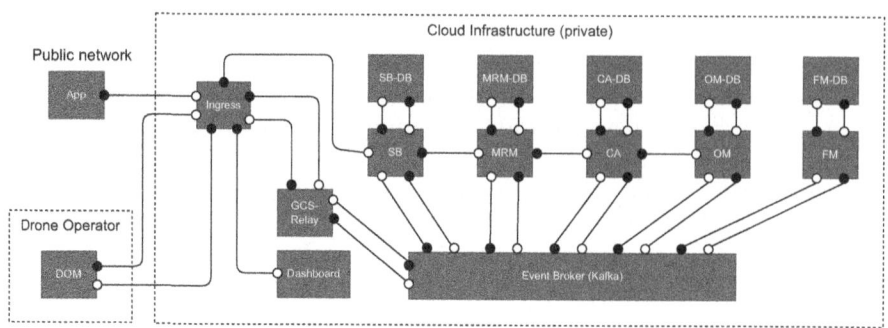

Fig. 1. Architecture of the RTAPHM digital platform.

Revision Model. We introduce a revision model for Cyberlog-claims to enable feasible reasoning. Knowledge revision is essential to prevent an uncontrolled increase in log data size at the distributed monitors.

We demonstrate the applicability of our approach on a prototypical cloud system from the avionics domain for booking transportation services with unmanned air vehicles. The implementation of CyberDS and case study artefacts are publicly available [22].

2 Motivating Example

We motivate our approach with a use case from the RTAPHM project[3]. Unmanned air vehicles (UAVs) have gained high popularity in the last few years, their applications range from cargo transportation, to SAR operations or environmental monitoring[4]. The RTAPHM project develops a cloud platform for orchestrating UAVs for such uses, i.e. for the transportation of transplant organs[5].

System Description. The platform comprises several micro-services (see Fig. 1): the Service Broker (SB) for request correctness and authorization; the Multi-Resource-Manager (MRM) for resource management; the Cognitive Assistant (CA) for resource allocation optimization; the Fleet Manager (FM) for UAV predictive maintenance; and the Operation Manager (OM) for UAV flight preparation and launch triggering. Each service has a local database and communicates via REST APIs and Kafka message broker[6].

The platform interfaces with Drone Operators (DOs) responsible for executing UAV missions. DOs are connected to the Drone Operation Manager (DOM), which executes UAV missions and sends status updates (e.g., position) to the

---
[3] https://www.fortiss.org/forschung/projekte/detail/rtaphm.
[4] https://droneii.com/237-ways-drone-applications-revolutionize-business.
[5] https://www.cbc.ca/news/canada/toronto/first-lung-transplant-drone-1.6208057.
[6] https://github.com/apache/kafka.

platform. Additionally, an ingress gateway connects the platform to a public web application (App) for service requests by users.

Typical Workflow. A user requests an organ transport from destination A to destination B. The SB validates the request and forwards it to the MRM and CA to select a UAV, transport assets, and freight-related constraints (e.g., delivery-time, organ weight). The user receives booking options with prices and selects one, converting the request into a mission. The OM triggers personnel for drone preparation and waypoint calculation. Upon completion, the DO receives a ready-to-fly (RTF) request with waypoints to launch the drone.

To assure the correct execution of the RTAPHM system for scheduling UAVs, multiple requirements have to be satisfied. We assume that these requirements are derived from a careful safety analysis of the system, e.g., using an STPA [18]:

1. (Process requirements): *All necessary steps before the takeoff of a drone have to be completed and documented (e.g., preparation of the drone by ground staff, service request is registered).*
2. (Temporal requirements): *The time between receiving and sending a message in the system shall not exceed a threshold. If an RTF-message arrives delayed at the DO, the organ will be damaged.*
3. (Data-related requirements): *The waypoints in the RTF message shall match the waypoints of one registered UAV base. Otherwise, an attacker could modify waypoints in the RTF message that are sent to DO.*

To monitor the discussed properties in the distributed RTAPHM system we are confronted with the following challenges:

- **Flexible Specification:** The implementation of the properties to be monitored shall not be coupled to the implementation of the underlying system to allow the modification of properties. For instance, the business logic in the system might change over time so that process requirements do change.
- **Verifiability and Documentation**: Data used for the monitoring of properties should be verifiable by third parties during as well as after the systems operation for documentation purposes. When e.g., regarding property 3, a monitor receives an RTF message, one should be able to verify that this message has been generated by DO. Otherwise, reasoning with *incorrect* claim becomes wrong. Also, applications such as RTAPHM, are subject to legal auditability requirements. E.g., §486.346 of the US Code of Federal Regulations requires organ transport operators to *"develop and follow a written protocol for packaging, labeling, handling, and shipping organs in a manner that ensures their arrival without compromise to the quality of the organ"*.
- **Distributed Setup:** Monitoring communication in a distributed setup and assuring at the same the auditability of system actions can lead to latencies if done not efficiently.

Our goal is to provide a monitoring mechanism so that requirements for a distributed system can be specified independently from the implementation of the components and monitoring results can be provided back to the system components. In the following sections we explain our approach to achieving this.

3 Monitoring Approach

To address the described challenges in Sect. 2, our proposed monitoring framework adds a verification layer to the system by appending security monitors to the distributed components of the system as shown in Fig. 2a. We consider having such a decentralized monitor setup to avoid a single point of failure and to reduce communication costs when monitoring properties [9]. In particular, the monitors collect and analyse the data to verify requirements during the systems' operation, and provide the results of their analysis back to the system components.

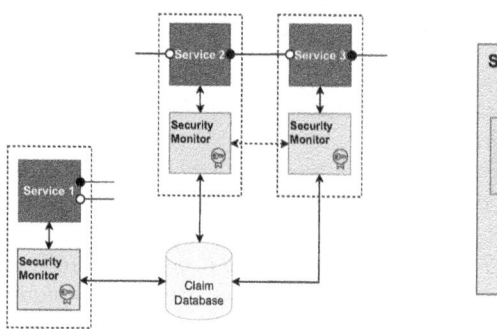
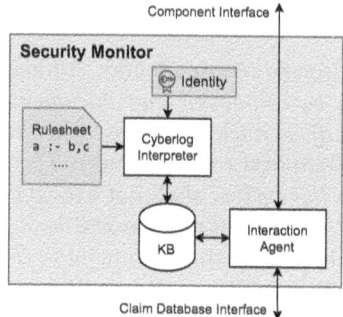

(a) Monitoring with security monitors   (b) Security monitor architecture

Fig. 2. Figure a) shows the deployed security monitors for each service communicating using a common database. Figure b) shows the architecture of a monitor.

The monitors can communicate claims about the process execution with each other, to observe the interactions between multiple system components. Further, each monitor has an assigned identity so that claims can be exchanged in the form of cryptographically signed attestations. That means, that the system which provides an attestation possesses a private and a public key. The private key is used to encrypt the hash (signature) of the attestation, while the public key can be used to decode the signature and verify the integrity of the message.

To enable the verifiability of processed claims as well as the documentation of processes, the monitors further communicate with a claim database that stores the exchanged claims in a tamper-proof manner.

In the following sections we outline in detail the three main components of our monitoring approach: the specification language for defining properties and specifying the behavior of the system, the monitoring architecture and the tamper-proof storage integration.

3.1 Specification Language

To monitor the dynamic behaviour of a system, we need to specify the required behaviour/properties in a specification language. For this, we propose an extension of Datalog [8], called *Cyberlog*. Our language is based on Datalog, because Datalog has a simple syntax and allows to specify an abstract view on the overall systems state by using *rules* applied over a dataset. Further, Datalog has a well-defined semantics and is often used as a query language for declarative databases, and has been already applied to specify the behaviour of distributed systems [1].

Cyberlog requires, such as Datalog an interpreter that operates on a *knowledge base (KB)* and uses the predefined rules to incrementally update the KB with new reasoned facts by applying rules on data in the KB. When new data, such as an API request in the web service is registered and logged in the KB, reasoning automatically takes place. This allows a continuous update of the view on the system's state when new events are stored in the KB. In the following, we describe briefly the syntax of Cyberlog:

Definition. A Cyberlog *rule* has the form a :- b_1, ..., b_n, where a, b_i are atoms. The atom a is called the *head* of the rule, while the b_i form the *body* of the rule. In Cyberlog, the atoms are attestations of the form principal attests symbol (t_1,...,t_n), where principal is a term denoting the identity of a component or person associated with the system, symbol is a *predicate* and the t_i are *terms*. A term can be a variable (starts with an uppercase character), or a constant, such as string literals 'text' or numbers.

A Cyberlog program, known as *rulesheet*, contains a set of rules designed for execution by a single principal. To simplify notation, principals can use symbol (...) to represent self attests symbol(...) where self refers to the principal executing the program. Rule heads are restricted to self-attestations to prevent principals from claiming things on behalf of others. Identity is established using X509 digital certificates (RFC 5280), with the rulesheet specifying certificate subject and issuer details.

Example. An example rulesheet for the principal 'SB' is given in Listing 1.1. It begins with a declaration of the identities of components, where we show the details only for SB.

```
1    'SB':  Subject: 'C=DE, ST=Hamburg, L=HH, O=ZAL, CN=SB'
2           Issuer:  'C=DE, O=Lets Encrypt, CN=R3'
3      ...
4    // interpretation of events
5    request(RequestId, Data, Time) :-
6      postRequest('/servicerequest', Time, Data),
7      get_param_int(Data, 'request_id', RequestId).
8
9    // workflow
10   good_rtf_exists(RequestId, AircraftId) :-
11     'SB' attests request(RequestId, Data, TimeRequest),
12     'MRM' attests feasible_config(RequestId, AircraftId),
13     'OM' attests tasks_done(RequestId, AircraftId),
14     'OM' attests ready_to_fly(RequestId, AircraftId, Data, TimeRTF).
15
```

```
16   // time
17   delayed_rtf(RequestId, DelayTime, SentTime) :-
18     'OM' attests ready_to_fly(RequestId, AircraftId, DataRTF,
         TimeRTF),
19     'CA' attests    mission_confirmed(RequestId, Data, SentTime),
20     DelayTime == TimeRTF - SentTime,
21     DelayTime > 1000.
```

Listing 1.1. Example Cyberlog program specifying workflow and timing related properties, as well as the interpretation of the systems' API messages.

The rule in line 10 formalizes property 1 from Sect. 2 and specifies that an RTF message is considered as *good* (modeled by `good_rtf_exist`) if SB has registered a Service Request (`request`), MRM has provided an appropriate booking option (`feasible_config`), and if all necessary UAV preparation tasks (line 13) have been done by OM (`tasks_done`). The claim `good_rtf_exist` is generated when all the claims in the body of the rule are available in the monitor's KB.

The rule in line 17 is an example of a time-related property that checks whether an RTF message issued by the DO is delayed (threshold is 1000 ms). The rule generates a `delayed_rtf` fact whenever a `ready_to_fly` fact exists, CA has confirmed the booking request, and the time between the sent RTF and confirmation exceeds the threshold.

Lines 4–7 demonstrate the definition of views on the available data using Cyberlog, which is in our example the modelling of a Service Request. The parameter `RequestId` is a unique identifier for a service request to distinguish between different requests, `Time` is the time when the request was made, and `Data` is the content of the message in the request (line 5). The body of the rule in lines 5–7 declares that a `postRequest` claim must exist, which represents an API request event. Cyberlog also provides predicates for working with JSON data, which is ubiquitous in cloud systems. In line 7 a request identifier is extracted from the request data `Data` (i.e. JSON data) and bound to the `RequestId` variable.

3.2 Security Monitor

Each monitor in `CyberDS` is configured with an identity and a rulesheet (s. Fig. 2b). The monitor has an integrated *Cyberlog-interpreter* for the evaluation of rules and a *local knowledge base* (KB), where Cyberlog claims are stored. The monitor continuously executes the rules from the rulesheet. Initially, the KB is empty. There are three ways in which claims can be added to the KB: a) A claim is derived by applying a Cyberlog rule. b) A claim is passed directly to the monitor, e.g., a claim representing an event. c) The monitor's interaction agent loads new claims from the claim database, a tamper-proof storage of claims. This is e.g., done when other monitors add claims to the claim database, which the monitor has not yet received. Also, a query interface exists where the services may request information from the knowledge base of a monitor. For instance, in our use-case, the DOM may query `good_rtf_exists` before starting a drone.

3.3 Tamper-Proof Claim Database and Auditability

When monitoring properties in the cloud, we need to assure that for each claim in the KB of a security monitor, the complete chain of how it was derived can be reconstructed later.

To achieve this, all Cyberlog rules and generated facts are stored in the tamper-proof claim database which provides the following guarantees: 1) *Immutability*: Once data has been stored in the database, it cannot be changed afterwards. 2) *Protection against manipulation*: Manipulation of the database can be detected. 3) *Proof of Inclusion*: The database provides cryptographic means to verify that a given claim is included in it[7]. The security monitors use inclusion proofs to make sure that only recorded claims are used in reasoning. Inclusion proofs are necessary, e.g., to cope with the situation where an attacker attacks the interface of the claim database so that submitted claims are not stored or stored in a modified form, but are reported to the monitors being stored in the original form.

Auditability of claims is then enabled by storing evidence for claims in the database. Different kind of evidence exist: (1) For claims derived using rules, the evidence is the instance of the rule used to derive the claim. E.g., for a claim `request(id,data,time)` the evidence would be the rule in line 9 with the claims `postRequest('/servicerequest', time, data)` and `get_param_int (data, 'request_id', id)` in Listing 1.1. (2) For claims loaded from the claim database, the evidence is an *inclusion proof* of the claim in the claim database. (3) Claims which have been submitted directly to the KB, such as `postRequest(...)` (i.e., API requests), are atomic assertions that cannot be verified further.

4 Claim Revision Control

We have outlined how Cyberlog allows us to track events and to automatically derive auditable claims about the systems execution. While flexible, the approach is naive with respect to the resource consumption of monitors: The KB of the individual monitors will only grow larger during the operation of the system. For the continuous operation of the security monitors, it is essential to bound the amount of data in the local KBs.

As there seems to be no general criterion to decide when a claim will not be needed anymore for reasoning, claims can not be removed fully automatically. Instead, we propose an approach where the user can control the local KB explicitly by retracting claims using *rulesheet* specifications.

Our approach can be seen as a generalization of Dedalus [2], an extension of Datalog with a discrete notion of time. Dedalus models logic programming with a discrete notion of times, by annotating claims with time and using rules such as `next a :- b, c`. The rule expresses that if b and c are true at some time t, then a will be true at time $t+1$. However, for monitoring applications, one would be interested in storing only the claims of the current time.

---
[7] https://transparency.dev/verifiable-data-structures/.

One of the main contributions of this paper is to generalize Dedalus to a workflow language for distributed claim revision control. In particular, we replace the time indices by revision in a revision system inspired by version control systems such as git or mercurial.

4.1 Revision Model

The revision model works as follows: new knowledge generated by the monitors in their local KBs is considered as a *staging* revision of claims. This revision is completed by *committing* it to the claim database. Once stored in the claim database, revisions cannot be modified. However, it is possible to revise a revision, e.g., to revoke claims, by committing a new revision that *supersedes* it.

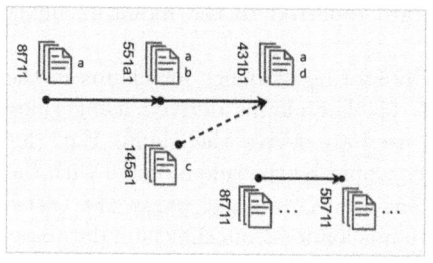

```
Revision 431b1:
 - owner: SB
 - supersedes: 551a1
 - includes: 145a1
 - rules: <reference to rules>
 - claims:
     - a (evidence: signed by SB)
     - d (evidence: derived by
           rule d :- b, c)
```

(a) Evolving revisions of claims over time.

(b) Revision information.

Fig. 3. Revision model with superseded and included revisions of claims.

The main concept in the revision model is shown in Fig. 3. A revision (black dot) stores a set of Cyberlog claims and is identified by a unique ID, typically a hash like 431b1. Our revision model accounts for two possible relations between revisions: a revision can *supersede* another revision (represented by a solid arrow, e.g., 551a1 supersedes 8f711) or it can *include* another revision (represented by a dashed arrow, e.g., between 145a1 and 431b1). Superseding another revision means that claims in the superseded revision are replaced by claims in the superseding revision. Only the owner of a revision is allowed to commit a superseding revision. The inclusion of one revision in another means that the included claims can be used as facts when applying rules.

To enable auditability, each claim in a revision is stored together with evidence to make it possible to reconstruct how it was generated. For instance, having the claim d in revision 431b1 requires that the claims b and c do exist. The inclusion relation between revisions allows us to include claims by other principals when applying rules, and all claims from an included revision may be used to generate new claims by applying rules.

4.2 Monitoring with Revisions

To integrate the revision model into our monitoring framework, we regard the local KB of each monitor as an incomplete revision, called `staging`. When the monitors start, `staging` is empty. As before, a monitor can add claims to `staging` or facts can be derived by applying rules. A `staging` revision is *transient*, and needs to be `committed` to be stored persistently in the claim database. When committed, the claims from `staging` become a new revision R in the claim database. After a commit, the monitor starts with a fresh `staging` revision that will supersede R when it is committed. The user can define how the fresh `staging` revision should be initialized from R by means of *next-rules* in Cyberlog rulesheets. A next-rule has the form `next a :- b1, ..., bn` and is applied after the commit of a `staging` revision. The claims `b1,...,bn` are claims of the committed revision. This rule has the effect of adding the claim `a` to the new staging revision after a commit.

As an example, recall the example from Listing 1.1. The following example for a Cyberlog next-rule specifies that a request should stay in the staging revision after a commit when it is still in process.

```
1    next request(Id, Data, TimeRequest) :-
2       request(Id, Data, TimeRequest),
3       in_process(Id).
```

The monitors regularly commit their revisions, creating a linear chain where the latest claims are found in the last revision. To access claims from other monitors, a monitor can include facts from specific revisions for reasoning by specifying their revision IDs. If the included revision is later superseded by another revision, then its claims and consequences will be removed and replaced with the claims of the *new* revision.

5 Implementation and Experimental Results

To evaluate the monitoring approach in practice, we have implemented CyberDS with Kubernetes[8] and the Istio service mesh[9].

Claim Database. Our implementation of the claim database is based on the tamper-proof log Trillian [12]. Trillian implements an append-only storage and uses a Merkle-tree data structure for preventing tampering of data and for providing inclusion proofs, as discussed in Sect. 3.3. It is used in large-scale applications, such as Certificate Transparency or sigStore[10].

Security Monitors. The security monitors implement a Cyberlog interpreter and the revision model. They are implemented in OCaml and use an efficient in-memory implementation of Datalog[11] for performing reasoning. In the current

---
[8] https://www.kubernetes.io.
[9] https://www.istio.io.
[10] https://www.sigstore.dev.
[11] https://github.com/c-cube/datalog.

prototype implementation, the monitors are compiled to JavaScript using the ReScript compiler and then executed using the NodeJS runtime to make it possible to use libraries not available in OCaml, e.g., for verifying Trillian inclusion proofs.

Cloud Integration. To apply the monitoring approach to a cloud system, the monitors must be placed into the system to observe events and communicate with the components under observation. Our implementation integrates transparently with an infrastructure built from Kubernetes and Istio. Kubernetes is a system for orchestrating containerized applications. Istio extends Kubernetes with a programmable software-defined network and is responsible for routing traffic, enforcing authorization policies, and similar tasks via so called proxy containers. We deploy the monitors together with the proxy containers with each application component and use the *external authorization service interface* of the proxy container to give the monitors access to the network traffic to perform authorization decisions in the monitors. Figure 4 shows the integration of the security monitor inside a Kubernetes pod (a logical host machine).

We describe the monitoring workflow: An HTTP request which goes to a service, is forwarded to the monitor. The monitor adds information about the request to its local KB. E.g., a POST request generates the fact postRequest ('/endpoint', body, timestamp) in the KB, where body is the message body. Recall that the monitor applies its Cyberlog rules continuously, so all consequences are also added to the KB.

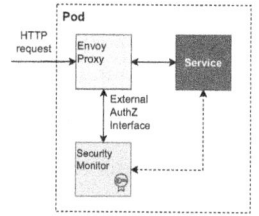

Fig. 4. Security monitor integration using Istio.

Prototypical Evaluation. To demonstrate the applicability of our approach, we deployed the RTAPHM system (s. Section 2) together with CyberDS monitors in an on-premise kubernetes cluster with k3s[12]. We specified the systems behaviour in the rulesheets of the monitors to be able to monitor multiple properties (s. Section 2). We initiated the scheduling process of a UAV through simulating requests by sending API requests to the SB. Outside the cluster a DOM application was deployed and connected to a UAV simulation (Gazebo). We evaluated our framework in two ways:

(1) First we evaluated whether our framework can detect different types of attacks: in the first scenario, we forged UAV base coordinates (i.e., monitoring requirement 3 from Sect. 2). For the evaluation, we have used a flaw in

---

[12] https://k3s.io.

the frontend (web application) to modify UAV coordinates of three bases which are stored in the SB component. In the second scenario, we forged a ready-to-fly message that contained an ID of a service request that does not exist and sent the message to the MRM (i.e., monitoring requirement 1 from Sect. 2). Using our framework, in both scenarios, all forged messages could have been detected and the launching of the UAV was prevented. The monitor implementation and definition of the rulesheets for the case study are available online [22].

(2) In the second part, we estimated the performance of our approach. In particular, we triggered multiple parallel booking processes and evaluated *processing delays* caused by forwarding API requests to security monitors. Further, we measured the amount of facts stored in each KB of a monitor. The minimal delay was 2ms (at SB), while the average delay was 29ms. The maximal introduced delay among all monitors was 113ms at MRM which is relatively high. This is not surprising, as in the rules of the MRM monitor mostly JSON arrays have to be processed and computation time is required to generate facts for each element of an array. Further, the minimal overall delay was 135ms, and maximal 157ms. We believe that the execution delays can further be reduced by parallelizing the fact reasoning in the monitor. Considering the number of facts, the minimal KB size was 163 (at SB), while the maximal was 259 (at CA). After one service request the KB size increased maximally by less than 184 facts (maximal fact size is 15KB). The monitors' processing required 1mCPU, which is negligible.

6 Discussion

One of the limitations of our approach is to deal with a threat, where all private keys for signing claims get stolen and an attacker generates all required evidence for a claim. Furthermore, organizational challenges have to be considered, for instance, trust is required that a party agrees launching a monitor next to its service, and communication is logged correctly. Information processed by a component could be confidential, and only accessible by authorized components. A fundamental assumption in implementing a claim database using centralized storage like Trillian is relying on a trusted operator. Redundant storage of Trillian logs can enhance resilience against database failures. Maintaining multiple copies of the log across different parties allows manual intervention in case of storage inconsistencies. To detect compromised claim databases that present conflicting records (split view attack), gossiping, as suggested by Meiklejohn et al. [19], could be employed. The performance of these solutions in conjunction with our monitoring approach requires further evaluation. Regarding the expressiveness of the specification language, we cannot express negated properties in Cyberlog. However, it is possible to specify rules with negation in a revision, since a revision is immutable. It is future work to extend our language to support this.

7 Related Work

We give a brief overview of related work for the monitoring of distributed applications and describe some work related to the auditability of system events.

7.1 Runtime Monitoring

Runtime monitoring of distributed applications has been exhaustively studied. Many approaches are based on generating automata from temporal logic specifications or using stream processing [6,9–11]. For instance, Sen et al. [21] introduce a distributed specification language to generate monitors. Bauer et al. [5] address the distributed synthesis of monitors from LTL specifications based on system execution traces. Gan et al. [11] present an approach to generate finite state automata for monitoring liveness/safety properties of web services.

However, these approaches are limited in expressing data-related properties compared to our approach. Also, the contribution of our paper lies not in proposing a distributed monitor synthesis algorithm, but in defining a framework for the transparent monitoring of properties in the cloud.

Cotroneo et al. [9] have proposed a stream processing approach for monitoring cloud systems. API requests are first logged in a fault-free system execution to learn the correct behaviour of the system using rules to observe later whether the system obeys the rules. In this approach, it is not possible to specify properties that consider data in the API requests, and monitoring results are not provided back to the components, to control the overall systems execution.

Our monitoring approach is related to the authorization framework Open Policy Agent (OPA)[13], which decouples the enforcement of authorization policies from the application code. Policies are specified in a declarative language called Rego which is similar to Cyberlog. While OPA has an interface to integrate external data when applying rules, processed and reasoned data is not auditable and no attentions are supported. Also, since OPA is only an authorization framework, no monitoring of workflow events is possible, compared to CyberDS.

7.2 Auditability of System Events

Google's Certificate Transparency (CT) is a prominent example for making decisions based on auditable data: the framework aims to detect falsely or maliciously issued certificates by recording all HTTPS certificates from publicly trusted Certificate Authorities in the tamper-proof log Trillian [12]. Trillian serves as a scalable alternative to distributed, tamper-proof data storage systems like blockchain. This concept has been adopted in various projects, including supply chain security in sigStore, identification of malicious firmware with Binary

---

[13] https://www.openpolicyagent.org.

Transparency [13], detection of malicious software updates from package managers [16], and verifiable auditing of (confidential) data accesses [15]. Our approach uses the underlying tamper-proof storage Trillian to store system events and uses its API for retrieving inclusion proofs of data.

An extension of standard SIEM approaches is provided by syslog-ng[14] ensuring secure transmission and tamper-proof storage of logged data for compliance. Nevertheless, it is not possible to monitor requirements which depend on this data, so that components can incorporate the monitoring results in their decisions.

Another work [20] has applied blockchain for the runtime monitoring of business processes. Recorded events are auditable, but the transaction time takes more than 7 min. This approach is therefore only applicable in a use case with long-running tasks, such as logistics or manufacturing. Further, a blockchain deployment is complex and requires a considerable amount of resources, in contrast to our approach.

Hoye et al. [23] have designed and implemented a logging approach based on Hyperledger to share auditable information between organizations. However, they build on different assumptions of trust between partners and the availability of system components.

8 Conclusion

We have presented, CyberDS, a novel approach for monitoring cloud applications with tamper-proof guarantees. Our flexible monitoring approach uses a declarative and rule-based specification language for the configuration of the monitoring system. Monitors are injected transparently into the cloud system to track the communication between the services and to continuously monitor properties. The monitors are connected to the verifiable storage Trillian to provide auditable information for the reasoning process as well as for documentation purposes to satisfy regulatory requirements. Further, we presented a revision model, to cope with the increasing local monitor storage of runtime events. We demonstrated CyberDS on a use case from the avionics domain, where our framework could identify different types of attacks and achieve good performance results regarding the application example. We are investigating the automated configuration of monitors, i.e., the generation of Cyberlog-programs from API specifications and use case information to ease the application of our approach. We assume that our approach can contribute to the security of cloud systems and promote the deployment of applications for safety-critical systems in the cloud.

Acknowledgment. This work was supported by the Federal Ministry for Economic Affairs and Climate Action in the RTAPHM project, grant 20X1736M (Lufo V-3). We thank also colleagues and reviewers for their valuable feedback on that work.

---

[14] https://www.syslog-ng.com/.

Disclosure of Interests. The authors have no competing interests to declare that are relevant to the content of this work.

References

1. Alvaro, P., Condie, T., Conway, N., Elmeleegy, K., Hellerstein, J.M., Sears, R.: Boom analytics: exploring data-centric, declarative programming for the cloud. In: European Conference on Computer Systems (EuroSys 2010), pp. 223–236. ACM (2010)
2. Alvaro, P., Marczak, W.R., Conway, N., Hellerstein, J.M., Maier, D., Sears, R.: DEDALUS: datalog in time and space. In: de Moor, O., Gottlob, G., Furche, T., Sellers, A. (eds.) Datalog 2.0 2010. LNCS, vol. 6702, pp. 262–281. Springer, Heidelberg (2011). https://doi.org/10.1007/978-3-642-24206-9_16
3. Bachane, I., Adsi, Y.I.K., Adsi, H.C.: Real time monitoring of security events for forensic purposes in cloud environments using SIEM. In: SysCo 2016, pp. 1–3. IEEE (2016)
4. Balta, D., et al.: Accountable federated machine learning in government: engineering and management insights. In: Edelmann, N., et al. (eds.) ePart 2021. LNCS, vol. 12849, pp. 125–138. Springer, Cham (2021). https://doi.org/10.1007/978-3-030-82824-0_10
5. Bauer, A., Falcone, Y.: Decentralized LTL monitoring. Form. Meth. Syst. Des. (2016)
6. Bratanis, K., Dranidis, D., Simons, A.J.H.: An extensible architecture for run-time monitoring of conversational web services. In: MONA 2010, pp. 9–16. ACM (2010)
7. Cassar, I., Francalanza, A., Aceto, L., Ingó lfsdóttir, A.: A survey of runtime monitoring instrumentation techniques. EPTCS **254**, 15–28 (2017)
8. Ceri, S., Gottlob, G., Tanca, L.: What you always wanted to know about datalog (and never dared to ask). IEEE Trans. Knowl. Data Eng. **1**, 146–166 (1989)
9. Cotroneo, D., De Simone, L., Liguori, P., Natella, R., Scibelli, A.: Towards runtime verification via event stream processing in cloud computing infrastructures. In: Hacid, H., et al. (eds.) ICSOC 2020. LNCS, vol. 12632, pp. 162–175. Springer, Cham (2021). https://doi.org/10.1007/978-3-030-76352-7_19
10. Decker, N., Kühn, F., Thoma, D.: Runtime verification of web services for interconnected medical devices. In: ISSRE 2014, pp. 235–244. IEEE (2014)
11. Gan, Y., Chechik, M., Nejati, S., Bennett, J., O'Farrell, B., Waterhouse, J.: Runtime monitoring of web service conversations. In: CASCON 2017, pp. 2–17. ACM (2017)
12. Google: Trillian. https://transparency.dev/#trillian
13. Google: Firmware transparency (2021). https://github.com/google/trillian-examples/tree/master/binary_transparency/firmware
14. Harper, A., VanDyke, S., Blask, C., Harris, S., Miller, D.: Security Information and Event Management (SIEM) Implementation. McGraw-Hill Osborne Media (2010). https://doi.org/10.1036/9780071701082
15. Hicks, A., Mavroudis, V., Al-Bassam, M., Meiklejohn, S., Murdoch, S.J.: VAMS: verifiable auditing of access to confidential data. CoRR abs/1805.04772 (2018)
16. Hof, B., Carle, G.: Software distribution transparency and auditability. CoRR abs/1711.07278 (2017)
17. Influxdata: Influxdata. https://www.influxdata.com/blog/will-kubernetes-collapse-under-the-weight-of-its-complexity/

18. Leveson, N.: A new accident model for engineering safer systems. Saf. Sci. 237–270 (2004)
19. Meiklejohn, S., et al.: Think global, act local: gossip and client audits in verifiable data structures. CoRR abs/2011.04551 (2020)
20. Prybila, C., Schulte, S., Hochreiner, C., Weber, I.: Runtime verification for business processes utilizing the bitcoin blockchain. Future Gener. Comput. Syst. (2020)
21. Sen, K., Vardhan, A., Agha, G., Rosu, G.: Efficient decentralized monitoring of safety in distributed systems. In: ICSE 2004, pp. 418–427. IEEE (2004)
22. Tool: Cyberlog-Monitoring. https://git.fortiss.org/sd-tools/cyberlog-monitoring
23. Van Hoye, L., Maenhaut, P.J., Wauters, T., Volckaert, B., De Turck, F.: Logging mechanism for cross-organizational collaborations using hyperledger fabric. In: 2019 IEEE International Conference on Blockchain and Cryptocurrency (ICBC), pp. 352–359. IEEE (2019)

Automated Driving Systems

Anatomy of a Robotaxi Crash: Lessons from the Cruise Pedestrian Dragging Mishap

Philip Koopman

Carnegie Mellon University, Pittsburgh, PA, USA
koopman@cmu.edu

Abstract. An October 2023 crash between a GM Cruise robotaxi and a pedestrian in San Francisco resulted not only in a severe injury, but also dramatic upheaval at that company that will likely have lasting effects throughout the industry. Issues stem not just from the loss events themselves, but also from how Cruise mishandled dealing with their robotaxi dragging a pedestrian under the vehicle after the initial post-crash stop. External investigation reports provide raw material describing the incident and critique the company's response from a regulatory point of view, but exclude safety engineering recommendations from scope. We highlight specific facts and relationships among events by tying together different pieces of the external report material. We then explore safety lessons that might be learned related to: recognizing and responding to nearby mishaps, building an accurate world model of a post-collision scenario, the inadequacy of a so-called "minimal risk condition" strategy in complex situations, poor organizational discipline in responding to a mishap, overly aggressive post-collision automation choices that made a bad situation worse, and a reluctance to admit to a mishap causing much worse organizational harm downstream.

Keywords: Autonomous vehicles · robotaxi crash · regulation · safety lessons

1 Introduction

On October 2, 2023, a Cruise robotaxi Autonomous Vehicle (AV) in San Francisco struck and then – as part of a subsequent maneuver – dragged a pedestrian under the vehicle as part of a complex mishap scenario. A different, human-driven vehicle struck the pedestrian first. Cruise failed to proactively disclose the pedestrian dragging portion of the mishap. Many Cruise leaders were sacked, followed by a 24% workforce cut.

We examine the events of the mishap and the aftermath based on information made public in an external investigation report commissioned by Cruise [6]. That report is a single document file that contains two stand-alone parts with independent page numbering. The majority of references in this paper are to these two documents, with the tag for the report as indicated and a relevant report page number:

- QER: The primary report by Quinn Emanuel Trial Lawyers, with an emphasis on resolving questions of regulatory compliance and potential culpability. Much of its content is about who knew what, and who said what to whom, when. (109 pages)

- EXPR: A redacted technical root cause analysis for the mishap events. (86 pages)

The QER and EXPR state safety engineering and operational safety practices are out of their scope (QER 6, EXPR 13). However, they contain enough information to reconstruct a sequence of events that yields safety insights beyond what might be evident from a straightforward reading of the narrative and conclusions of those reports. A primary goal of this paper is to present a clarified view of those events.

At a high level, the QER portrays the crash itself as follows (QER 9–10). A pedestrian crossing against a "Do Not Walk" pedestrian signal stopped in traffic, and was first hit by a human driver in a different lane, who fled the scene. The pedestrian was "launched" into the path of the Cruise robotaxi AV by that first impact. The AV braked hard but was unable to avoid hitting the pedestrian. The subsequent pedestrian dragging was due to the AV doing something designed to enhance safety (achieving a so-called Minimal Risk Condition[1]) that went wrong due to the AV not detecting the pedestrian under the vehicle. None of this would have happened if a hit-and-run human driver of an adjacent vehicle had not hit the pedestrian first (QER 1,10).

The QER portrays the crash response and regulatory interfaces as due to "poor leadership, mistakes in judgment, lack of coordination, an 'us versus them' mentality with regulators, and a fundamental misapprehension of Cruise's obligations of accountability and transparency to the government and the public" (QER 7). The QER calls the ultimate regulatory operational suspension order "a direct result of a proverbial self-inflicted wound" due to mishandled interaction with regulators (QER 7).

A closer look at the material in the reports, however, reveals there is much to understand beyond that narrative. Indeed, the human driver in the other vehicle acted badly, and since-sacked Cruise management mishandled the crisis. However, there is much more texture to the situation, including significant technical shortcomings in the AV's design as well as significant room for improvement in operational safety procedures.

The AV potentially violated a California road rule by accelerating toward a pedestrian in a crosswalk. The AV had trajectory information on the human-driven vehicle and pedestrian a few feet away, but failed to recognize it as a pending collision. The AV could have stopped much sooner, even if only reacting to the pedestrian intrusion into the AV's lane, avoiding or mitigating initial impact injuries. The pedestrian was not thrown completely onto the ground at impact, but rather was at least partly on the hood of the AV before being run over. Multiple AV sensors indicated the pedestrian's presence under the vehicle, but the AV failed to recognize the scenario. The AV proceeded with a post-crash maneuver essentially instantly without giving a remote assistance team time to assess the situation. And, the contractor-staffed remote assistance team seemed to know of the dragging essentially immediately, but the Cruise crisis response team had to figure that out for themselves early the next morning.

We agree with the QER that cultural issues contributed to the regulatory mess created. But, we go further and say that fundamental changes are needed in the technical systems, operational procedures, and crisis response approach to address deep safety concerns beyond just the compliance topics within the scope of the QER.

---

[1] Despite the name, an MRC does not guarantee global minimal risk, but rather is merely a "stable stopped" condition "to reduce the risk of crash." The risk need not be objectively "minimal", and the defining SAE J3016 recommended practice disclaims safety scope [7].

2 Background

2.1 Terminology

Below are some key terms, with Cruise-specific terms defined based on the QER.

- CIRT: Cruise Incident Response team: Cruise employees in San Francisco.
- CMT: Crisis Management Team: Cruise employees in San Francisco. This seems to be a follow-on to the CIRT, although the relationship is not explained in the QER.
- CPUC: California Public Utilities Commission: state government regulator of paid commercial transport permits in California.
- CRAC: Cruise Remote Assistance Center: contractors in Arizona.
- Cruise: a subsidiary of General Motors that is in the robotaxi business. Cruise is headquartered in San Francisco, with some relevant staff working remotely.
- DSS: Driverless Support Specialist contractors who provided on-scene support.
- DMV: California Department of Motor Vehicles: state government safety regulator.
- NHTSA: National Highway Traffic Safety Administration: US Federal government vehicle equipment safety regulator.
- City: Generic term encompassing San Francisco Metropolitan Transportation Authority (SF MTA), San Francisco Fire Department (SFFD), and the San Francisco Police Department (SFPD).

2.2 Crash Context and Overview

Cruise had been previously granted permission to operate without a safety driver by DMV, and granted permission to charge for robotaxi rides by CPUC. Numerous on-road incidents had occurred that established tension between Cruise and the City, especially involving interference with emergency responders and public road congestion. The CPUC public hearings were particularly contentious. A collision with a fire truck that resulted in an AV passenger injury occurred a week after the most recent CPUC approval, after which DMV requested that Cruise reduce its operational tempo [4].

There was no human driver in the mishap AV, nor was there a continuous real-time remote safety supervisor. The mishap AV had no passengers on board. The CRAC operations center located in Arizona (perhaps 700–800 miles distant) oversaw remote operations. The DSS support team provided local on-road support in San Francisco, for example, to recover disabled vehicles. Engineering activities and the CIRT/CMT crisis response teams were centered in San Francisco, but drew upon staff in other states.

A mishap overview starts at EXPR 14. The mishap occurred starting at 9:29 PM US PT on October 2, 2023. Environmental factors are not mentioned as contributing issues, although at that time it would have been dark with some streetlight illumination. The speed limit was 25 mph (EXPR 18), with a maximum AV mishap speed of 19.1 mph.

Figure 1 shows a simplified diagram of the mishap geometry. At the start of the mishap sequence, two vehicles were stopped at a red traffic light next to each other at an intersection on a 4-lane surface road: the Cruise AV on the curb lane, and a human-driven Nissan in the medial lane, adjacent to the two-way street center-line.

After the vehicles' light had changed to green, a pedestrian entered the crosswalk on the far side of the intersection. The pedestrian first crossed the AV's lane, and then the

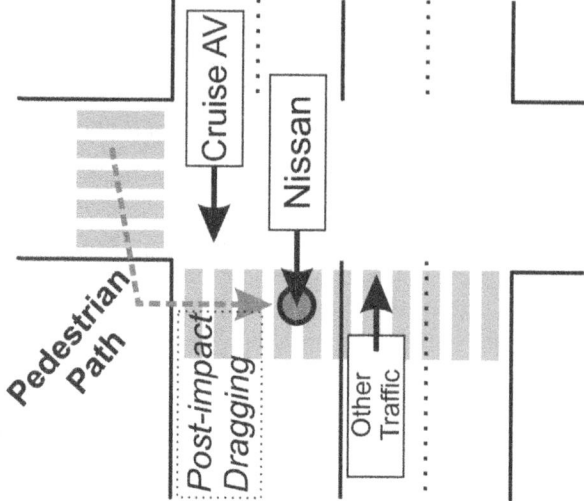

Fig. 1. Simplified diagram of mishap; not to scale.

Nissan's lane. Both vehicles accelerated straight through the intersection toward the far crosswalk as the pedestrian traversed their paths. The pedestrian stopped in front of the Nissan, with her path blocked by oncoming traffic in the other direction. The pedestrian was likely distracted from the approaching Nissan ("did not look left") while attempting to signal opposing traffic to permit continuing the crossing (interpretation of description at EXPR 47). The Nissan struck the pedestrian in its lane, braking only after impact.

After an impact interaction with the Nissan, the pedestrian separated from the Nissan and entered the AV's travel lane. The AV started braking just before impact, hitting the pedestrian at close to its maximum pre-impact speed, coming to a quick stop. The AV's front wheel ran over the pedestrian, leaving the pedestrian under the AV.

A split-second after coming to a stop, the AV initiated a pullover maneuver to achieve a so-called "Minimal Risk Condition" (MRC), incorrectly assessing the situation as a side impact rather than a run-over scenario (EXPR 16). CRAC remote operators connected with the vehicle while it was partway through this maneuver.

The pedestrian was entrapped under the vehicle, and was dragged along the pavement for 20 feet at speeds up to 7.7 mph. The AV might have continued this dragging situation for up to 100 feet (QER 14). However, the AV recognized a vehicle motion anomaly and stopped the motion prematurely without recognizing a trapped pedestrian.

We consider details in three phases: the day of the mishap (crash and immediate response), the day after the mishap (post-crash response), and the longer-term regulatory interactions. The regulatory interactions lasted for several weeks, with Cruise ultimately shutting down their public robotaxi operations for an extended period. Each phase presents lessons to be learned not only by Cruise, but by any company testing or deploying automated vehicle technology on public roads.

All times given in this paper are in the US PT (local) time zone. Relative times are from the time of the initial AV impact with the pedestrian, rounded to the nearest time unit available in the source material. While the QER and EXPR have several timelines and additional information in various places, we integrate the timelines to show some relationships and considerations not readily apparent in the source material.

3 Crash Details

The QER states "But for the human driver of the Nissan hitting the pedestrian, the October 2 Accident would not have occurred" (QER 10). While narrowly true, this is not the whole story. Adverse events not caused by a robotaxi can be expected, and to achieve acceptable safety it is a near certainty that an AV will need to be able to do something reasonable in response to the vast majority of such events. As a counterfactual example, if the pedestrian had seen the Nissan coming, she might have suddenly run back across the AV's path to reach the sidewalk. Given that the AV was tracking the pedestrian being caught by traffic in an adjacent lane, information regarding an emergent high-risk traffic situation was available to, but not recognized by, the AV.

Regardless of what might have been, more relevant to improving future safety are: What went right? What went wrong? And what lessons might be learned?

3.1 Crash Timeline

The following timeline highlights points especially relevant for our discussion. More detailed timelines and time series data graphs are available, but require some integration on the part of the reader (QER 10 *et seq.*, 18; EXPR 43–45, 50, 54, 68, 83, 85). Figure 2 shows a portion of the velocity and acceleration graphs during the mishap sequence.

Items in this timeline are sourced from EXPR 44 unless otherwise noted.

- −38.3 s: Pedestrian tracking begins. Pedestrian crosses street parallel to later vehicle motion, then turns left to cross in front of the vehicles on the far side of intersection. Pedestrian remains in clear sight of AV the entire time. Whether the AV tracked pedestrian gaze as attempting "eye contact" with AV or not before starting to cross at −7.9 s is not mentioned (EXPR 19, 48–49).
- −10 s.: Traffic light changes; both vehicles are motionless.
- −9.2 s: By this time, both Nissan and AV are accelerating straight toward an empty crosswalk on the far side of the intersection. The Nissan leads slightly.
- −7.9 s: Pedestrian enters crosswalk. AV traveling approximately 5.5 mph (Fig. 2).
- −7.7 s: AV predicts pedestrian will cross AV's lane. AV continues accelerating.
- −5.3 s: Pedestrian leaves AV travel lane after 2.6 s. AV traveling approximately 13.5 mph, and continues to accelerate (Fig. 2).
- −4.8 s: AV predicted paths for Nissan and Pedestrian "are consistent with a potential collision", but "the ADS did not consider the potential of a collision between the Nissan and the pedestrian" (EXPR 15, 50, 53, 77).
- −4.7 s: Pedestrian stops/pauses in Nissan's travel lane, blocked by traffic in the opposing direction, and remains in crosswalk (EXPR 47).
- −4.6 s: AV biases right in its lane, possibly due to the presence of the pedestrian in adjacent lane (EXPR 51).

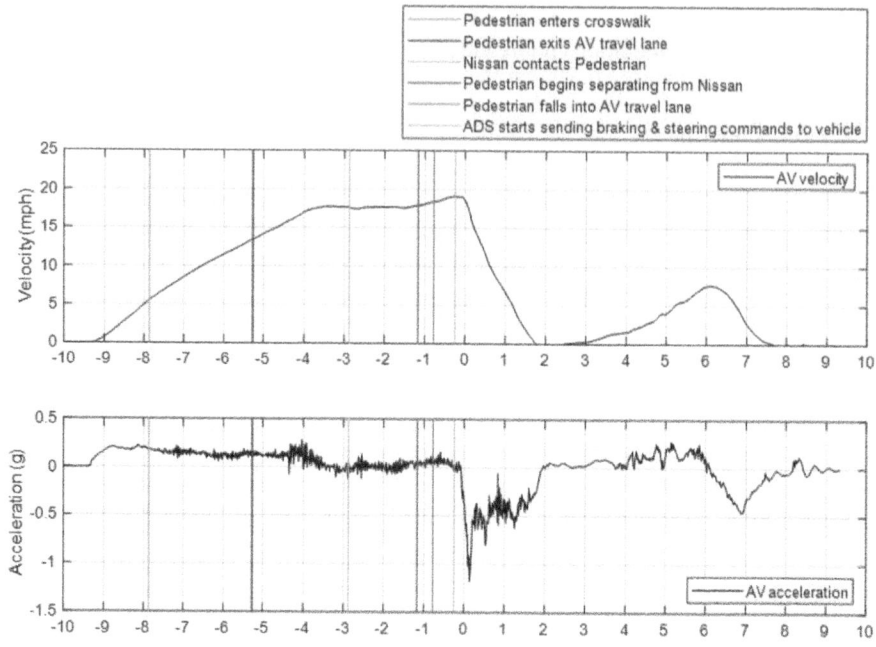

Fig. 2. AV velocity and acceleration profiles from EXPR 43.

- −2.9 s: Nissan impacts pedestrian at 21.7 mph without prior braking (EXPR 15, 52). AV is now at 17.6 mph (EXPR 53) and maintains this speed through the next second (EXPR 55). The pedestrian is moving at 2.6 mph at the time of collision, but the direction of motion is redacted (EXPR 51–52). The AV onboard camera captures a frame of that collision event, which was in view of the AV (EXPR 78).
- −2.0 s: AV Pedestrian tracking is dropped. However, intermittent classification and tracking continues until −0.3 s (EXPR 79).
- −1.5 s (approx.): AV resumes gentle acceleration (Fig. 2)
- −1.17 s: Visual separation of Pedestrian from Nissan. AV at 17.9 mph (EXPR 15).
- −1.07 s: Nissan initiates hard braking (brake lights) (EXPR 78).
- −0.78 s: Pedestrian enters AV's travel lane approximately 21.5 feet in front of AV. AV at 18.4 mph (EXPR 66).
- −0.78 s: Initiating braking now would have completely avoided pedestrian contact, but AV did not brake, instead increasing speed slightly (EXPR 66, Fig. 2).
- −0.41: imminent collision predicted by AV "collision checker".
- −0.3 s: AV last correct classification as a pedestrian (EXPR 15, 79).
- Intermittent classification and tracking of pedestrian. However, AV detecting "occupied space", leading to steering and braking commands (EXPR 15, 79).
- −0.25 s: AV sends steering and braking commands. AV at 19.1 mph (EXPR 15).
- −0.2 s: Nissan has completely stopped[2]

---

[2] There is a small discrepancy with the timing information given, with possible values of ranging from -0.2 to + 0.4 s (EXPR 56–57).

Anatomy of a Robotaxi Crash   125

- 0 s: AV initial impact with pedestrian. AV has slowed by 0.5 mph to 18.6 mph (EXPR 15). "Front bumper first contacted pedestrian" (EXPR 15). Collision detection system incorrectly identifies a side impact (EXPR 16). CRAC saw the pedestrian initially on the AV hood in a time-delayed video (QER 21).
- 0.23 s: AV left front wheel runs over pedestrian (EXPR 16, 81).
- 1.78 s: AV braking achieves a zero net speed (EXPR 16) for a fractional second.
- 1.83 s: AV begins post-impact acceleration toward MRC. AV starts dragging the pedestrian approximately 20 feet at speeds of up to 7.7 mph (EXPR 16, 82–83).
- "within seconds" after impact, AV sends a 3-s collision video to CRAC that includes just the collision event (QER at 18).
- 3.8 s: Left front wheel spins at perhaps 20 mph in a "traction control event" due to entrapped pedestrian "physically resisting the motion of the vehicle" (EXPR 16, 84, Fig. 64).
- Within 5 s: CRAC connected to AV video and audio. CRAC saw "ped flung onto hood of AV. You could see and hear the bumps" and the AV "was already pulling over to the side." (QER 21, 61)
- 5.8 s: Degraded mode entered due to wheel slip caused by pedestrian's legs (EXPR 16,84) is said to trigger an "immediate stop." In reality this mode triggers "gradually slowing to a stop" (EXPR 36), consistent with timeline. AV might instead have dragged the pedestrian up to 100 feet or one full block (QER at 34).
- 8.8 s: Final point of rest for AV reached (EXPR 16, 84). Pedestrian is largely under rear of vehicle. Legs protrude from left rear, with tire on top of at least one pedestrian leg (QER 33). The pedestrian's feet and lower legs were visible in the wide-angle left side camera view throughout the event (EXPR 83). Legs were briefly detected but neither classified nor tracked after the collision (EXPR 16).

3.2 Crash Analysis

What Went Right: An undifferentiated obstacle was seen in front of the AV, prompting an emergency stop due to a presumption that it might be a vulnerable road user (EXPR 17,32). The AV noticed an impaired ability to move accurately and entered degraded states accordingly. These degradations played a role in preventing the outcome from being worse, avoiding potentially dragging the pedestrian up to 100 feet.

Accelerating into Pedestrian: The AV accelerated the entire time that it detected a pedestrian in a crosswalk in its travel lane, with speed more than doubling (8 mph increase) during the 2.6 s the pedestrian was in its travel lane. This was due to the AV's machine learning-based (EXPR 29) prediction that the pedestrian would be out of the way by the time the AV arrived (EXPR 44).

Accelerating into a pedestrian in a crosswalk is inconsistent with California Rules of the Road, which state: "(c) The driver of a vehicle approaching a pedestrian within any marked or unmarked crosswalk shall exercise all due care and <u>*shall reduce the speed of the vehicle*</u> or take any other action relating to the operation of the vehicle as necessary to safeguard the safety of the pedestrian." ([1] with emphasis added) Other California Vehicle Code sections relevant to the mishap are listed on QER 9 fn 12.

Delay in Braking: The laws of physics did not preclude stopping in time to avoid AV/pedestrian impact. The AV had enough time to totally avoid impact if it had immediately responded to the pedestrian presence in its lane, even accounting for braking latency (EXPR 66). Even a delayed but earlier braking "would have potentially mitigated severity of the initial collision" (EXPR 66). Due to a slower-than-possible response, AV aggressive braking beyond 0.5 g only occurred after impact time 0 (Fig. 2).

Brittle Pedestrian Position Prediction: The AV tracked the pedestrian coming to a stop in a traffic lane and to the point of being hit by an adjacent vehicle, which then braked aggressively. Any of these factors could have been treated as a red flag that a high-risk situation was developing, but were not.

The QER recounts statements regarding the impracticality of tracking a pedestrian in this particular impact incident, with staff using terms such as "unrealistic" and "insane hypothetical" (QER 66). However, at least some drivers would have braked or performed an evasive maneuver in this type of situation (EXPR 66).

Vanishing Pedestrian: The AV acted in a way consistent with "forgetting" a pedestrian was in the vicinity. The AV did not stop because of a tracked pedestrian in its lane, but rather because there was occupied road space immediately in front of it that it assumed could be a vulnerable road user. It seems inappropriate to "forget" a pedestrian who has just been hit by a vehicle a few feet away. Moreover, there was sensor information showing the pedestrian, but the system was not up to the tracking challenge.

Near-Impact Sensor Self-occlusions. Impact diagnosis depended on available object tracking information immediately prior to impact. However, this impact occurred in an area in which the pedestrian was substantially vehicle-self-occluded from lidar sensors, contributing to an incorrect assessment of the position of the pedestrian during and after impact (EXPR 80–81). Sensor self-occlusions near and under the vehicle contributed to AV misdiagnosis of the immediate pre-crash and post-crash situations.

Moving with an Entrapped Pedestrian: The vehicle had recent historical information available that a pedestrian was likely to have been struck and then lost to tracking. This situation should have required at least an override from CRAC before moving the vehicle further. Instead, the vehicle moved again, entirely on its own, approximately $1/20^{th}$ of a second after stopping. While a mapping error contributed to initiating movement (EXPR 17), any such system will need to be robust in the face of inevitable mapping errors. EXPR 17,76 admit: "After the AV contacted the pedestrian, an alert and attentive human driver would be aware that an impact of some sort had occurred and would not have continued driving without further investigating the situation."

Functional Insufficiencies. The EXPR asserts no sources of hardware or software failure were identified (EXPR 14). This emphasizes the need to account for functional insufficiencies [3] beyond more narrowly defined implementation defects.

The Sensor-Based Safety Narrative. The AV industry safety narrative commonly features an emphasis on superior sensor coverage combined with superhuman reaction time. Yet in this mishap the pedestrian was not tracked accurately, and braking was not initiated particularly quickly for a computer driver. EXPR 17, 85 admit: "The AV's lack of anticipation of a potential future incursion of the pedestrian into its travel lane was a contributing factor to this incident."

Sensors and fast reactions only present the possibility of safety, which must be complemented by robust classification and tracking capabilities. The reasoning ability of the vehicle did not include consideration of a disappearing pedestrian who had just been hit. Despite multiple kinesthetic clues and AV camera imagery showing the legs of a pedestrian being dragged under the vehicle, the AV proceeded with its pullover maneuver, continuing until time 8.8 s.

Safety Comparison Baseline. The EXPR narrative switches apparent standards of comparison from a super-proficient robot (accelerating into a pedestrian against California Rules of the Road because it is confident the road will be clear when it arrives), to a naïve robot (failing to predict a collision based on intersecting tracks between a pedestrian and a neighboring vehicle), to an immature robot that cannot deal with the unexpected (not tracking a pedestrian deflected into its travel lane; forgetting that a pedestrian a few feet away is likely still somewhere close when tracking is lost), to a reasonable human driver (who could not have reacted that fast, even while saying a robot could have), and back to an immature robot that had no way to know the thing it saw in its camera and was driving over with ample sensory clues was in fact that same nearby untracked person. (See especially EXPR Sects. 3.2.3 and 3.3.4.)

Off-Nominal Situations and Safety. At a higher level, the scenario points out that robotaxis will need to deal with severe off-nominal scenarios with potentially high consequences. Lacking a human safety driver, the opportunities for disastrously incorrect decisions in complex, messy situations based on incomplete sensor data in scenarios the design team feels are unforeseeable seem numerous.

Cruise internal discussion statements such as "the pedestrian is well past our lane of travel into the other lane" (QER 39) and "it would not be reasonable to expect that the other vehicle would speed up and proceed to hit the pedestrian, and then for the pedestrian to flip over the adjacent car and wind up in our lane" (QER 39) do not change the fact that the mishap occurred, and will not prevent future similar mishaps.

3.3 Potential Lessons

Below are some potential lessons that could be helpful for Cruise and other designers of automated vehicle functions:

- High prediction confidence of pedestrian intent can be a risky basis for significant acceleration. Accelerating toward a vulnerable road user in the own-vehicle path reduces reaction time to surprise events, and can be contrary to rules of the road.
- A vulnerable road user mishap in an adjacent lane can present substantial risk due to inherently unpredictable outcomes, and should not be ignored.
- Arguments that superior sensor capability and fast reaction time will necessarily produce safer-than-human-driver outcomes overlook the more difficult areas of perception and prediction. Post-crash vulnerable road user detection can be expected to be especially challenging.

- Having another road user initiate a mishap does not absolve the AV from a responsibility to react in a reasonable way to inherently unpredictable events.
- Subjective judgements of the reasonableness of scenarios by developers should not override methodical safety engineering practices.
- Automating post-collision actions requires robust sensing and somewhat different prediction capabilities during and after collision.
- A so-called "Minimal Risk Condition" is, at best, a relative statement. At the very least, the position of all vulnerable road users – including potentially under the AV – should be unambiguous before any vehicle movement is attempted.

4 The Immediate Response

4.1 Immediate Response Timeline

1. (within + 5 s) 9:29 PM: Remote Assistance at CRAC connected to AV, apparently with slightly delayed video and audio (QER at 21, 61). The pullover maneuver and pedestrian dragging was still in progress.
2. (+2 min) 9:31 PM: 911 emergency phone call dispatchers were alerted to the crash by a bystander calling in (QER 24). There is no indication that Cruise or their contractors ever contacted City emergency dispatchers.
3. (+3 min) 9:32 PM: The AV sent a 14-s video showing the collision to CRAC, but not the pullover maneuver/dragging segment (QER at 11).
4. (+8 min) 9:37 PM: City emergency responders are on scene. They ask CRAC via AV connection to keep AV in place pending extrication (QER 69).
5. (+10 to + 15 min): Cruise Driverless Support Specialists ("DSS") arrive physically at the scene (QER 20).
6. (+20 min) 9:49 PM: CIRT labels accident "Sev-1" (minor collision). (QER 11) based on input from CRAC.[3]
7. (within + 20 min): National pager system is activated to notify Cruise employees of a "minor" mishap (QER at 22).
8. (+45 min) 10:17 PM: Cruise contractors arrive on scene, noting post-crash AV movement and victim blood and skin patches on the ground (QER 11, 21).
9. (+2 h) 11:31 PM: Cruise raises accident to "Sev-0" (major with injury) and initiates a War Room response, notifying more employees (QER 11).
10. (+2.5 h) 11:55 PM: Cruise CEO joins the War Room Slack channel and shortly thereafter the War Room Google Meet (QER 22).
11. The QER cites a lack of "conclusive evidence" that Cruise employees including senior leadership had knowledge of the pedestrian dragging on Oct. 2 (QER at 23). Some participants recall a discussion about the pedestrian having been dragged before a shift end at 4:00 AM Oct 3 (QER 22 fn 21).

---

[3] The Sev-1 characterization by Cruise is based on "the fact that Cruise's Remote Assistance had characterized the accident a 'minor collision...'" (QER 41). On the other hand, "there is no indication Cruise spoke with the Remote Assistance contractors who were interacting with the AV" until weeks later (QER 103). Perhaps there was messaging rather than voice communications that are not described. In any event, communications and entries in any communication records between CRAC and Cruise seem significantly under-reported in the QER.

4.2 Post-Crash Analysis

What Went Right: The remote assistance channel opened essentially immediately, giving CRAC the ability to see and hear what had happened, including both the AV/pedestrian collision and the dragging sequence. This channel was successfully used to communicate with on-scene responders who, for example, advised CRAC to keep the vehicle stationary. A CIRT initial response occurred relatively quickly after the mishap, providing Cruise with the potential to coordinate a company-wide response.

Failure to Stay on Plan: There was a response playbook, but it was not followed because it was "too manually intensive" (QER 22). Although the War Room was supposed to address accident causation and next steps, there was a quick focus on media narrative "almost exclusively" (QER 23). CIRT arguably did not know about the dragging, and was concerned that the media and an SFFD statement emphasized a pedestrian pinned under the vehicle rather than the initial Nissan collision (QER 24). While this reaction might be attributable to the equivalent of an inevitable Fog of War, deviating from the procedural playbook was a missed opportunity to stay on plan.

Failure to Request Timely Remote Intervention: One of the challenges of immature vehicle automation is ensuring that the AV asks for help when it needs to instead of doing something dangerous without additional support. The AV in this case got that decision wrong. The connection to CRAC was established quickly, but the AV had already initiated the pullover maneuver that undoubtedly led to a severe increase in pedestrian injury, rather than waiting for CRAC guidance.

One can imagine a series of design decisions that lead to such an outcome, especially motivated by intense scrutiny and criticism of previous robotaxi strandings blocking traffic in highly publicized events [5]. Nonetheless, designing a vehicle that all on its own decides to move after striking an unclassified object (presumed to be a vulnerable road user (EXPR 32)) seems a poor system-level safety approach.

Failure to Alert Emergency Responders. There is no indication that Cruise or its contractors initiated contact to City emergency services. Similarly, there is no indication CRAC or any other Cruise operational team considered emergency service contact. The first mention of emergency responders is when they were already at the scene contacting CRAC via physically accessing the vehicle communications feature to ask for the vehicle to remain stationary. By all accounts, CRAC knew a pedestrian injury had occurred. CRAC should have reached out to City emergency responders immediately.

Failure of Coordination Between Operations and Engineering/Management: CRAC and on-the-ground response teams both knew a serious pedestrian injury had occurred. The initial misclassification as Sev-1 is unexplained. According to the QER, CIRT did not realize a pedestrian dragging had occurred in a timely manner. A failure to coordinate the CIRT response with CRAC at the time of an incident, and especially when the incident was upgraded to Sev-0 (major) seems a major process failure. (See [2] for additional perspective.)

Premature Removal of Safety Supervisor: A physically present safety supervisor could have prevented the pedestrian dragging, and arguably initiated a safety stop well before the first pedestrian impact based on a developing high-risk situation. Indeed, many other incidents such as mass AV strandings would likely have been mitigated or avoided entirely by a physically present backup driver in the AV. Removal of the safety supervisor seems motivated by business and publicity interests rather than a primary focus on safety. There were ample warnings that safety supervisors had been removed prematurely in the form of numerous incidents of emergency responder interference.

4.3 Potential Lessons

- While quick movement of an AV out of travel lanes is desirable, doing so after a detected collision seems high risk due to the high variability and uncertainty of possible post-collision scenarios. A brief wait for remote assistance seems justified by safety considerations regardless of public pressure to minimize road disruptions.
- It is difficult to manage a crisis in general. Deviating from procedures is all too easy, but makes things worse. Periodic practice of crisis responses is required for effective execution when the real thing happens. Such training is not mentioned.
- Remote assistance operators should routinely and immediately alert emergency responders if they have any reason to believe an injury collision has or might have occurred. In this incident, Cruise got lucky that a bystander made that call.
- There is no explanation given for the apparent communication gap between CRAC and CIRT. Incident response communications should be scripted in the response plan, with specific handoffs and information to be transferred. CRAC should keep a contemporaneous log of such communications. Potential bad news, including especially road user injuries, should be highlighted to the CIRT and logged.
- There is no mention of a methodical approach to preserve evidence such as the vehicle configuration, communication logs, and other technical data beyond doing a data download from the vehicle. Post-incident procedures should address this area.
- In retrospect, the human in-vehicle safety supervisor for Cruise vehicles was likely removed prematurely. Having one might have prevented the initial AV pedestrian collision by reducing vehicle speed in response to a dangerous road situation. Even if that had not happened, an in-vehicle safety supervisor could be expected to notice and intervene with running over a pedestrian followed by dragging, and would be a backup method for contacting emergency services.

5 Organizational Response

We treat everything happening starting the next day (October 3, 2023) as the organizational response. CMT activity lasted until approximately 24 h after the crash, but changed in tenor to a public relations and regulatory response for these activities within a few hours of the crash. Here is a timeline, abbreviated due to space constraints:

1. (+6.3 h) 3:45 AM: A War Room Slack message unambiguously communicates the AV pulled forward with the pedestrian under the vehicle. This was done via CMT accessing data from CRAC (QER 27) rather than via a proactive hand-off process from CRAC to CMT. The QER says that no other "Cruise employee" had accessed this data previously (QER 28), but we note that CRAC staff are contractors, so the QER statement can easily be misunderstood.

2. Oct. 3: Cruise meets with DMV. DMV claims that Cruise did not disclose the pullover maneuver and dragging, only learning about it from another government agency later (QER 1–2). Internet connectivity issues and lack of affirmative discussion regarding the dragging are said to have contributed to the situation (QER 2). Interviews with DMV personnel were out of scope for the QER.
3. Oct. 3: NHTSA was said to have received the full video (QER 2). However, NHTSA was not shown the full video during the meeting for various reasons and the pedestrian dragging was not discussed (QER 41–42).
4. Oct. 3: City officials (SF MTA, SFPD, SFFD) shown the full video with discussion about the pedestrian dragging; no apparent internet issues (QER 2–3).
5. Oct. 3: Media outlets were shown a video that ended at the moment of impact and omitted the pedestrian dragging (QER 2). Cruise leadership believed they did not have an obligation to disclose the full details to the press (QER 3).
6. Oct. 3: A required "1-Day Report" to NHTSA failed to mention the pedestrian dragging (QER 15). This is said to have been due to a paralegal with "little oversight" drafting and filing (QER 98). However, Slack messages document Cruise's Deputy General Counsel and Communications Director both affirmatively approving a draft narrative missing that information (QER 49).
7. Oct. 11: A 10-Day report to NHTSA filed by the same paralegal also omits mention of the pedestrian dragging (QER 16).
8. Oct. 13–16: Cruise meets with DMV and shares full video (QER 69).
9. Oct. 24: DMV issues suspension order of Cruise's California operating permits. Cruise suspends California driverless operations (QER 87–88).
10. Nov. 2: Cruise submits a 30-Day report to NHTSA that includes mention of pedestrian dragging, along with a recall of 950 vehicles (QER 88).
11. Aftermath: Eleven Cruise employees involved with briefings to government regulators depart, including resignation of the CEO, followed by a 24% Reduction in Force (QER 91).

5.1 Organizational Response Analysis

What Went Well: Cruise proactively contacted local government and regulators to set up meetings. A War Room activation plan was carried out, ensuring that relevant Cruise stakeholders had a means of communication. Videos of the crash were created, including a video of the full mishap sequence, before external stakeholder meetings.

Narrow Focus on Narrative: The Cruise management team immediately became focused on asserting control of a public narrative they saw as unfair and felt "under siege" (QER 24,26). In part, this was apparently due to incomplete knowledge on both sides. For example, SFFD was making statements based on having been on the scene in which grievous injuries to a pedestrian had occurred, but without first-hand information about the initial Nissan impact. On the other hand, Cruise knew about the Nissan impact, but apparently did not know about the pedestrian dragging when establishing their public narrative, with that momentum carried forward into the following day.

Failure to Stand Down Fleet: The Cruise decision not to stand down fleet operations after a dramatic, severe pedestrian injury was justified by the CEO labeling the event an "extremely rare event" "edge case" and considering the event in the context of "the overall driving and safety records" (QER 35).

Failure To Proactively Disclose Material Bad News: By the time regulatory and media discussions were happening, more than 100 people in Cruise, including the CEO, knew of the pedestrian dragging (QER 28). Nonetheless, this fact was not proactively disclosed, even though Cruise knew it was their "biggest issue" (QER 40).

Regardless of intentions, later revelations/clarifications that pedestrian dragging had been under-disclosed caused a furor (e.g., QER 73). Arguably this communication mistake (rather than the mishap itself) was the main reason Cruise operations eventually had to be shut down. A common industry attitude is not volunteering information not explicitly asked for by government regulators. That attitude might avoid initiating negative news cycles, but ultimately resulted in a dramatically bad outcome for Cruise.

5.2 Potential Lessons

Ultimately, as the saying goes, the cover-up is always worse.

- While it is natural to be defensive after a mishap involving an organization's technology, lack of transparency and lack of affirmative disclosure of bad news can bring serious negative consequences, including loss of trust and stakeholder backlash. A key misstep was deciding to double down on not proactively revealing the pedestrian dragging in various ways. This included not ensuring regulators noticed the pedestrian dragging issue, and not correcting overly favorable media reports.
- The fear of triggering a negative media cycle (QER 31) ultimately resulted in increased harm to the company. If it becomes clear that stakeholders have incorrect or inadequate information (1) acknowledge as quickly as possible to those stakeholders that a revision will be necessary and (2) use extra effort to disclose fully and completely what is known to the organization as soon as possible.
- Do a safety stand down in response to a major event. This could have given Cruise more credibility with regulators, and blunted the public perception of their emphasis on scaling up operations regardless of cost or harm.
- Organizations should consider having a communications specialist who is not invested in the organization's narrative coordinate crisis communications to avoid narrative capture by stakeholders heavily incentivized to continue operations.

6 Conclusions

Another major mishap involving an AV is inevitable. Road travel is not perfectly safe, and such a mishap occurring should not all on its own be a reason to discontinue development of AV technology. However, the wrong mishap occurring in the wrong way, which is handled poorly, can potentially pose an existential threat to a company – and perhaps to the entire industry. This Cruise mishap has unquestionably reverberated throughout the industry, in large part because of the mishandled regulatory response.

Technical issues related to the crash identified include weaknesses in recognizing and responding to nearby mishaps, challenges in building an accurate world model of a postcollision scenario, and the inadequacy of a so-called "minimal risk condition" strategy in complex situations. Post-collision issues center around poor organizational discipline in responding to a mishap as well as overly aggressive post-collision automation that made a bad situation worse. Organizational issues identified center around a reluctance to admit to and address a mishap, ultimately causing organizational harm.

Cruise had previously spent significant effort to spin the narrative as a bad human driver in the Nissan causing a mishap, with their robotaxi doing the best it could in a situation nobody could have reasonably predicted. However, casting blame on the driver who initially hit the pedestrian (even if deserved) does not serve the purpose of identifying safety issues for the robotaxi that should be improved.

With the availability of the QER and EXPR, we have the opportunity to learn more about other aspects of the mishap and its aftermath, with key lessons listed in Sects. 3.3, 4.3, and 5.2. We present potential lessons regarding the vehicle's handling of a pre-crash scenario and post-crash scenario. We also present potential lessons regarding both the immediate and longer-term organizational responses to such a mishap.

Rather than attempting to create an accident report to parallel the one presented by the EXPR, this paper seeks to make available information in the reports more accessible to a broader audience. A potential threat to the validity of these findings stems from incomplete information in the reports. The QER is based on interviews limited to available employees and contractors but not outsiders (QER 4–5), and the EXPR is heavily redacted in places. We optimistically assume that redacted information would not undermine or contradict visible information. Nonetheless, given the situation we recommend assuming the information presented in both the QER and EXPR are crafted to present the most favorable picture possible toward Cruise's stated goal of regaining regulatory and public trust.

No external support funded the preparation of this paper.

References

1. California Vehicle Code 21950. https://leginfo.legislature.ca.gov/faces/codes_displaySection.xhtml?sectionNum=21950.&lawCode=VEH
2. Cummings, M.: Commentary on the January 24, 2023 Quinn Emanuel Report (2023). https://doi.org/10.13140/RG.2.2.28821.09448
3. INT'L ORG. FOR STANDARDIZATION [ISO]. (2022). ISO 21448:2022 Road vehicles — Safety of the intended functionality. https://www.iso.org/standard/77490.html
4. Korosec, K.: Cruise told by regulators to 'immediately' reduce robotaxi fleet 50% following crash (2023). https://techcrunch.com/2023/08/18/cruise-told-by-regulators-to-immediately-reduce-robotaxi-fleet-50-following-crash/
5. Mitchell, R.: San Francisco's North Beach streets clogged as long line of Cruise robotaxis come to a standstill. LA Times (2023). https://www.latimes.com/california/story/2023-08-12/cruise-robotaxis-come-to-a-standstill
6. Quinn Emanuel Trial Lawyers. Report to the Boards of Directors of Cruise LLC, GM Cruise Holdings LLC, and General Motors Holdings LLC Regarding the October 2, 2023 Accident in San Francisco (2024). https://bit.ly/49tmcKd
7. SAE, Taxonomy and Definitions for Terms Related to Driving Automation Systems for On-Road Motor Vehicles, SAE J3016_202104 (2021)

Comprehensive Change Impact Analysis Applied to Advanced Automotive Systems

Nicholas Annable[1](✉), Mehrnoosh Askarpour[2], Thomas Chiang[1], Sahar Kokaly[2], Mark Lawford[1], Richard F. Paige[1], Ramesh Sethu[3], and Alan Wassyng[1]

[1] McMaster Centre for Software Certification, Department of Computing and Software, McMaster University, Hamilton, Canada
{annablnm,chiangte,lawford,paigeri,wassyng}@mcmaster.ca
[2] General Motors Canada, Markham, Canada
{mehrnoosh.askarpour,sahar.kokaly}@gm.com
[3] General Motors Corp, Detroit, USA
ramesh.s@gm.com

Abstract. Like many manufacturers of complex cyber physical systems, automotive OEMS depend on incremental development. When changes are made to vehicles that were previously assured to be safe, it can be difficult to understand the impact of a change on the overall safety of the vehicle. In previous work we introduced Workflow$^+$, a model-based framework for modelling development and safety processes and their outputs for safety critical systems, and then generating safety assurance from the models. In this work, we demonstrate how the extensive traceability inherent in Workflow$^+$ can be leveraged to enable comprehensive safety-related change impact analyses. This facilitates sound incremental safety assurance to complement incremental development already in use.

1 Introduction

"Incremental assurance" is a significant challenge for automotive OEMS. The challenge lies in how to assure the safety of a family of vehicles that was previously shown to be safe, but has been changed, without going through the entire safety assurance activity. Incremental assurance is difficult primarily because safety is a global property and subject to emergent (mis)behaviours [16], and treating safety in a compositional way is considered an open problem [15].

Safety Assurance Cases (ACs) are mandated by the automotive functional safety standard ISO 26262 [10]. ACs are also used in many other safety-critical application domains. A comprehensive report on assurance cases by Rushby et al., can be found in [17]. In the context of safety ACs, there is considerable

---

Partially supported by the Natural Sciences and Engineering Research Council of Canada.

© The Author(s), under exclusive license to Springer Nature Switzerland AG 2024
A. Ceccarelli et al. (Eds.): SAFECOMP 2024, LNCS 14988, pp. 134–149, 2024.
https://doi.org/10.1007/978-3-031-68606-1_9

research on the topic of incremental assurance. For example, it has been studied by exploring the concept of safety case maintenance [7,12], where the main idea is to ensure safety by maintaining an existing AC in response to small (incremental) updates to the system or its assurance. While this area of research has made progress towards efficient and effective incremental assurance, there is still room to improve the traceability of the effect of changes throughout the safety and development process, and the data guiding the possible required changes in the vehicle and analyses. This process is commonly known as "Change Impact Analysis" (CIA).

Our previous attempt to tackle this challenge can be found in [19]. However, with comprehensive CIA as a goal, we found that Assurance Case Templates (ACTs) had no significant advantage over other ACs that are based on a tree structure – like Goal Structuring Notation (GSN) [11]. The Workflow$^+$ (WF$^+$) [1–5] modelling framework solves this problem in any application domain, such as the automotive domain we used for our case studies. It was developed with comprehensive CIA as a main objective from the beginning, and in this work we demonstrate how it can be used for CIA.

One of the defining attributes of WF$^+$ is the extensive traceability inherent in the models, tightly integrating models of the process, data and a generated AC. This AC is currently structured like GSN, but it does not need to be. We have used the extensive traceability of WF$^+$ to develop a novel approach to comprehensive CIA that facilitates incremental safety assurance. At the metamodel level, the CIA traces effects of modifications throughout the generic processes, data and "AC template" for vehicle families. At the instance level, the CIA traces effects of modifications throughout the development and AC of the actual vehicles.

2 Background on CIA for Incremental Safety Assurance

This section provides background information as to why effective CIA is vital in understanding the scope and scale of changes to a vehicle and its environment in the quest for incremental safety assurance. It also puts WF$^+$ in context, and explains our motivation for pursuing this approach.

CIA in complex software has been an active area of research for many years. We focus here only on directly related work. One example is the Boundary Diagram Tool for impact analysis in Simulink models [14]. Several approaches have emerged, leveraging MBSE methods to support CIA specifically related to safety assurance. Two examples are Kokaly et al. [13] and Cârlan et al., [7,8], which present model-based approaches for CIA in GSN ACs. Francis et al., [9] show how to treat spreadsheets as models, facilitating traceability.

To the best of our knowledge, existing approaches perform CIA for safety assurance by focusing on rules for propagating changes through the safety argument itself, which is often documented using GSN. A high quality GSN-like AC will have comprehensive links to its system of interest. Theses links are primarily through the evidence nodes that support terminal claims, although links in other nodes may exist. While these links may be extensive, they are not complete in that there are many artifacts in the system, related analyses, development and safety processes, and even the system's environment that are not directly relevant to the AC. If the CIA process is driven from the AC it is unlikely that we will find all the impacts that indirectly affect the AC.

Another challenge is that it can be difficult to trace claims and affected data in an argument structure to the processes that produce them. Two data items may not appear to be connected whereas if they are both used in a process, a change in one may affect the other. This is visible only in the process model.

In WF^+ we approach CIA by leveraging the extensive traceability between process and data models to propagate changes through the system. When an initial change is made the processes that used the changed information can be readily identified, as can all associated data. The AC is generated from constraints in the model, and each time one of those constraints is identified as potentially impacted, this will result in a terminal claim in the AC being identified as potentially impacted. In this way the WF^+ approach to predicting the impact of a change on the assurance of a system is unique, and has the potential to avoid some of the challenges faced by existing approaches.

Figure 1 shows stages in using WF^+ metamodels to perform CIA on an existing vehicle family, to facilitate incremental safety assurance. Stage 0 represents the starting point (the original vehicle family and its assurance). Stages 1–5 are the CIA, which establishes the safety-related impacts of a change to the vehicles by leveraging traceability. Stage 1 begins by identifying the direct changes being made to the system, for example based on a change request. Stage 2 identifies the elements potentially impacted by the direct change, and stage 3 confirms that these are actual impacts in the system. Stage 4 identifies how the confirmed impacts affect the safety argument for the system, and stage 5 confirms which parts of the safety argument are actually impacted. Stage 6 is a simplified view of the stage where the results of the CIA are used to indicate changes to the system. Stage 7 represents updating the assurance argument, where some of the impacted claims have been modified and some did not need to be modified. Stages 6 and 7 are often iterative. The story is the same for the instance models but much more difficult to show in a simple diagram.

3 Running Example

We use a simplified automotive example throughout Sects. 4 and 5 to illustrate how we use WF^+ for CIA. The example is an electric propulsion system driving the rear wheels of a vehicle. A block diagram of the system is shown in Fig. 2.

The propulsion system consists of one controller and a motor for each rear wheel. The torque requested by the driver is input to the motor controller through the accelerator pedal position. The battery status from the battery management system is also an input to the motor controller, and is used to determine how much power is available for use by the motors. Each motor receives a torque request from the motor controller along with power from the battery,

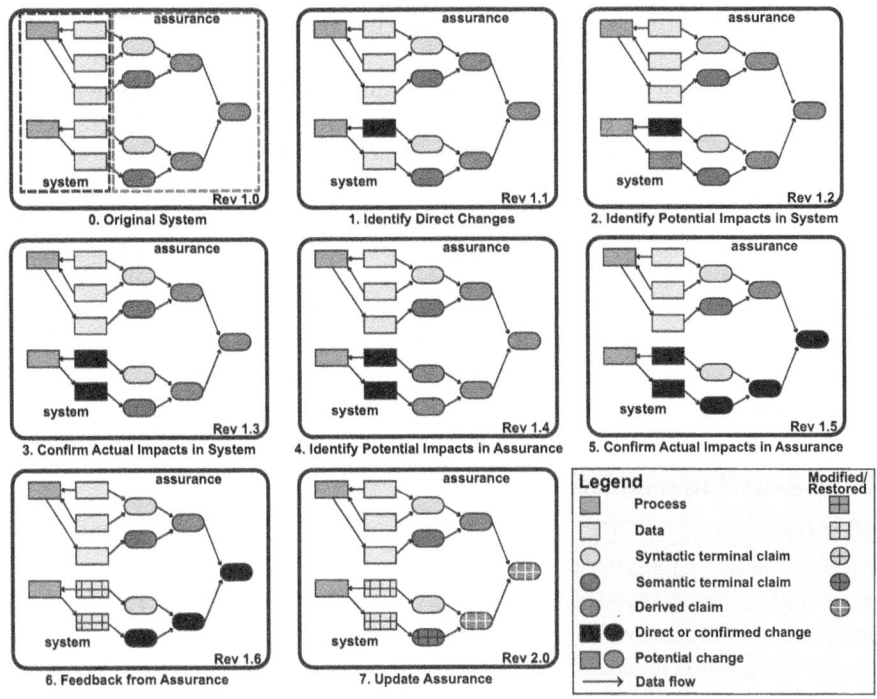

Fig. 1. Stages of WF$^+$ CIA Resulting in Incremental Assurance

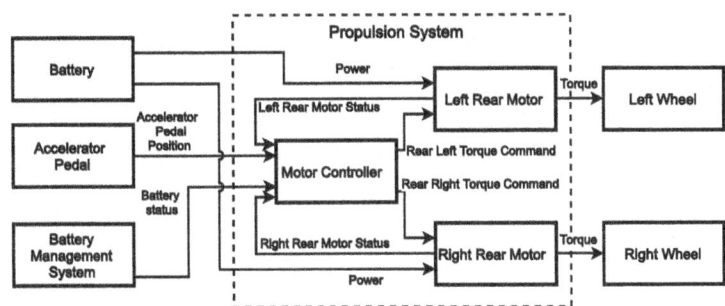

Fig. 2. Running Example: An Electric Vehicle Propulsion System

and produces torque applied to the rear wheels along with a feedback signal on the motor's status back to the motor controller.

This type of electric propulsion system is seen in hybrid-electric vehicles where the rear wheels are driven by electric motors and the front is driven by an internal combustion engine. For the sake of simplicity we omit the front motor and other details from the example.

We now use this example to illustrate how we use WF$^+$ in an effective CIA, taking us through Stages 0–5 in Fig. 1.

4 Original Vehicle Family Models Stage 0

Suppose that the WF+ models for a vehicle family exist, including the generated AC. Based on guidance in ISO 26262 [10], we show simplified extracts of the processes used to assure the safety of these vehicles. These processes were selected to be representative of what is done in practice: a Hazard Analysis and Risk Assessment (HARA) representing the concept phase of ISO 26262, a Hazard and Operability Analysis (HAZOP) representing the development phase, and a Verification Planning process representing the later stages. A detailed example illustrating the semantics of WF+ models and related arguments can be found in [3]. Below we present the WF+ metamodels and instance models used for this assurance, where only claims necessary to illustrate the impact analysis are included. Figure 3 also shows a legend for all the WF+ models.

4.1 HARA Metamodels

The HARA process used here is a simplified version of that described in ISO 26262-3 and SAE-J2980 [18]. The hazard analysis portion of HARA (Fig. 3) begins with an analysis of each function provided by the system to determine how it may malfunction. Following this, each malfunction is mapped to the vehicle-level hazard it may cause, along with assumptions regarding the hazard caused and the worst-case mishap associated with it.

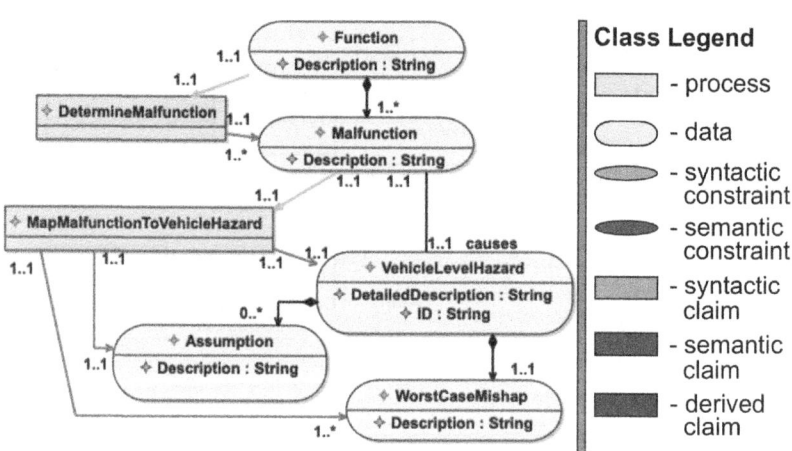

Fig. 3. Stage 0 – Hazard Analysis

The risk assessment portion in Fig. 4 uses a vehicle-level hazard, its related assumptions and worst-case mishap to determine severity, exposure, and controllability. These determine the ASIL associated with that hazard.

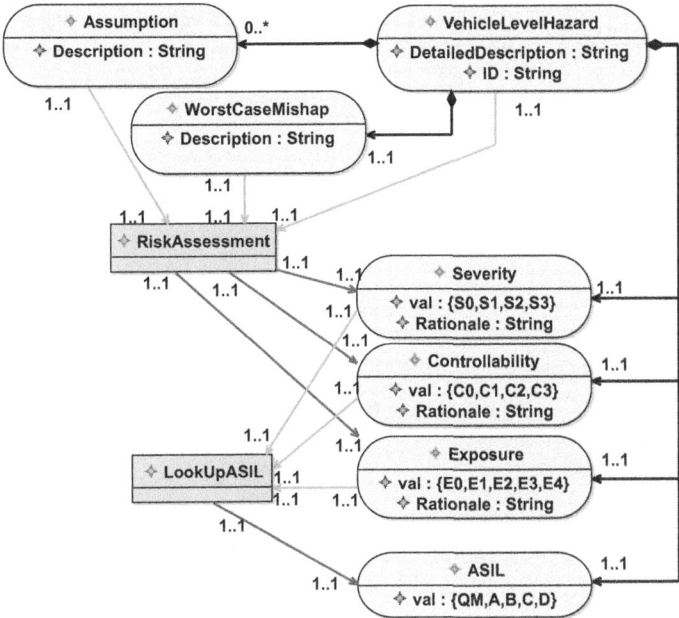

Fig. 4. Stage 0 – Risk Assessment

4.2 HAZOP Metamodels

ISO 26262-4 (section 6.4.4) recommends use of qualitative inductive safety analyses on the architectural design. In accordance with ISO 26262-9 (section 8.2) we selected HAZOPs. The model of HAZOP used here is based on [6].

The objective of the HAZOP process is to determine the effect of deviations from the intent of signals being passed between subsystems. In the first process, shown in Fig. 5, each signal passed between subsystems is compared to a set of guidewords to determine the consequence of the deviation described by the guideword. Note that the Component and Signal classes and their associations are the metamodel for the block diagram shown in Fig. 2, and that figure is actually a partial instantiation of this metamodel.

Following this, the consequence of each deviation is used to determine the vehicle-level hazard(s) (previously identified in the HARA) that the consequence causes and a causality mapping is created, shown in Fig. 6. Here we include several sample claims relating to multiplicities and a review of the causality link. The argument is by no means complete, but it can be used to illustrate traceability between system data and assurance for our CIA example.

Finally, safety requirements for mitigating the effect of each deviation are derived based on the consequence, the hazard(s) caused and the ASIL of that hazard (Fig. 7). Another example claim related to a multiplicity constraint is shown, along with one relating to the adequacy of the safety requirement.

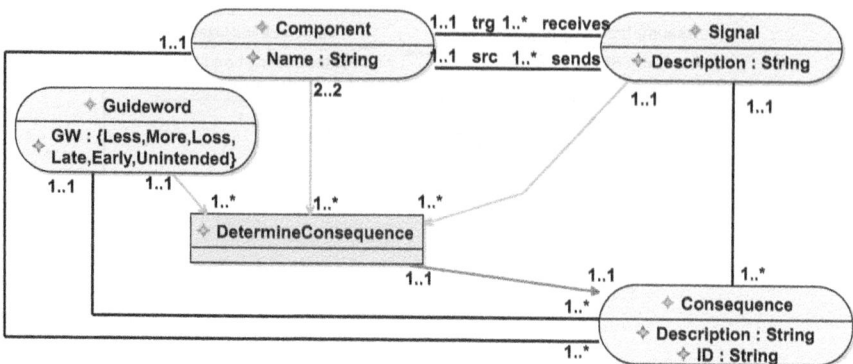

Fig. 5. Stage 0 – HAZOP Part 1

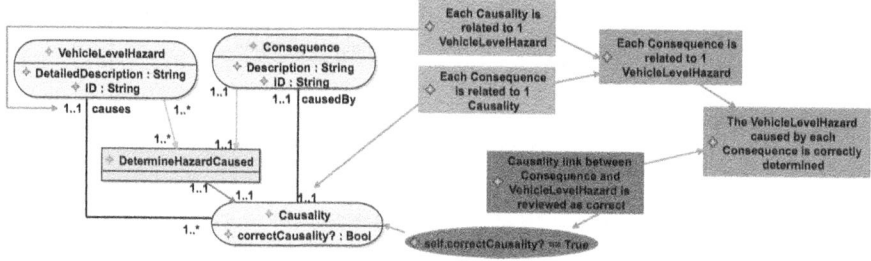

Fig. 6. Stage 0 – HAZOP Part 2

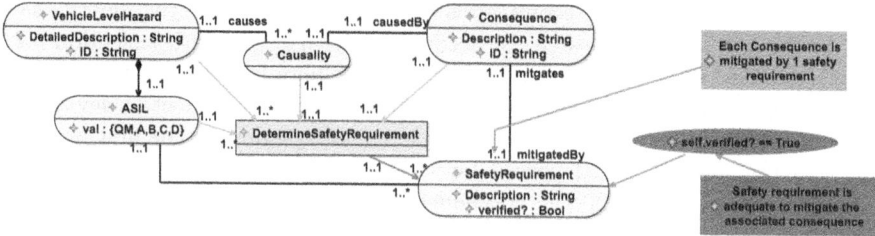

Fig. 7. Stage 0 – HAZOP Part 3

4.3 Verification Metamodel

ISO 26262-4 section 7.4 provides requirements and recommendations for checking compliance with technical safety requirements such as those derived during the previous processes. For instance, section 7.4.3.2.2 recommends that different types of testing be used, with the types and the strength of the recommendation varying based on the ASIL associated with the safety requirement.

The same section states that the verification should be carried out in accordance with ISO 26262-8, clause 9. In our example we use a greatly simplified version of the Verification Planning described in ISO 26262-8 section 9.4. In this process, the safety requirements derived in the previous process are input to a Verification Planning process, from which requirements-based and fault injection tests are derived in line with the methods outlined in 7.4.3.2.2. Notice from the multiplicities that requirements based tests are required for each safety requirement since the standard highly recommends them, but fault injection tests are not required for each safety requirement since the standard only recommends them for ASILs A and B, and highly recommends them for higher ASILs. This is shown in Fig. 8, along with an example constraint and claim that requires safety requirements with an ASIL greater than B to have fault injection testing.

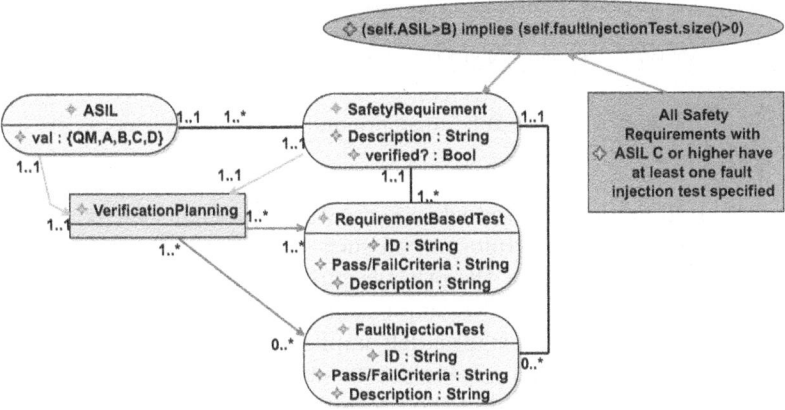

Fig. 8. Verification Planning

4.4 Instance Models

In practice, parts of instances are created in third-party tools that need to be interfaced with such as Medini Analyze, Simulink, and very often Excel tables. Here, we use tables to represent instances as they are an intuitive and compact representation that many engineers already are familiar with. We use the metamodel to derive a WF^+ example instance and from this design a table template that users can fill in to describe the instance data. After this there is a bi-directional mapping between a WF^+ instance and its associated table.

The top part of Fig. 9 shows an instance of an execution of DetermineHazardCaused from Fig. 6. Below that is the derived tabular format. Columns in the table correspond to classes or attributes from the WF^+ metamodel. Each row contains the instance data for each of the classes in the metamodel. For example, in Fig. 9 each Consequence (class) has an attribute ID, which is used to relate to the specific Hazard Caused (class), through that Consequence. Note that the name of a column may sometimes be altered to improve the readability of a table or satisfy a preference, but the mapping back to the WF^+ model is always

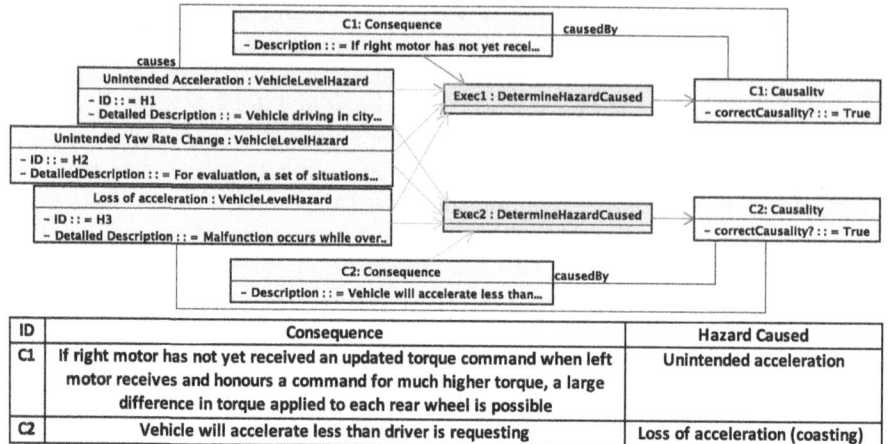

Fig. 9. Example WF$^+$ Instance and its Tabular View

maintained. Information omitted when generating a tabular view is maintained in the WF$^+$ instance model. Information such as the Causality class that connects the data in the table, or the other hazards considered when determining the effect of a consequence, are left implicit in the view.

The fact that information is often left implicit in these common representations is part of the reason impact analysis has been so challenging without any detailed traceability maintained in the background.

4.5 Instances for Stage 0

We show the instances resulting from the execution of the previous processes. Figure 10 first shows an instance of the execution of the HARA based on the HARA Metamodel. The text is an extract from Table D3 in [18]. The figure also shows an instance of HAZOP, followed by an instance of Verification Planning. Note that safety requirements and test cases are intentionally left as placeholders. Together these instances, or rather the detailed WF$^+$ models and arguments behind these spreadsheet views, form the content of Stage 0 in Fig. 1.

5 CIA After a Change to the Battery Management System

Suppose that after some time the designers of the battery management system determine that the limitations on peak power drawn from the battery are too conservative and could be increased. In this section we walk through the WF$^+$ approach to performing a CIA using the example described in Sect. 3. Stages 5–7 are explained, but updated models are not included in the example.

ID	Function	Malfunction	Vehicle Level Hazard	Assumption	Detailed Description	Worst Case Mishap	S	Rationale	E	Rationale	C	Rationale	ASIL
H1	Provide requested Drive Torque	Provide more drive torque than requested	Unintended Acceleration	Unintended acceleration does not cause destabilization	Vehicle driving in city or on country roads behind another car	Front/rear collision with the vehicle in front	S2	...	E4	...	C2	...	B
H2	...	...	...	...	...	...	S3	...	E4	...	C2	...	C
H3	Provide requested drive torque	provide less drive torque than requested	Loss of acceleration (coasting)	If driving situation is almost at limit of stability, the sudden loss of torque possibly can lead to some destabilization. (side force is affected due to change in longitudinal force)	Malfunction occurs while overtaking another vehicle on country road with oncoming traffic	frontal collision with oncoming traffic	S3	...	E2	...	C1	...	QM

ID	Source	Target	Signal	Description	Guideword	Consequence	Hazard Caused	ASIL	Safety Requirement(s)
C1	Motor Controller	Right Rear Motor	Right Torque command	Torque to be applied to rear right wheel by right rear motor	Late	If right motor has not yet received an updated torque command when left motor receives and honours a command for much higher torque, a large difference in torque applied to each rear wheel is possible	Unintended acceleration	B	SR1 - Motor Controller shall...
C2	Pedal	Motor Controller	Pedal Position	Position of accelerator pedal	Less	Vehicle will accelerate less than driver is requesting	Loss of acceleration (coasting)	QM	N/A

Safety Requiremnt	Requirement-Based Test	Pass/Fail Criteria	Fault Injection Test	Pass/Fail Criteria
SR1	(RBT1) Test by...	Some criteria...	N/A	N/A
	(RBT2) Test by...	Some criteria...	N/A	N/A

Fig. 10. Excel Tabular Instances of HARA (top), HAZOP (middle) and Verification Planning (bottom) for Stage 0

5.1 Identifying Direct Changes – Stage 1

In Stage 1 we note the direct changes in the instance models that this causes. By direct change, we mean the parts of the system that are the source of the change and are thus the starting point of the CIA. In our example there is only one direct change, which is to the battery management system, corresponding to the model element appearing in Fig. 2, so this element in our model should be "flagged" as changed and is shown as black in Fig. 11.

5.2 Identifying Potential Impacts in the System – Stage 2

There are two possibilities to consider when identifying potential impacts. The first is in the design of the system itself, where changes in the behaviour of one part of the system may affect the behaviour of other parts. This part of the CIA can be done effectively on detailed system models, such as by using the Boundary Diagram Tool [14] on Simulink models to trace dependencies through a model. We manually apply this step by tracing dataflow through Fig. 2. The Motor Controller, Left Rear Motor and Right Rear Motor in the propulsion system may be affected by the direct change, so these elements should be flagged as potentially impacted by the change and are shown as purple in Fig. 11.

The next possibility is that the outputs of processes previously executed using data that is impacted may no longer be valid. We can identify these types of impacts by following input and output trace links through the instance model and flagging outputs when necessary. This requires impact propagation rules for each process that specify conditions in which an impact is propagated. These rules are defined at the metamodel level and are thus reusable for any execution of that particular process. The rules vary in complexity and depend upon the

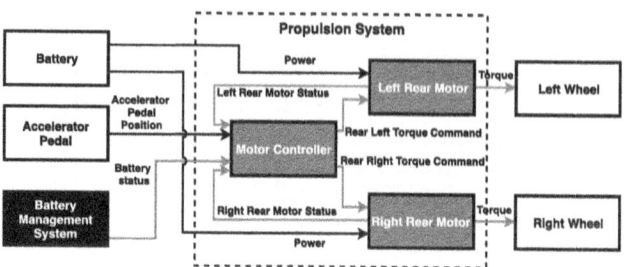

Fig. 11. Stages 1 and 2 of CIA

level of detail in the model. In this example we use only the simplest rule, that if any inputs are flagged, all outputs should be flagged as well. These impact analysis rules are applied iteratively until no new impacts are found.

Applying this to our example, we follow the trace links from the directly changed and potentially impacted system elements, highlighted as black and purple in Fig. 11, into the WF$^+$ models in our example. System classes Component and Signal in Fig. 5 are used by process DetermineConsequence, and this results in the table shown in Fig. 10. Figure 12 shows the impacts into the two HAZOP executions. Both IDs C1 and C2 in that table thus contain direct and potential system impacts, annotated as "1". These can be traced in Fig. 5 by an input arrow to DetemineConsequence, whose outputs are also potentially impacted and annotated as "2" in Fig. 12. This logic continues, following the trace links defined in Figs. 6, 7. This results in the Hazard Caused being potentially impacted and annotated as "3" from Fig. 6, and Safety Requirement, annotated as "4" from Fig. 7. It also leads to the execution of the Verification Planning process through the link from SafetyRequirement to VerificationPlanning (Fig. 10), resulting in a potential impact, "5", in Fig. 12.

ID	Source	Target	Signal	Description	Guideword	Consequence	Hazard Caused	ASIL	Safety Requirement (s)
C1	Motor Controller [4]	Right Rear Motor [1]	Right Torque command [1]	Torque to be applied to rear right wheel by right rear motor [1]	Late	If right motor has not yet received an updated torque command when left motor receives and honours a command for much higher torque, a large difference in torque applied to each rear wheel is possible [2]	Unintended acceleration [3]	B [4]	SR1 - Motor Controller shall... [4]
C2	Pedal	Motor Controller [1]	Pedal Position	Position of accelerator pedal	Less	Vehicle will accelerate less than driver is requesting [2]	Loss of acceleration (coasting) [3]	QM [4]	N/A

Safety Requiremnt	Requirement-Based Test	Pass/Fail Criteria	Fault Injection Test	Pass/Fail Criteria
SR1 [5]	(RBT1) Test by... [5]	Some criteria... [5]	N/A	N/A
	(RBT2) Test by... [5]	Some criteria... [5]	N/A	N/A

Fig. 12. Propagating Changes into Instances of HAZOP & Verification Planning

5.3 Confirming Actual Impacts in the System – Stage 3

Next we confirm that changes are not false positives. Since impacts are found by following traceability through the model less detail results in more false positives.

Identifying false positives requires understanding of the semantics of the system. Currently, we cannot achieve this automatically. We can use a systematic approach to identify false positives. We start with the direct change and manually review the changes it causes. In our example, the first impacts to consider are those on the propulsion system, which are found to be real impacts. Next, we evaluate the impacts labelled "2" in Fig. 12. We confirm the impact in row C1, but find that in row C2, the power limitation on the battery has no impact on the less-than-intended pedal position signal resulting in lower torque than the driver requests. As a result, the impact designated as "2" in row C2 is a false positive. Impacts that follow ("3" and "4" in the same row) are also false positives. We continue with the same method, and find that all remaining impacts in this example are not false positives.

5.4 Identifying Potential Impacts in the Assurance Case – Stage 4

Once impacts have been confirmed, the next Stage is to identify the impact on assurance arguments. Each terminal claim in the argument depends on a single constraint in the model, and the only way the satisfaction of a constraint can be changed is if a model element in its scope is changed. We can thus determine which terminal claims are potentially impacted, by checking if any data used by the corresponding constraint are flagged as impacted.

Figure 13 demonstrates this for the same instance shown in Fig. 9. The black elements are confirmed changes from Stage 3. From there we can trace to constraints and terminal claims. This results in flagging the upper claim shown in purple, while the lower red claim is not. There are no higher-level arguments shown in this example but the links between arguments can be followed up the argument tree to find those potentially impacted by the change.

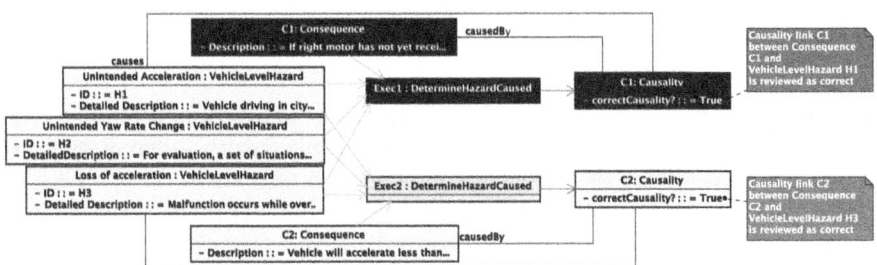

Fig. 13. Propagating Changes into the Assurance Argument

5.5 Confirming Actual Impacts in the Assurance – Stage 5

Once potentially impacted claims in the argument have been identified, the next Stage is to identify whether or not changes in evidence could actually necessitate a change to the associated claims. For example, consider the potentially impacted (purple) claim Fig. 13. Whether or not this claim is actually impacted depends on if the hazard caused by a consequence may actually change. Similarly to Stage 4, this requires understanding of the semantics of the system. It is possible that we will be unable to answer this question with reasonable confidence until the resolution for changes in prior processes are better understood – it would be very difficult to say if the hazard caused by a consequence will change until we understand how the consequence itself will change. Because of this Stage 5 often needs to be carried out iteratively as the design is updated.

5.6 Incremental Assurance – Stages 6 and 7

Incremental safety assurance is challenging and its details are beyond the scope of this work. We briefly clarify the role the CIA plays in incremental assurance.

After a CIA the team implementing the change must systematically "repair" the model by making changes to the system design and re-executing various processes as required (Stage 6), with the goal of reaching a point where the system is adequately safe and its safety is soundly argued (Stage 7). The change impact highlighted by the CIA provides a starting point for Stage 6. Since impacted claims and supporting data are traceable to the processes that produced that data, engineers can be guided to processes that need to be (re-)executed. Suppose the team is pointed to HARA documented in Fig. 10 and they determine that the increased power changes the controllability of H1 from C2 to C3, and its ASIL becomes C. SR1 would then be associated with ASIL C, and the constraint in Fig. 8 would be invalidated since SR1 now requires fault injection testing. This makes the impact on assurance more concrete, where we see exactly how the claim is invalidated. With the model in this new state, iterations of earlier stages may be needed to reassess the impact of the change.

6 Discussion

Our running example is relatively simple, where part of a system is modified but its structure remains the same. We have experimented with different types of changes, such as adding or removing components or signals in the system. We found that the approach to using CIA for incremental assurance is the same in general, and is usually implemented iteratively. Real world examples as to why we need incremental assurance abound. One of these is over-the-air updates for automotive vehicles. We are now applying WF^+ to this challenge.

The anticipated upfront cost and effort of using a framework such as WF^+ to enable such comprehensive CIA is high. We predict that the investment includes the construction of models of automotive vehicle families that are useful over many years, and some of these models would exist in some form anyway. The cost benefit hinges on the fact that we anticipate that the initial WF$^+$ metamodels will be used many times, with few modifications. This is borne out by the examples we have constructed over the past few years.

The effectiveness of the CIA with respect to incremental safety assurance depends on a number of factors.

1. *Models on which the CIA is based need to be accurate and faithfully represent the abstractions we choose.* The benefit of WF$^+$ in this regard is that it models processes and data emphasizing constraints, and reviews that target relevant process and data attributes. These are built on experience/knowledge of what is needed for safety.
2. *Models need to be detailed with enough traceability to link changes to "all" their potential impacts.* This is a feature of WF$^+$. Without this we cannot expect to avoid false negatives, which means that our assurance argument may not be sound. In WF$^+$ the link to an impacted terminal claim is very specific, and determined directly by the WF$^+$ CIA. Changes may impact multiple terminal claims that are then easy to trace *upwards* through the tree-structured assurance argument. Also, this mechanism results in a much finer granularity of the evidence that supports the argument, strengthening the link between evidence and terminal claim.
3. *The CIA algorithms and tools must cope with highly complex models to be usable in real world situations.* We have been applying WF$^+$ CIA to realistic problems in the automotive domain. However, we cannot yet claim that this has been demonstrated conclusively in practice.

7 Conclusion

We have shown that the extensive traceability in WF$^+$ models facilitates a comprehensive CIA. It can highlight dependencies between items in the development process and its data, the safety process and its data, the environment, and the assurance case. It is limited only by the detail included in the WF$^+$ models.

A primary objective in the CIA is to avoid false negatives. If there is a dependency we want to find it and report it. What this means in practice is that we need to model the data that is output or updated by a process in as much detail as possible. This will protect against false negatives. However, CIAs are notorious for producing false positives, and these adversely affect the way users treat the results from these analyses. We can save time and effort by not including substantial detail in the models of the processes and their data. However, our observation is that lack of sufficient detail in the data often leads to false negatives. Lack of detail in the processes can cause false positives, since we have to assume that all outputs of a process are affected by all its inputs – unless there is detail that shows that this is not true. Detail in the process is something that we need to balance against the prevalence of false positives.

With the aid of interface tools we can perform seamless CIA that traces dependencies throughout the WF$^+$ models and into third-party applications that are used for instantiation of WF$^+$ metamodels. We have interfaced WF$^+$ to Excel, Simulink and Medini Analyze. All are used in practice to provide evidence for the assurance case. Comprehensively modelling development and assurance together is a valuable way of integrating assurance into the development process, and of creating a comprehensive and effective change impact analysis that facilitates sound and effective incremental assurance.

References

1. Annable, N.: A model-based approach to formal assurance cases. Master's thesis, McSCert, Department of Computing and Software, McMaster University (2020)
2. Annable, N., Bayzat, A., Diskin, Z., Lawford, M., Paige, R., Wassyng, A.: Model-driven safety of autonomous vehicles. In: Proceedings of CSER (2020)
3. Annable, N., Chiang, T., Lawford, M., Paige, R.F., Wassyng, A.: Generating assurance cases using workflow$^+$ models. In: Trapp, M., Saglietti, F., Spisländer, M., Bitsch, F. (eds.) SAFECOMP 2022. LNCS, vol. 13414, pp. 97–110. Springer, Cham (2022). https://doi.org/10.1007/978-3-031-14835-4_7
4. Annable, N., Chiang, T., Lawford, M., Paige, R.F., Wassyng, A.: Lessons learned building a tool for workflow+. In: 2023 ACM/IEEE 26th International Conference on Model Driven Engineering Languages and Systems (MODELS), pp. 140–150. IEEE (2023)
5. Chiang, T.: Creating an editor for the implementation of WorkFlow+: a framework for developing assurance cases. Master's thesis, McSCert, Department of Computing and Software, McMaster University (2021)
6. Ericson, C.A.: Hazard and Operability Analysis, chap. 21, pp. 365–381. Wiley (2005)
7. Cârlan, C., Gallina, B., Soima, L.: Safety case maintenance: a systematic literature review. In: Habli, I., Sujan, M., Bitsch, F. (eds.) SAFECOMP 2021. LNCS, vol. 12852, pp. 115–129. Springer, Cham (2021). https://doi.org/10.1007/978-3-030-83903-1_8
8. Cârlan, C., Gauerhof, L., Gallina, B., Burton, S.: Automating safety argument change impact analysis for machine learning components. In: 2022 IEEE 27th Pacific Rim International Symposium on Dependable Computing (PRDC), pp. 43–53 (2022). https://doi.org/10.1109/PRDC55274.2022.00019
9. Francis, M., Kolovos, D.S., Matragkas, N., Paige, R.F.: Adding spreadsheets to the MDE toolkit. In: Moreira, A., Schätz, B., Gray, J., Vallecillo, A., Clarke, P. (eds.) MODELS 2013. LNCS, vol. 8107, pp. 35–51. Springer, Heidelberg (2013). https://doi.org/10.1007/978-3-642-41533-3_3
10. ISO 26262: Road vehicles – Functional safety. Int. Organization for Standardization, Geneva, Switzerland (2018)
11. Kelly, T.: Arguing safety – a systematic approach to managing safety cases. Ph.D. thesis, University of York (1998)
12. Kelly, T.P., McDermid, J.A.: A systematic approach to safety case maintenance. Reliab. Eng. Syst. Saf. **71**(3), 271–284 (2001)
13. Kokaly, S., Salay, R., Chechik, M., Lawford, M., Maibaum, T.: Safety case impact assessment in automotive software systems: an improved model-based approach. In: Tonetta, S., Schoitsch, E., Bitsch, F. (eds.) SAFECOMP 2017. LNCS, vol. 10488, pp. 69–85. Springer, Cham (2017). https://doi.org/10.1007/978-3-319-66266-4_5

14. Mackenzie, B., et al.: Change impact analysis in simulink designs of embedded systems. In: In European Software Engineering Conference and Symposium on the Foundations of Software Engineering, pp. 1274–1284 (2020)
15. Rushby, J.: Composing safe systems. In: Arbab, F., Ölveczky, P.C. (eds.) FACS 2011. LNCS, vol. 7253, pp. 3–11. Springer, Heidelberg (2012). https://doi.org/10.1007/978-3-642-35743-5_2
16. Rushby, J.: On emergent misbehavior (2015). https://www.csl.sri.com/~rushby/slides/emergentm12.pdf. Accessed 14 Feb 2024
17. Rushby, J., Xu, X., Rangarajan, M., Weaver, T.L.: Understanding and evaluating assurance cases. Technical report, SRI International (2015)
18. SAE: Considerations for ISO 26262 ASIL Hazard Classification (2018)
19. Wassyng, A., Joannou, P., Lawford, M., Maibaum, T.S., Singh, N.K.: New standards for trustworthy cyber-physical systems. In: Romanovsky, A., Ishikawa, F. (eds.) Trustworthy Cyber-Physical Systems Engineering, pp. 337–371 (2016)

A Case Study of Continuous Assurance Argument for Level 4 Automated Driving

Hideaki Kodama[1,4], Yutaka Matsuno[1(✉)], Toshinori Takai[2], Hiroshi Ota[3], Manabu Okada[4], and Tomoyuki Tsuchiya[4]

[1] Graduate School of Science and Technology, Nihon University, Tokyo, Japan
`cshd23008@g.nihon-u.ac.jp, matsuno.yutaka@nihon-u.ac.jp`
[2] Change Vision, Inc., Fukui, Japan
`toshinori.takai@change-vision.com`
[3] Hitachi Industry & Control Solutions, Ltd., Tokyo, Japan
`hiroshi.ota.fn@hitachi.com`
[4] TIER IV, Inc., Nagoya, Japan
`manabu.okada@tier4.jp, tomoyuki.tsuchiya@tier4.jp`

Abstract. Automated driving has a wide-ranging impact on our lives, making it crucial to form a consensus among various stakeholders for the dependability, including safety and availability. Moreover, a *continuous assurance argument* is essential to cope with changes and uncertainties. In this paper, based on existing development and operational documents and logs in an automated driving system company, we create assurance cases for level 4 automated driving safety in a Japanese city using Goal Structuring Notation (GSN). The GSN diagram includes results from safety analyses such as STAMP/STPA and test results. Additionally, as a step toward continuous assurance argument, we have prototyped a system for monitoring operational parameters such as speed of oncoming vehicles, ensuring the values remain within the assumed ranges. The GSN diagram with other documents such as safety analysis documents and operational monitoring data will be available via a web-based GSN tool.

Keywords: automated driving system · assurance cases · continuous assurance argument · monitoring

1 Introduction

Automated driving has a wide-ranging impact on our lives, making it crucial to form a consensus among various stakeholders for the dependability, including safety and availability.

The most challenging issue is that automated driving systems are *open systems* [5]. An open system operates in an open environment, and these systems are always faced constant changes and uncertainties. To cope with such changes and uncertainties, we proposed an abstract process model in [15] (Fig. 1). The abstract process consists of a development cycle and an operation phase. The

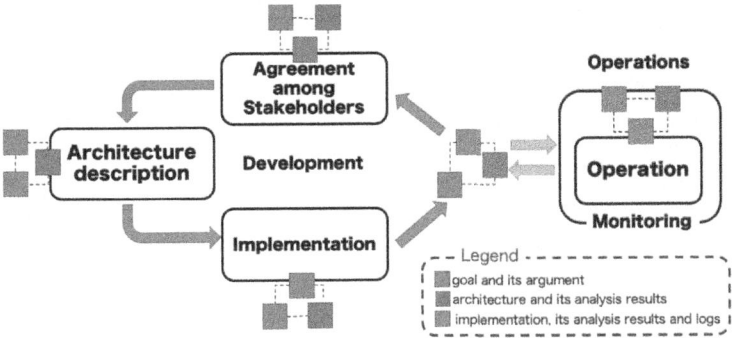

Fig. 1. An Abstract Process for Assuring Automated Driving Systems

development cycle has three phases: agreement among stakeholders, architecture description, and implementation. In the operation phase, the system and environment are always monitored to address changes and uncertainties. Throughout the process, assurance cases, architecture, analysis results, and system logs are continually updated and managed. The rationale of the abstract process is as follows: To cope with changes and uncertainties, the stakeholders discuss their best possible solutions by agreement; based on the agreement, the architecture is designed and implemented in an iterative process; changes and uncertainties in operational phases are monitored and the information and logs are feed-backed to the development phase. Following the abstract process, a *continuous assurance argument*-i.e., "The safety case is a 'living document' that evolves over the safety life-cycle" [3]-will be achieved.

TIER IV, an automated driving startup, envisions "The Art of Open-Source - Reimagine Intelligent Vehicles" and aims to build an ecosystem where anyone can contribute to technology development by leading the world's first open-source automated driving OS, "Autoware[1]". Aiming to solve social issues like reducing traffic accidents and addressing labor shortages, the company has partnered with many to develop the technology, accumulating over 100 automated driving demonstrations in Japan and abroad. The Autoware Foundation manages Autoware, providing a foundation for rapid research and development. This platform enables developers worldwide to enhance Autoware's functionality. At TIER IV, Autoware serves as the basis for their product, Pilot.Auto, which offers evidence for quality assurance and safety and dependability arguments. This activity leverages Web.Auto, a software platform for development support. Developers using Autoware can also utilize Web.Auto, aiming to create an environment where development activities similar to those at TIER IV are achievable throughout the entire Autoware ecosystem.

Based on the abstract process, this paper summarizes TIER IV's activities for Level 4 automated driving safety using Goal Structuring Notation (GSN) [2]. The GSN diagram includes results from safety analyses such as STAMP/STPA

---

[1] https://github.com/autowarefoundation/autoware.

[12] and Use Cases. The target automated driving systems have already been developed and tested in several local cities in Japan by TIER IV. Currently, this development and testing have been conducted using conventional methods without assurance cases. The goal of the paper is to demonstrate what scale and content of assurance cases can be constructed from the development of actual automated driving systems and to obtain insights for future development and operation of automated driving systems based on continuous assurance arguments. To achieve this goal, we present a case study in Shiojiri, a local city in Nagano prefecture, Japan, where TIER IV has conducted field trials of automated driving systems on actual roads.

Contributions of the paper are as follows.

- We summarize TIER IV's efforts towards Level 4 automated driving safety, showcasing the use of GSN as an example of actual development efforts in automated vehicles by a company, presented as a practical experience report.
- In a case study, we detail an assurance argument for a predefined route in Shiojiri, a local city in Japan. Unlike previous works on creating safety cases for automated vehicles, detailed technical information appears in our GSN diagram and the GSN diagram will be publicly available via a web-based GSN editor called D-Case Communicator [13,14].
- We have implemented a system atop the city hall to monitor vehicles, including those making a right turn into the city hall, and developed a toolchain with D-Case Communicator. Monitored operational values, such as the average speed of vehicles, sometimes appear in context or assumption nodes in a GSN diagram for automated vehicles. By using the toolchain, we can continuously check whether the operational phase parameter values are within the range assumed in the assurance case, which enables dynamic updating of the argument and system. This is a step toward achieving a continuous assurance argument.

The paper's structure is as follows. Section 2 presents related work. Section 3 introduces a case study on creating assurance cases for Level 4 automated driving safety in a local Japanese city. Section 4 describes a prototype implementation of the proposed toolchain, including an assurance case tool and a monitoring system, to verify if context information, such as "velocity of oncoming vehicles is within 60 km/h", is met in the operational phase. Section 5 discusses lessons obtained form the case study. Section 6 concludes the paper.

2 Related Work

There are several works on creating assurance cases for automated driving systems, such as [4,18]. Unlike our work, these papers do not target specific automated driving systems but present general assurance case structures. These general structures will aid our efforts in creating GSN diagrams for actual automated systems. Aurora presents a self-driving safety case framework on their website to address the safety of both autonomous trucks and passenger cars. The main goals include being proficient, fail-safe, continuously improving, resilient, and trustworthy. The goal of continuous improvement aligns with our focus on the continuous assurance argument.

In [1], a framework for providing legitimate trust in the functionality of automated systems incorporating machine learning-based components is proposed through dynamic assurance cases (DAC). A DAC offers a dynamic means of assurance that recognizes operational situations and continuously monitors the system's performance. The literature demonstrates how a DAC was created for an unmanned aerial system with general-purpose automated taxiing capabilities for airport ground movement tasks. A DAC is created by initially forming a static assurance case. This approach is similar to ours, which ensures the safety of automated driving systems by using assurance cases combined with monitoring data.

In UL4600 [17], an international standard for autonomous vehicles, Safety Performance Indicators (SPIs) are introduced for quantifiable safety measurements. SPIs provide the means to detect whether any claim within the safety case is falsified during design, simulation, testing, and deployment [11]. Our toolchain with a monitoring system, as shown in Sect. 4 could serve as an example of an SPI mechanism.

3 A Case Study in a Local City

TIER IV has been conducting demonstrations of Level 4 automated driving in various cities across Japan. In this paper, we create a GSN diagram for a case study using the round-trip route from Shiojiri train station to the Shiojiri city hall (about 2 km) in Nagano Prefecture, Japan (see Fig. 2).

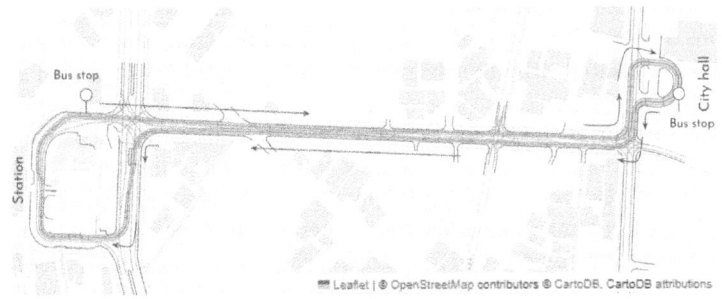

Fig. 2. A round-trip route from Shiojiri train station to the city hall

Specifically, we examine the use case where the automated shuttle bus turns right to enter the city hall (Fig. 3). The outlined approach can be applied to the entire route to achieve full coverage.

Fig. 3. The automated shuttle bus and a scene of making a right turn and enter the city hall passing through the sidewalk

The automated shuttle bus faces various dependability concerns beyond safety. For instance, overly conservative safety margins for objects on the road could prevent vehicles from making right turns indefinitely, thereby reducing the overall traffic system's availability. This paper mainly discusses safety of automated driving systems and other issues are briefly discussed. In practice, it is crucial to balance safety with other considerations such as availability. We are currently engaging with stakeholders to discuss the dependability of automated systems.

3.1 Top Level of Assurance Case for Level 4 Automated Driving

At TIER IV, most documents are maintained in cloud systems, and several tools, including Jira, are used to share development and operational information of the systems. From these numerous artifacts, we have collected those related to the shuttle bus system being tested in Shiojiri city and created an assurance case for Level 4 automated driving safety. Figure 4 illustrates the top-level structure of the assurance case.

In Fig. 4, the top goal, G1, states: "The automated driving shuttle bus satisfies the absence of unreasonable risk". We refer to the term "unreasonable risk" from the UN/WP29/R157 regulation and define it as "the overall level of risk for the driver, vehicle occupants and other road users which is increased compared to a competently and carefully driven manual vehicle". For the top goal, there are four contexts (C1 through C4): The system is operated as level 4 automated driving (C1), a LSAD system according to ISO 22737:2021 [7] (C2), comply with Japanese Road Traffic Act[2], ISO Guide 51:2014 [10], ISO 26262:2018 [6], ISO 21448:2022 [8], and ISO 34502:2022 [9] (C3), and the route and the intersections are predefined in the round trip between Shiojiri train station and the city hall (C4). Automation level 4 is defined as "high driving automation". A level 4 ADS performs the dynamic driving tasks (DDT), including DDT fallback; There is no human driver

---

[2] https://www.japaneselawtranslation.go.jp/en/laws/view/2962/en.

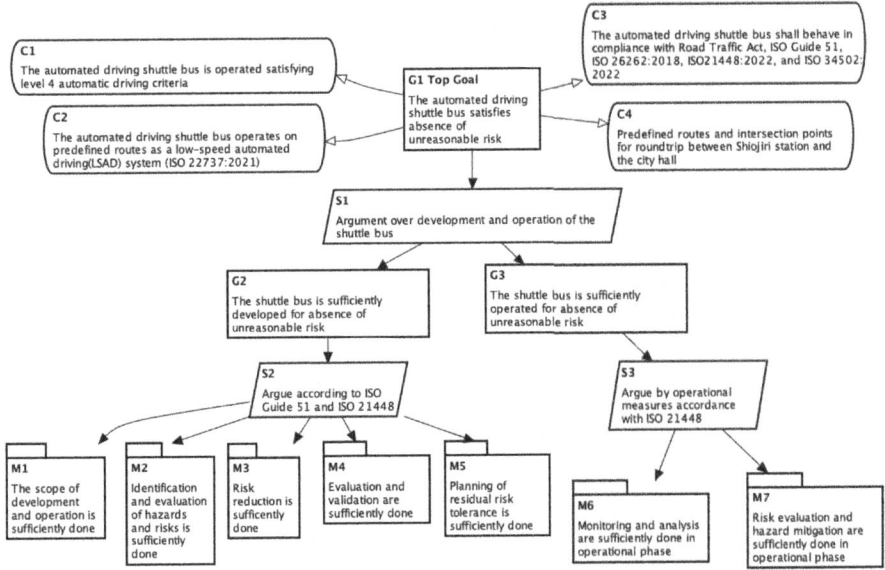

Fig. 4. Top level structure of the assurance case

or request to intervene in fallback. Automation level 5 is defined as "full driving automation". The only difference from level 4 is "unlimited operational design domain (ODD)" [16]. For arguing the top goal, we use strategy S1: Argument over the development and operation of the shuttle bus, and state sub goals G2 and G3 for development and operational argument for mitigating unreasonable risk.

For goal G2: "The shuttle bus is sufficiently developed to ensure the absence of unreasonable risk", we adopt strategy S2: "Argumentation based on ISO Guide 51 and ISO 21448", outlining five GSN modules (M1 through M5) as subgoals. These modules align with the structure of ISO Guide 51, covering: Step 1: Scope of development and operation; Step 2: Identification and evaluation of hazards and risks; Step 3: Risk reduction; Step 4: Evaluation and validation; and Step 5: Planning of residual risk tolerance. This approach is pertinent for limited ODD automated driving, such as an automated driving shuttle bus service on a predefined route. GSN module M1 focuses on identifying users, intended use, and reasonably foreseeable misuse as required by ISO Guide 51. It aims to define the safety scope and goals, identify system boundaries, and outline strategies for safety achievement. At TIER IV, activities in Step 1 are implemented through the Concept of Operations (ConOps) and Operation Concepts (OpsCon).

For goal G3: "The shuttle bus is operated with sufficient measures to ensure the absence of unreasonable risk", we implement strategy S3: "Argue by operational measures in accordance with ISO 21448", identifying two GSN Modules (M6 and M7) as sub goals. GSN Module M6 focuses on monitoring and analysis during the operational phase, while M7 addresses risk evaluation and hazard mitigation in the operational phase.

In the following two subsections, we will specifically discuss GSN Module M2: Identification and evaluation of hazards and risks, and M4: Evaluation and validation.

3.2 GSN Module M2 for Identification of Risk and Hazard

Module M2 encompasses the outcomes of risk and hazard identification efforts. We employ STAMP/STPA for identifying risks and hazards, as ensuring dependability is crucial for delivering automated driving services. This necessitates considering various perspectives and conducting trade-off studies on risks, including aspects beyond safety. Next, we will focus on a specific scenario where the ego-vehicle makes a right turn at an intersection without traffic lights, presenting the findings from STEP1 and STEP2 of the STAMP/STPA process.

STEP1 in STPA (Setting the goal of the analysis): In this step, we recognize the risk in the broadest sense, i.e. including safety, availability, laws and regulations, etc. In the following, we describe some accidents as undesirable consequences for each aspect of risk that is dealt with in this paper. For the safety risk, we define personal injury or damage and property damage according to the severity aspect in ISO 26262:2018. The availability risk is considered time loss of shuttle bus users and a traffic block of other road users. For the aspect of the risk with respect to laws and regulations, we define one undesirable consequence, i.e. a traffic law violation. Next, we define three hazards for those undesirable consequences (1) ego vehicles enter other road users' trajectories; (2) other road users enter ego vehicle trajectories; and (3) approaching road structures.

STEP2 in STPA (Building a control structure): For constructing a control structure, we first recognize that the intersection with the relationship to its traffic participants including the ego-vehicle in the target scenario as a "system". We consider the intersection as a system because to identify hazardous scenarios, we address the distance between the ego-vehicle, in this case a shuttle bus controlled by an automated driving system and other traffic participants as a critical parameter. We show the accidents for the case study in Table 1.

Table 1. Defined accidents for the case study

Accident ID	Accident
A1	Another vehicle is damaged
A2	Any object on the road is damaged
A3	The ego-vehicle is damaged
A4	A user of the ego-vehicle is killed or injured
A5	A user of the other-vehicle is killed or injured
A6	A pedestrian is killed or injured (including feeling fear)
A7	A user of the ego-vehicle cannot move on schedule
A8	Traffic for other road user is disrupted
A9	The ego-vehicle violates the road traffic laws at the intersection

The set of initial hazards is defined as shown in Table 2. Note that the other road users include vehicles and users of the intersection such as pedestrians and bicycles.

Table 2. Defined hazards for the case study

Hazard ID	Hazard
H1	The ego-vehicle cannot stop within the defined safety distance m**(varies according to the velocity) from the ego-vehicle to other road users
H2	The ego-vehicle cannot stop within the defined safety distance m**(varies according to the velocity) from other road users to the ego-vehicle
H3	The behavior of the ego-vehicle violates legal requirements(e.g. Article 37 of the Road Traffic Act, the structural requirement of one-person-operated bus)
H4	The behavior of the ego-vehicle exacerbates the traffic condition
H5	The ego-vehicle cannot stop within the defined safety distance m**(varies according to the velocity) from the ego-vehicle to other road users
H6	The behavior of the ego-vehicle is not within the safety maximum acceleration m/s^2**
H7	A user of the ego-vehicle fails in her/his duty to ensure her/his safety in the shuttle bus (i.e. getting out of your seat, not holding onto the handrail)

A part of safety constraints is shown in Table 3. There may be other safety constraints, but in this paper we only show some examples.

Table 3. Example safety constraints

safetyconstraints ID	safety constraints
SC1	The ego-vehicle shall keep the safety distance from the other road users
SC2	The ego-vehicle shall maintain clear paths for other road users
SC3	The behaviors of the ego-vehicle shall comply with all relevant laws
SC4	The ego-vehicle shall keep the safety distance from any object other than road users
SC5	The behaviors of the ego-vehicle shall keep within the maximum acceleration m/s^2**
SC6	The behaviors of the ego-vehicle shall actuate braking within the maximum deceleration m/s^2**
SC7	The users of the ego-vehicle(a shuttle bus) shall comply with their duty to ensure their own safety

STEP3 in STPA (Identifying UCA): To characterize UCAs, we first define 4 decision points during the turn right as (1) deciding to enter the intersection, (2) deciding to enter the oncoming lane, (3) deciding to enter the sidewalk, and (4) deciding the turning right operation had been completed. The result of the subsequent steps of STPA will not be discussed in this manuscript, but we will explain an analysis for identifying UCAs that were important in this case study.

A trade-off relation exists between the distance from the ego-vehicle to the other road users. To successfully turn right at an intersection without causing accidents, vehicles must keep a safety distance (Fig. 5). The safety distance at an intersection can be considered with the distance traveled by the oncoming vehicle for the time it takes for the ego-vehicle to pass through the opposite lane as a base. Estimating the time required to pass an intersection should also consider the width of the road and road users other than the oncoming vehicles. We define the safety distance by classifying three categories, "crash", "scared", and "usually(Safe)". Figure 5 shows the image of those categories. The category "crash" is when the ego-vehicle fails to appropriately keep the safety distance from the oncoming vehicle. A distance classified in the category "scared" is defined as the one at which the collision can be avoided by the brake operation of the oncoming vehicle, but the driver of the oncoming vehicle feels fear. A distance in "usually(Safe)" is defined as when the ego-vehicle can pass the intersection when it turns right without any concern, and even when the speed of the ego-vehicle is too high, the behavior to reduce the speed to the legal one is acceptable. Additionally, when slowing down the oncoming vehicle to below the speed limit, it is considered unacceptable as it constitutes obstruction of traffic. The actual safety distance in the automated driving service in Shiojiri city is defined as several different values for each situation. For example, for the situation that the ego-vehicle turns right and enters a privately owned area, the least recognition distance is set to 82.8 m with the assumption that the maximum speed of the oncoming vehicle is defined as 85% of the actual average speed of that point (that is, 48km/h). The safety distance is not a fixed value but is determined by the predicted time the ego-vehicle passes the intersection and the speed of the oncoming vehicle. On the other hand, we compare the calculated safety distance to the one at which the typical human driver decides to turn right, and the result is that the latter is shorter than the automated driving case. We consider the backgrounds of the phenomenon include (1) the length of the shuttle bus is longer than the typical cars and (2) the specification of the automated driving vehicle prohibits sudden acceleration.

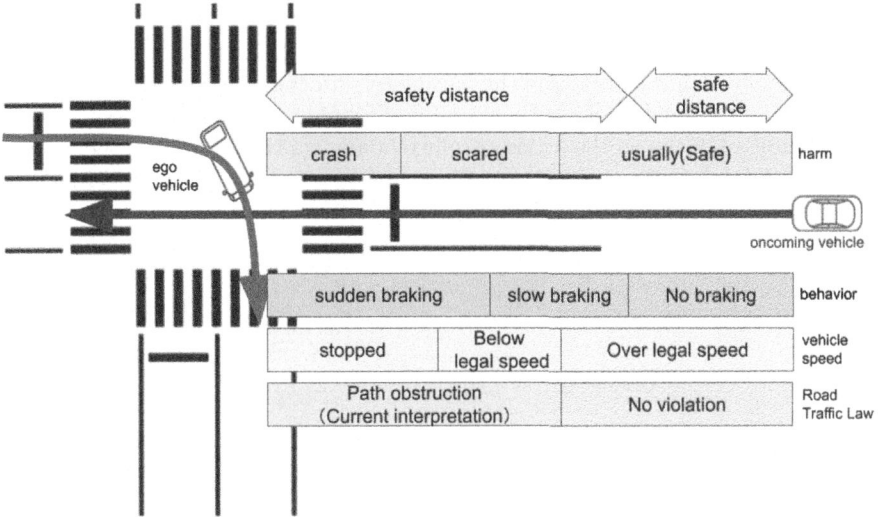

Fig. 5. Defined categories of the safety distance

Based on the obtained UCAs considering safety distances, a set of loss scenarios is derived and evaluated, and safety measures are determined. To implement effective safety measures, safety engineers need to consider existing technologies available as safety mechanisms, the use of external measures, their limitations, and system specification/configuration. Additionally, the safety measures require appropriate safety processes and methods to avoid systematic failures in safety mechanism development phase. Referring to ISO/IEC Guide 51, safety engineers follow a three-step method to propose safety measures.

3.3 GSN Module M4 for Evaluation and Validation

Module M4 includes evaluation and validation of the safety of the automated shuttle bus system. In this section, we focus on argument for the validation (argument for the evaluation is omitted due to space limitation). The safety validation argument is mainly divided by driving scenes (e.g., departure from a bus stop, right turn at an intersection, T-junction, etc.), which is a typical argument structure. Figure 6 shows the GSN diagram for the driving scene of making a right turn into the Shiojiri city hall.

The driving route in Shiojiri city (Fig. 2) is divided into 32 driving scenes, and the requirements necessary for autonomous driving in these divided scenes are listed. These listed requirements are called the "RequirementList" and are classified into the following seven categories. Currently, the driving scenes used in Shiojiri city hall into four categories: A: Normal Driving, C: Intersection, E: Signal Recognition, and X: Emergency Vehicles. There are two types of requirements: main requirements and sub-requirements. If the behavior of the main requirement changes depending on the surrounding situation, sub-requirements are defined.

Table 4 shows a portion of the RequirementList. The main and sub-requirements are composed of one or more use cases. Use cases are organized by the requirements (main and sub-requirements) and the parameters for those requirements (e.g., the speed of an autonomous vehicle that can handle sudden pedestrian appearances). In this study, we focus on the driving scene of making a right turn into the Shiojiri city hall, and detailed arguments for the goal G8 on safety are conducted at the use case level, which is the smallest unit constituting the scene (Fig. 6).

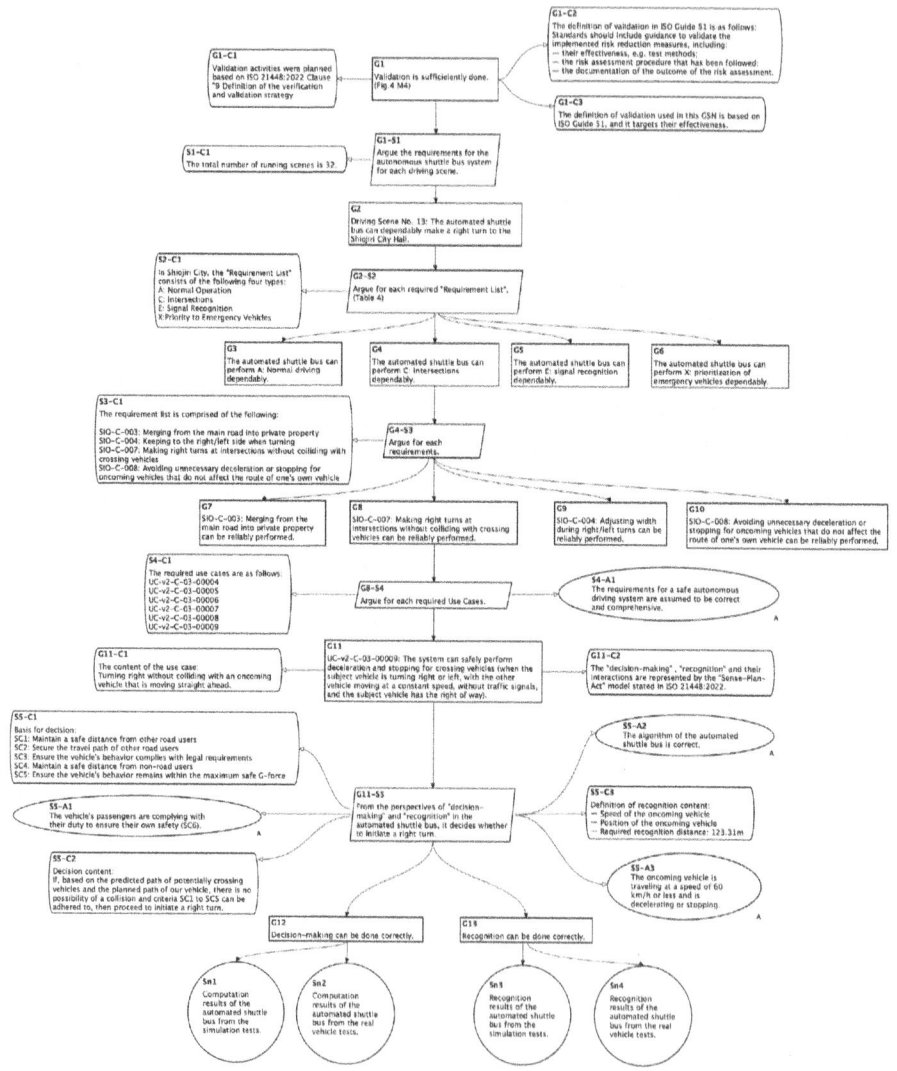

Fig. 6. A GSN diagram for validating a use case in Shiojiri city (a part of GSN Module 4 in Fig. 4)

Table 4. Part of the Requirement List

ID	classification	Abstraction	Requirements
SIO-A-006	A: Normal Operation	Stop for Objects on the Route	If there is a stationary object on the route, stop with a gradual deceleration while maintaining a safe distance between vehicles (UC-v2-A-03-00001, UC-v2-A-03-00002, UC-v2-A-03-00003, UC-v2-A-03-00004, UC-v2-A-03-00005) Continue to stop even if pedestrians approach from the side while stopped for obstacles on the route. (UC-v2-A-03-00010) Start moving again if the obstacle on the route is no longer present. (UC-v2-A-03-00012)
SIO-A-006-1			When encountering a moving object, such as an animal, approaching the vehicle, always decelerate to maintain a safe distance from other vehicles, allowing for a safe stop If gradual deceleration is insufficient, perform an emergency deceleration.(UC-v2-A-03-00008, UC-v2-A-03-00009)
SIO-A-006-2			In cases such as a pedestrian suddenly stepping out, where sufficient stopping distance cannot be assured, perform an emergency deceleration to reduce speed as much as possible and come to a stop.(UC-v2-A-03-00011)
SIO-C-007	C: Intersections	Make a right turn at the intersection without colliding with crossing	Proceed through the intersection with a right turn, ensuring not to obstruct the movement of crossing vehicles and confirming no collision will occur with them. If there is a possibility of colliding with a crossing vehicle, stop. (UC-v2-C-03-00004, UC-v2-C-03-00005, UC-v2-C-03-00006, UC-v2-C-03-00009) Safely make a right turn at the intersection even when crossing vehicles are accelerating and entering the intersection (UC-v2-C-03-00007, UC-v2-C-03-00008). Signal with the turn indicator at least 3 seconds before changing lanes and continue signaling until the lane change is complete.
SIO-C-007-1			Even in the presence of blind spots, safely make a right turn at the intersection (UC-v2-C-03-00013).
SIO-C-007-2			If an oncoming vehicle turning right deviates from its lane and enters the path of your vehicle, detect this and come to a stop. (UC-v2-C-03-00203).
SIO-E-002	E: Signal Recognition	Follow hand signals while driving	On the driving route, if a police officer is giving instructions through hand signals, follow those instructions (Use case under development)
SIO-X-001		Priority to Emergency Vehicles.	When an emergency vehicle is approaching, yield the road to the emergency vehicle (Use case under development)

In Fig. 6, the goal G8 focuses on one of the use cases in category C: Intersection within the driving scene of making a right turn into the Shiojiri city hall. The content is "UC-v2-C-03-00009: Making a right turn without colliding with oncoming vehicles". In UC-v2-C-03-00009, attention is paid only to the movement of oncoming vehicles, and obstacles or pedestrians are not considered. Handling obstacles and pedestrian darting are argued in separate use cases (UC-v2-A-03-00011: Braking in case of failure to maintain a safe following distance, UC-v2-A-03-00001: Recognizing stationary objects on the road and stopping while maintaining a following distance). This study introduces one use case that constitutes the scene, and validation requires combining 43 use cases as a scene. In practice, the decision to start making a right turn is determined from the perspectives of "recognition" and "decision-making" of the automated shuttle bus. The following explains the "recognition" and "decision-making" of the autonomous driving system, based on predictions from the available development materials. The system is under development and improvement, and it may differ from the actual system.

In this use case, the "recognition" elements are the type, speed, and position of the oncoming vehicle. The speed limit in the driving scene is 40 km/h, and based on a survey of actual speeds, the highest actual speed of oncoming vehicles is assumed to be 60 km/h. Under the following conditions, the required recognition distance was calculated to be 123.31 m.

- Proceeding without stopping at a slow acceleration to 10 km/h from a stopped state
- Assuming that oncoming vehicles traveling at 60 km/h, begin to decelerate gently (Acc-2.5, Jerk±0.5) one second after the self-driving car starts turning right

As a premise for "decision-making", it is assumed that the safety constraints SC1 to SC5 in Table 3 can be adhered to, and SC6 is described as an assumption. Since the behavior of oncoming vehicles is uncontrollable, it is described as an assumption. From the perspective of judgment, based on the data of the oncoming vehicle received from recognition, the predicted path of potentially intersecting vehicles is assumed. If it is determined that there is no possibility of collision with the planned path of the subject vehicle and SC1 to SC5 can be adhered to, a right turn is initiated.

4 A Toolchain of an Assurance Case Tool and a Monitoring System

We have been developing a toolchain consisting of D-Case Communicator [14] and a monitoring system. The monitoring system is provided by Hitachi Industry & Control Solutions, Ltd. Fig. 7 is a screenshot of the monitoring system on the city hall rooftop to monitor vehicles turning right into the city hall and other vehicles. By the toolchain, such logs are sent to the D-Case Communicator and the D-Case Communicator can automatically check the context and assumption nodes of the GSN diagram. Figure 8 is a screenshot of the D-Case Communicator in which the same GSN diagram of Fig. 6 is shown. In Fig. 8, assumption node G2-A3 is marked as "Fail", indicating that "The oncoming vehicles are traveling at a speed of 60 km/h or less" is not satisfied. This program allows embedding codes and conditional statements in a GSN diagram, with an example of a conditional statement ($velocity \leq 60$ km/h) embedded in the G2-A3 node.

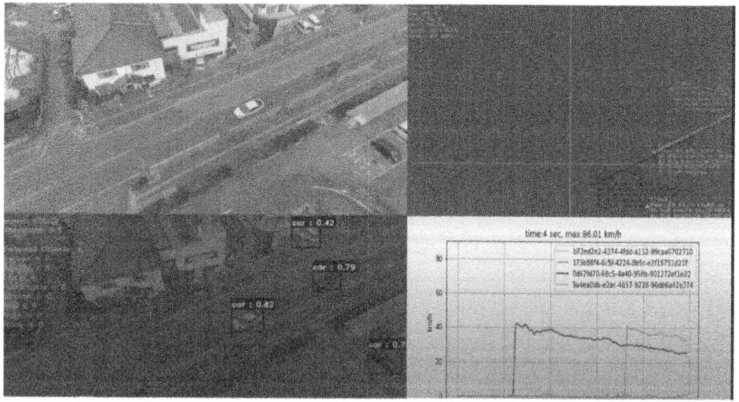

Fig. 7. A screen shot of the monitoring system

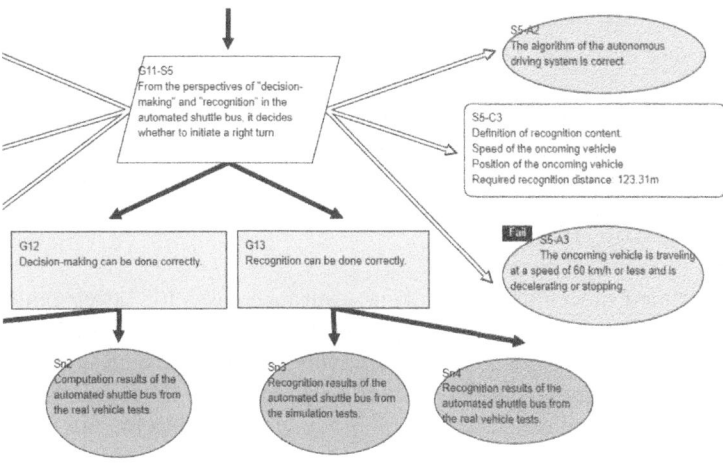

Fig. 8. A screenshot of D-Case Communicator

5 Lessons Obtained from the Case Study

From the case study in Shiojiri, the following insights were obtained. These insights are considered applicable to other fields as well.

- The description of the assurance case was a first attempt at TIER IV, and it was significant to share and visualize how the safety of level 4 automated shuttle bus is being achieved through a goal-oriented approach.
- Assurance cases for a Level 4 automated driving system can become extensive; for just a 2 km round trip between Shiojiri train station and the city hall, 32 driving scenarios are required for validation, and a detailed STAMP/STPA analysis is necessary even for a vehicle's general right-turn maneuver. We observe that the size and granularity of assurance cases require consensus formation in the field of system safety and dependability.
- The volume of documents related to the development and operation of automated driving systems is vast and sometimes disorganized. It seems difficult to perfectly manage all documents related to the system safety and dependability. We believe some feasible management methods should be considered.
- It is important to share safety and dependability information among stakeholders for system assurance. Our experience shows that GSN diagrams are not necessarily easy for some stakeholders to understand. We believe that safety and dependability experts should create GSN diagrams and then use techniques such as generative AI to translate the information into formats and content that are easier for various stakeholders to understand.
- System safety cannot be analyzed independently because it is related to availability and other system attributes. Therefore, an analytical method for various system attributes, i.e., dependability, is required. Currently, we are developing a dependability analysis method based on STAMP/STPA.

- In this paper, we presented a prototype toolchain of a monitoring system and an assurance case tool. To apply it in the actual development and operational phases of an automated driving system and realize a continuous assurance argument, specific monitored parameters and more real-world examples based on actual systems should be publicly shared and discussed.

6 Concluding Remarks

We have outlined our efforts in creating assurance cases for Level 4 automated driving, utilizing development and operational documents at TIER IV. Additionally, we have introduced a toolchain involving a monitoring system and the D-Case Communicator, designed to verify conditional statements in the GSN diagram. We have been continuing to create assurance cases for the Shiojiri case study, and it will be publicly available in the near future.

Acknowledgements. We thank the anonymous reviewers for their helpful comments. We also thank Michio Hayashi for proofreading of the paper and discussions, Yoshikazu Nojiri, Ikuhiro Mori, and Kou Toriyabe for their discussions. This research is based on results obtained from a project subsidized by the New Energy and Industrial Technology Development Organization (NEDO).

References

1. Asaadi, E., Denney, E., Menzies, J., Pai, G.J., Petroff, D.: Dynamic assurance cases: a pathway to trusted autonomy. Computer **53**(12), 35–46 (2020)
2. Assurance Case Working Group: Goal structuring notation community standard version 3 (2021). https://scsc.uk/r141C:1?t=1
3. Bishop, P., Bloomfield, R.: A methodology for safety case development. In: Safety-Critical Systems Symposium (SSS 1998) (1998)
4. Gauerhof, L., Munk, P., Burton, S.: Structuring validation targets of a machine learning function applied to automated driving. In: SAFECOMP 2018, pp. 45–58 (2018)
5. IEC: IEC 62853:2018 Open systems dependability (2018)
6. ISO: ISO 26262:2018 road vehicle - functional safety, part 1 to part 10 (2018)
7. ISO: ISO 22737:2021 Intelligent transport systems - Low-speed automated driving (LSAD) systems for predefined routes - Performance requirements, system requirements and performance test procedures (2021)
8. ISO: ISO 21448:2022 Road vehicles - Safety of the intended functionality (2022)
9. ISO: ISO 34502:2022 Road vehicles Test scenarios for automated driving systems - Scenario based safety evaluation framework (2022)
10. ISO/IEC: ISO/IEC Guide 51:2014 Safety aspects - Guidelines for their inclusion in standards (2014)
11. Koopman, P.: UL 4600: what to include in an autonomous vehicle safety case. Computer **56**(5), 101–104 (2023)
12. Leveson, N.G., Thomas, J.P.: STPA Handbook (2018)
13. Matsuno, Y.: D-Case Communicator Web Page. https://www.matsulab.org/dcase/

14. Matsuno, Y.: D-case communicator: a web based GSN editor for multiple stakeholders. In: Tonetta, S., Schoitsch, E., Bitsch, F. (eds.) SAFECOMP 2017. LNCS, vol. 10489, pp. 64–69. Springer, Cham (2017). https://doi.org/10.1007/978-3-319-66284-8_6
15. Matsuno, Y., Takai, T., Okada, M., Tsuchiya, T.: Toward dependability assurance framework for automated driving systems. In: Guiochet, J., Tonetta, S., Schoitsch, E., Roy, M., Bitsch, F. (eds.) SAFECOMP 2023. LNCS, vol. 14182, pp. 32–37. Springer, Cham (2023). https://doi.org/10.1007/978-3-031-40953-0_4
16. SAE International: SAEJ3016: Taxonomy and definitions for terms related to driving automation systems for on-road motor vehicles (2021)
17. UL: UL 4600 Evaluation of Autonomous Products, 3rd edn. (2023)
18. Wardziński, A.: Safety assurance strategies for autonomous vehicles. In: Harrison, M.D., Sujan, M.-A. (eds.) SAFECOMP 2008. LNCS, vol. 5219, pp. 277–290. Springer, Heidelberg (2008). https://doi.org/10.1007/978-3-540-87698-4_24

Security of Safety-Critical Systems

TitanSSL: Towards Accelerating OpenSSL in a Full RISC-V Architecture Using OpenTitan Root-of-Trust

Alberto Musa[1]($^{\boxtimes}$), Franco Volante[2], Emanuele Parisi[1], Luca Barbierato[2], Edoardo Patti[2], Andrea Bartolini[1], Andrea Acquaviva[1], and Francesco Barchi[1]

[1] Università di Bologna, Via Zamboni, 33, 40126 Bologna, Italy
{Alberto.Musa,Emanuele.Parisi,Andrea.Bartolini,Andrea.Acquaviva,
Francesco.Barchi}@unibo.it
[2] Politecnico di Torino, Corso Duca degli Abruzzi, 24, 10129 Torino, Italy
{Franco.Volante,Luca.Barbierato,Edoardo.Patti}@polito.it

Abstract. RISC-V open-hardware designs are emerging in cyber-physical systems and security-critical embedded platforms. Among them, OpenTitan emerged as an open-source silicon Root-of-Trust, which provides secure-boot and execution-integrity functionalities, exploiting its internal hardware accelerators. In this paper, we explore a novel exploitation of OpenTitan as a secure cryptographic accelerator. To this purpose, we designed TitanSSL, a secure software stack that offloads cryptographic tasks to OpenTitan, and we study the trade-offs between offloading overhead through the stack and the obtained computation speed-up. TitanSSL includes an OpenSSL backend, a Linux driver for communications, and an OpenTitan firmware. We executed TitanSSL on a cycle-accurate simulator of a RISC-V CVA6 application processor integrated with OpenTitan on the same System-on-Chip. We compared our implementation with a pure software version across different cryptographic payloads. Finally, we provide guidelines for the use of OpenTitan as a coprocessor in secure cyber-physical systems designs based on open-hardware architectures.

Keywords: RISC-V Software Stack · Secure Systems · Software Stack

1 Introduction

Open-hardware System-on-Chip (SoC) technology is currently emerging across various application domains, including safety-critical and security-critical Cyber-Physical Systems (CPS). These systems are characterized by their highly dynamic operations and fast responses, enabling them to function effectively in critical environments but raising challenges to ensure safety and security against cyber threats. To address this challenge, modern SoCs incorporate hardware components to enhance security. For instance, [21] integrates an application processor and a silicon Root-of-Trust (RoT) within the same SoC to facilitate secure boot functions. Similarly, a recent work [4] combines the CVA6 RISC-V application processor [26] with OpenTitan as a RoT [5].

Integrating a RoT such as Open Titan within an SoC provides a secure execution environment isolated from the rest of the SoC but incurs a substantial area cost. For instance, [4] demonstrates that the OpenTitan Root-of-Trust is more than 75% larger than the host platform. Making the most of OpenTitan functionalities is highly desirable to pay off this cost. For instance, if properly interfaced, OpenTitan hardware support for key cryptographic algorithms can accelerate security libraries adopted by user applications.

In this work, we address the problem of providing a secure backend for OpenSSL exploiting memory isolation and cryptographic acceleration of OpenTitan-featured SoC.

Providing OpenSSL with a secure backend eliminates the need for additional dedicated security modules and enables the use of OpenTitan's private and protected memory for a wide range of software applications, avoiding the storage of cryptographic secrets within the main memory. Enabling OpenTitan for applications based on OpenSSL requires the development of an end-to-end software stack comprising *i)* an OpenSSL engine, *ii)* a driver for managing communication with OpenTitan, and *iii)* a firmware for the Security Controller (SC), the RISC-V core present in OpenTitan.

TitanSSL stack has been designed to match the following functional requirements: *i)* the cryptographic accelerators must be exploitable; *ii)* the functionalities for secret management and cryptographic key generation must be ensured; *iii)* the application's virtual memory must be usable transparently by the SC; and, finally, *iv)* the application processors must communicate with the security controller through a dedicated channel.

To evaluate TitanSSL functionalities and associated overheads on a real-life case, we executed TitanSSL on a RISC-V System-on-Chip platform based on CVA6 application processor [26]. The considered platform is representative of state-of-art RISC-V based systems for secure autonomous systems such as Micro-Aerial Vehicles. The platform is implemented on FPGA integrating OpenTitan Root-of-Trust and running Linux [4]. To extract performance profiles, we characterized OpenTitan execution on the QuestaSim cycle accurate simulator.

The conducted experiments highlight the communication overhead between the host and OpenTitan systems, the speedup gained with respect to the pure software implementation, and providing a set of design guidelines to boost the performance of the resulting systems in future designs.

We conduct a comparative analysis, revealing that, for typical cryptographic workloads, the overhead introduced by the TitanSSL software stack is compensated by the speed-up achieved with OpenTitan cryptographic accelerators. For example, when using AES with a 256-bit key in CBC mode on a 1024-byte payload, we observe a 2.4x speed-up compared to pure software. Similarly, SHA-256 on a 1024-byte payload gets a 3.4x speed-up, and RSA operations with a 1024-bit private key result in a 3.0x speed-up. This characterization highlights OpenTitan viability as a Security Controller (SC). Additionally, we have identified hardware and software bottlenecks that present opportunities for future optimization.

The structure of the manuscript is described as follows. Section 3 describes the hardware architecture used for our experimental setup. Section 4 reports the software framework and the technologies used. Section 5 explains the characterization methodology adopted to evaluate the solution and it presents experimental results on the proposed software stack compared to the default OpenSSL implementation. Section 6 defines the analysis conducted on the system to establish and ensure its security measures. Finally, Sect. 7 reports concluding remarks and future works.

2 Background and Related Works

Our work envisions a scenario in which OpenTitan is integrated into the same System-on-Chip (SoC) alongside a RISC-V application core lacking built-in cryptographic hardware support or other execution isolation mechanisms, as discussed in [26]. In this configuration, OpenTitan provides extensive security features through its hardware accelerators and memory isolations, as detailed in [5]. The OpenTitan core acts as a Security Controller (SC). It is responsible for executing cryptographic primitives, utilizing cryptographic accelerators, and ensuring the isolation of secrets from the application processor. In literature, similar implementations are attributable to *EdgeLock Secure Enclave* [20], *Keystone* [16], and *Timber-V* [25].

The EdgeLock Secure Enclave serves as an integrated security subsystem designed to enhance the security of NXP Processors. It utilizes hardware-level isolation and cryptographic safeguards to ensure the integrity and confidentiality of data, boasting robust security features. However, being a cutting-edge proprietary solution, it poses challenges when it comes to porting or replication in custom architectures.

Keystone, a framework for creating secure software environments (Enclaves) on RISC-V architectures, ensures physical isolation through RISC-V Physical Memory Protection (PMP) [3]. It isolates specific memory areas, restricting access to privileged users. While Keystone is versatile and suitable for diverse use cases and architectures, its flexibility comes with the drawback of incorporating a relatively expensive software Secure Monitor. This addition may lead to performance issues for the entire architecture, given that Enclaves share the same processor as the rest of the system.

TIMBER-V instead exploits the concept of tagged memory [14] to implement secure enclaves. More in-depth, TIMBER-V uses a 2-bit tag for every 32-bit word memory that indicates the privilege, namely U-mode for user privileges and S-mode for supervisor privileges, and the allowable accesses that the memory block has. By combining the hardware concepts of tagged memory and the privileges modes introduced by the RISC-V architecture, the objective of TIMBER-V is mainly focused on low memory and resource consumption while providing an acceptable level of portability.

Several other works propose different RISC-V architectures and analyze their security constraints and requirements [9]. In particular, T. Lu's survey [18] highlights the growing popularity of RISC-V architectures in the research community,

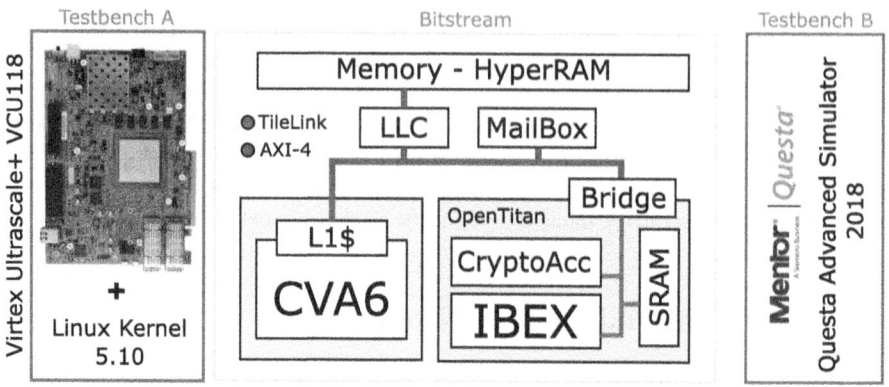

Fig. 1. Experimental setup and the testbeds used to evaluate the software stack.

particularly in security domains. This survey emphasizes how the openness of the RISC-V architecture can enhance hardware security against cyber threats through rigorous audits. These proposed RISC-V solutions encompass both software frameworks [1,17] and hardware approaches [24]. The latter introduces a solution optimizing context switches and effectively isolating the RISC-V architecture.

Programmable Root-of-Trust (RoT) IPs, such as OpenTitan, have been introduced to support secure boot and security services but have not been utilized as Security Controllers (SC) previously. One notable advantage is that isolation is not tied to a specific Operating System (OS) or architectural feature in the main processor, unlike popular commercial solutions such as *Intel SGX* [6] or *ARM Trustzone* [22]. This flexibility enables the integration of various processors that can function as Security Controllers, featuring cryptographic hardware accelerators capable of significantly enhancing the performance of cryptographic tasks while maintaining a high level of both performance and portability. Lastly, to the best of the author's knowledge, this work represents the first attempt to characterize the trade-offs and microarchitectural implications of the proposed architecture while supporting cryptographic tasks using OpenSSL.

3 Hardware Architecture

The reference platform considered in this work is shown in Fig. 1. It is representative of state-of-art RISC-V based systems, such as the one presented in [4] for secure autonomous Micro-Aerial Vehicles. The considered platform features three main components: *i*) a RV64GC Application Core, namely CVA6, *ii*) a Programmable Multi-Core Accelerator, and *iii*) OpenTitan, the hardware Root-of-Trust. CVA6 [26] is a Linux-capable RV64GC core featuring a six-stage, single-issue, in-order pipeline. It is meant to run general-purpose platform configuration tasks and to manage the system peripherals. The Programmable Multi-Core Accelerator consists of a cluster of eight CV32E40-based cores [11] with custom ISA extensions designed to accelerate computationally intensive DSP and machine learning tasks. OpenTitan is an open-source silicon Root-of-Trust inspired by Titan [15]. It consists of a set of hardware accelerators to enable efficient computation of common cryptographic primitives, plus the OpenTitan Big Number core (OTBN), a programmable coprocessor for asymmetric public-key cryptography. Sensitive data are stored in the OpenTitan internal ROM and scratchpad, two tamper-proof memories featuring scrambling functionalities. All the operations of the Root-of-Trust are coordinated by Ibex [7], it is a RV32IMC microcontroller optimized for embedded control applications. Additionally, OpenTitan features a range of protection against physical attacks, including power analysis countermeasures and tamper detection circuits. As mentioned, the target design integrates OpenTitan within the same silicon as the Application Processor and the Programmable Multi-Core Accelerator. A MailBox mediates the communication between the CVA6 Application Core and OpenTitan, as detailed in Sect. 4; it consists of a shared memory region meant for data sharing, plus two specialized registers, namely Doorbell and Completion, whose LSB is connected to the interrupt controller of Ibex and CVA6 respectively, to implement an easy asynchronous acknowledging system between the two cores. The memory Hierarchy employed by the CVA6 Application Processor and the RoT controller consists of three levels: private cache, Last-Level Cache (LLC), and main memory. The CVA6 core's private memory encompasses two L1 set-associative caches, a 16 KiB of L1 I-cache and a 32 KiB write-through L1 D-cache. The IBEX core lacks a cache memory but can access a 32 KiB SRAM scratchpad exclusive to the RoT system. Both systems share access to the LLC, an eight-way set-associative cache with 128 KiB and 256 lines. Finally, the SoC utilizes a 32 MiB main memory composed of four HyperRAM chips, each with a 64 Mbit capacity, situated off-chip.

4 TitanSSL Software Architecture

The TitanSSL software stack is illustrated in Fig. 2. As detailed in Sect. 3, it redirects OpenSSL cryptographic tasks needed by applications running on the application processor, such as SHA-256, RSA, and AES, to the Security Controller (the OpenTitan core). Consequently, the TitanSSL software stack is divided between the portion running on the Application Processor and the portion running on the Security Controller.

4.1 Application Processor

The software stack within the Application Processor plays a crucial role in facilitating communication from an untrusted environment, where applications utilize the OpenSSL library [10], to the Security Controller. The primary component is the TitanSSL Security Controller Engine (TitanSSL-SCE). This component seamlessly integrates OpenSSL functionalities with the Global Platform [13] standard through the implementation of the Global Platform Client

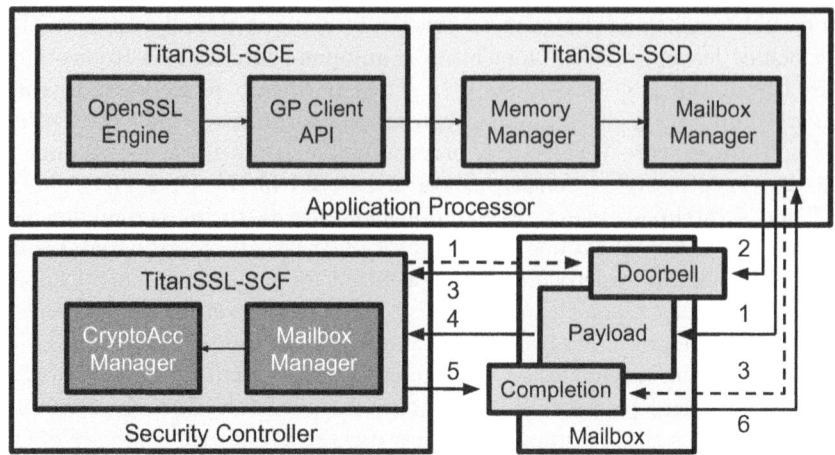

Fig. 2. Software stack and communication between Mailbox Managers.

API [12]. It establishes communication with the TitanSSL Security Controller Driver (TitanSSL-SCD) and subsequently with the TitanSSL Security Controller Firmware (TitanSSL-SCF). The TitanSSL-SCD represents the final software stack component residing in the Application Processor. Its responsibility is to ensure that the TitanSSL-SCF receives data from the TitanSSL-SCE via a communication mailbox and vice-versa. The software components, depicted in Fig. 2, running on the Application Processor, execute either in the *User Space* (illustrated in green) or in the *Kernel Space* (depicted in orange).

The software stack located in the user space includes the two key components of the TitanlSSL-SCE. These components intercept cryptographic operations, such as encryption and digital signatures, needed by a user-defined software (Application) and are activated by the application using the OpenSSL library.

The first component of the TitanSSL Security Controller Engine is an OpenSSL Engine, which serves as a library-defined extension mechanism of OpenSSL. Its purpose is to redirect the execution of cryptographic operations towards the Security Controller. This engine is linked to a Global Platform Client API implementation, a library specification designed to facilitate the management of security devices [12]. The OpenSSL engine implementation has two primary roles: initializing a communication context, TEEC_Context, and a communication session, TEEC_Session, as outlined by the GP Client API, and redirecting cryptographic requests (such as SHA-256, AES, or RSA) to the lower layers of the proposed architecture.

The second component of the TitanSSL Security Controller Engine is a Global Platform Client API implementation. This component abstracts the underlying Security Controller to our OpenSSL Engine, allowing the establishment of a secure connection with the TitanSSL Security Controller Driver (TitanSSL-SCD). Specifically, it ensures correct and secure communication through TEEC_Context and TEEC_Session. It transmits all the data needed to

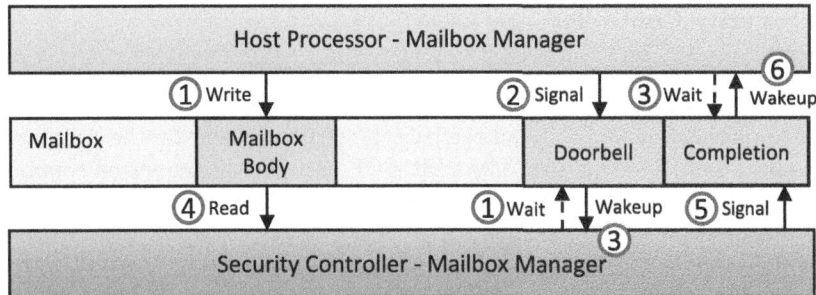

Fig. 3. Representation of the communication between the two Mailbox Managers and the Mailbox.

perform the operations required by the application to the Kernel Space by populating the TEEC_Operation structure (Fig. 3).

The TitanSSL-SCD is a software-stack component residing in the kernel space. It manages requests directed to the Security Controller and facilitates communication between the Application Processor and the Security Controller through a mailbox. Its functionalities are exposed exclusively through a character device accessible through the TitanSSL-SCE. When a command is invoked, the Device Driver processes the data by translating virtual addresses from User Space to physical addresses for the Security Controller. This process involves two steps: i) Pinning the page containing the address in memory to prevent data corruption and ii) Translate the virtual address to a physical one the Security Controller uses. The TitanSSL-SCD incorporates a Mailbox Manager to facilitate data exchange between the Application Processor and the Security Controller. If references to memory areas provided by the application occupy more than one page (4KiB in the worst case), the Mailbox Manager can handle memory fragmentation through a Multipage Management System. A new memory page, called "ReferencesPage", is allocated and then filled with the physical address information of all involved memory pages. By sending the ReferencesPage physical address to the Security Controller, the Security Controller will be able to access, in the best case, about 1GiB of virtual memory and, in the worst case, about 1MiB of virtual memory, depending on the size of the pages. If a single ReferencesPage is insufficient, it will be necessary to iteratively create additional ReferencesPages and provide them to the Security Controller.

The Mailbox Manager handles the mailbox as depicted in Fig. 2. This mailbox has a 32-bit field for the message payload and Doorbell and Completion registers designed to synchronize communication at both ends. The Mailbox Manager of the Application Processor ① composes a message in the mailbox, specifying the task and the physical addresses of involved memory pages. Subsequently, it ② activates the Doorbell register, thereby triggering the awakening of the Security Controller Mailbox Manager. Finally, the Mailbox Manager enters an idle state ③, awaiting activation through the Completion register ⑥. Once awakened, it delivers cryptographic operation results to the upper software layers.

4.2 Security Controller

TitanSSL Security Controller Firmware (TitanSSL-SCF) executes cryptographic operations within the secure space provided by OpenTitan. Leveraging its direct access to physical memory, it can retrieve the structure sent by the Application Processor. From this structure, TitanSSL-SCF extracts the requested command, the operation parameters, and the addresses for reading data and writing operation outputs.

As depicted in Fig. 2, the Mailbox Manager within the Security Controller remains in a waiting state ① until it receives a signal from the Doorbell register ③. Upon awakening, it ④ retrieves the requested cryptographic task and the physical address of the structure containing inputs and parameters from the Mailbox. Then, the Cryptographic Accelerator Manager carries out the demanded operations, such as generating a hidden ephemeral key, initiating a signature or signature verification procedure, computing a cryptographic hash, or encrypting a data sequence. Upon the completion of the cryptographic operation, the Mailbox Manager ⑤ sends a signal to the Completion register, awakening the Mailbox Manager of the Application Processor.

5 Experimental Results

To assess the efficacy of the suggested software stack, we evaluate the performance of three commonly utilized cryptographic algorithms: SHA-256, AES-256-CBC, and RSA. Section 5.1 compares the runtime achieved by offloading computation to OpenTitan in comparison to executing a purely software-based implementation on the host core. Then, Sect. 5.2 reports a precise, cycle-accurate, analysis of the OpenTitan firmware to identify possible inefficiencies and limitations in its current platform design. To run OpenSSL software implementations in Linux, we synthesized the reference platform on a Xilinx Virtex UltraScale+ VCU118 FPGA due to limited simulation speed in RTL simulators like QuestaSim. All experiments ran on Linux kernel v5.10.7, built with a standard RISC-V toolchain.

5.1 Comparison with Software Implementation

Table 1 and Fig. 4 present a comprehensive performance evaluation of the cost of offloading cryptographic operations to OpenTitan using the described software stack for various algorithms and payload sizes, comparing their effectiveness with the same algorithms executed on the host processor without our engine. Experiment runs are compared measuring the number of cycles required to process the specified payload size. SHA-256 and AES are run on four payload sizes: 16, 64, 256, and 1024 bytes, while RSA is evaluated on 512, 1024 and 2048 bits key length. Each experiment generated three measurements: *OpenSSL*, *TitanSSL*, and *SCF*. *OpenSSL* illustrates the cost of running the selected algorithm using the default SSL software implementation on the CVA6 host processor without

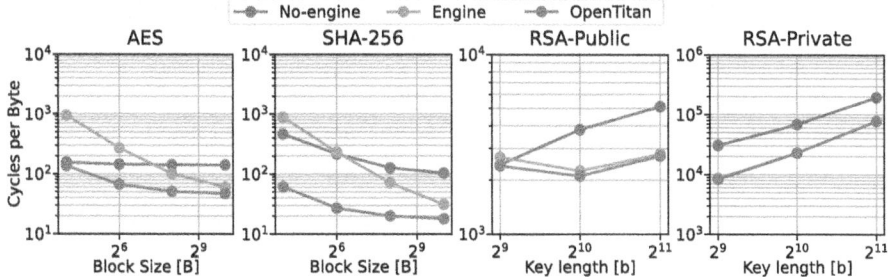

Fig. 4. The results are presented in cycles per byte for AES and SHA-256, as well as for different key lengths for RSA.

harnessing the cryptographic accelerators within OpenTitan. *TitanSSL* represents the clock cycles required to execute the proposed software stack in this study, encompassing the code running on the CVA6 core and the OpenTitan firmware. *SCF* accounts for the cycles needed by the TitanSSL-SCF only. The *Overhead* column emphasizes the overhead introduced by the SSL engine and OpenTitan device driver in comparison to the actual computation within the RoT. The *Speedup* value summarizes the advantage of accelerating cryptographic workloads using OpenTitan in terms of clock cycles compared to a pure software implementation on CVA6.

The first two sections of Table 1 present the performance results for the SHA-256 and AES algorithms, exhibiting similar behaviour. In both cases, OpenTitan demonstrates slower performance than the default software implementation for block sizes of 16 and 64 bytes. However, it outperforms the default software implementation for block sizes greater than or equal to 256 bytes, yielding speedups ranging from 1.4x to 3.4x. One contributing factor to this phenomenon is the lack of coherence between the IBEX private memory and the L1 data cache of the CVA6 core. Consequently, OpenTitan reads input data either from the Last Level Cache (LLC), with a latency of 15–20 cycles per word, or, in the event of an LLC miss, from the off-chip HyperRAM, incurring a latency of approximately 100 cycles per word. This imposes a notable overhead at the commencement of any computation within the RoT, necessitating prefetching data from the LLC memory into the OpenTitan scratchpad. Furthermore, any operation that triggers OpenTitan results in overhead due to the software layers that the user-space application must navigate to interface with the SSL engine and the OpenTitan device driver. These two side effects are quantified by the significant overhead observed for block sizes of 16 and 64 bytes, ranging from 75.2% for AES on a 64-byte block to 93.1% for SHA-256 on a 16-byte block.

The last two sections of Table 1 present the performance of the RSA algorithm when used for encryption (*rsa-public*) or decryption (*rsa-private*). Notably, the computational operations carried out by OpenTitan incur low overheads, ranging from 0.6% to 10.5%. Furthermore, in the best-case scenario, OpenTitan provides a speedup of 3.5x over software implementation, while in the worst case, it is

Table 1. TitanSSL results in terms of cycles per byte.

Test	Cycles per Byte			Speedup	Overhead
	OpenSSL	TitanSSL	SCF		
16 B	462	877	61	0.5	93.1 %
64 B	216	231	27	0.9	88.2 %
256 B	127	71	20	1.8	72.1 %
1024 B	104	31	18	3.4	41.6 %
aes					
16 B	155	947	136	0.2	85.6 %
64 B	145	270	67	0.5	75.2 %
256 B	142	101	51	1.4	50.1 %
1024 B	141	60	47	2.4	21.2 %
rsa-public					
512 b	2 430	2 678	2 396	0.9	10.5 %
1024 b	3 815	2 250	2 109	1.7	6.3 %
2048 b	5 126	2 790	2 720	1.8	2.5 %
rsa-private					
512 b	30 518	8 618	8 337	3.5	3.3 %
1024 b	68 531	23 016	22 875	3.0	0.6 %
2048 b	193 955	78 140	78 070	2.5	0.1 %

marginally slower than CVA6, offering a speedup of 0.9x. This discrepancy can be attributed to the fact that RSA involves complex mathematical computations on large integer numbers that cannot be efficiently executed on the CVA6 core. In contrast, the OpenTitan Big Number Accelerator (OTBN) accelerator within OpenTitan is specifically designed to accelerate arithmetic operations on large integers, resulting in significantly faster exponentiation operations required for RSA. Additionally, the greater speedup of OpenTitan in *rsa-private* compared to *rsa-public* can be explained by the fact that the private key in an RSA public-private key pair typically being a large number, resulting in a heavier computational burden during the exponentiation process compared to the exponentiation performed when encrypting with the public key, which usually involves a smaller number.

5.2 OpenTitan Firmware Analysis

We highlight the potential bottlenecks of the designed OpenTitan firmware by simulating the computation of a 1500 bytes packet digest in QuestaSim. Table 2 shows the details of the simulation outcome, reporting the number of instructions executed and the corresponding cycle fraction for each opcode executed, in every stage of the algorithm. The OpCodes are categorized according to their

Table 2. OpenTitan SHA-256 firmware analysis.

Op.	OpCodes [#]						OpCodes [cycles]					
	ALU	CF	M.OT	M.RAM	M.MBOX	TOT	ALU	CF	M.OT	M.RAM	M.MBOX	TOT
Prologue	2	0	10	0	3	15	0.0%	0.0%	0.2%	0.0%	0.3%	0.5%
Init	19	12	7	0	0	38	0.2%	0.1%	0.2%	0.0%	0.0%	0.4%
MultiPage	5	4	0	2	0	11	0.0%	0.1%	0.0%	0.2%	0.0%	0.3%
Push	1891	1507	1127	375	0	4900	15.2%	6.0%	34.6%	40.6%	0.0%	96.4%
Wait	0	0	0	0	0	0	0.0%	0.0%	0.0%	0.0%	0.0%	0.0%
Process	9	9	10	0	0	28	0.1%	0.1%	0.3%	0.0%	0.0%	0.5%
Final	36	13	24	9	0	82	0.2%	0.1%	0.7%	0.8%	0.0%	1.8%
Epilogue	1	0	10	0	0	11	0.0%	0.0%	0.2%	0.0%	0.0%	0.2%
TOT	1963	1545	1188	386	3	5085	15.7%	6.4%	36.2%	41.6%	0.3%	24948

type (ALU/$Control$-$Flow$), and memory accesses are additionally discriminated by the memory location they target: main memory ($m.RAM$), OpenTitan private scratchpad ($m.OT$) or mailboxes ($m.MBOX$). The *Op.* column represents the different steps involved in digest computation. The function *Prologue* and *Epilogue* save/restore registers from the stack. *Init* refers to accelerator initialization. *Push*, *Wait*, and *MultiPage* specify the operation involved in feeding the accelerator FIFO and busy waiting when it is full, respectively, while taking into account the multi-page protocol. Process and Final refers to the operations of loading the last 512 bits block of data to the accelerator, and loading the final computed digest. The analysis highlights that over 96% of the 24,948 cycles required to run the computation are spent in the Push phase. The reason is that while, according to the OpenTitan datasheet, processing a 512 bit data block requires 80 cycles, moving 32 4-bytes words from the main memory to the accelerator FIFO requires approximately 25 cycles per word, around 20 cycles for an LLC-hit, plus 5/6 cycles to access the OpenTitan private scratchpad. While the results indicate that the memory subsystem of the reference platform is a bottleneck, they also demonstrate the potential of using this software stack to speed up cryptographic operations in the Root-of-Trust. This would improve computational efficiency and security guarantees simultaneously.

6 Security Assumptions and Implications

The security assumptions for our system are rooted in the threat model discussed in the Opentitan analysis for SCRAMBLE-CFI [19]. In that analysis, the authors explore how the OpenTitan chip withstands an attack when an adversary gains physical access to the system. They outline how OpenTitan is inherently equipped with countermeasures against such attackers. These countermeasures are designed to prevent the manipulation of stored and transmitted data, as well as the instructions within the chip. As a result, the attacker considered in the threat model of this section has only the ability to manipulate input data with the aim of gaining privileges for executing cryptographic tasks or extracting sensitive data stored within OpenTitan storage-all without necessitating physical access to the system.

Our analysis combines STRIDE [23] with OWASP [2] and CWE [8] classifications for software vulnerabilities. Certain OWASP vulnerabilities that are not applicable to the analyzed system have been excluded. The remaining vulnerabilities are aligned with STRIDE categories based on potential attacks when exploited. The threat model encompasses the entire architecture and assigns criticality levels ranging from 0 (not analyzed) to 9 (involving sensitive data). Sensitive data categories include secrets and crucial physical addresses that are pivotal for system security.

Our assessment employs the STRIDE approach to evaluate both software blocks and data flows: i) Software blocks are vulnerable to potential threats such as Spoofing, Repudiation, and Elevation of Privilege. ii) Data flows are susceptible to Tampering, Information Disclosure, and Denial of Service. It's important to highlight that the Application and its data flow are excluded from our analysis. In this context, the OpenSSL Engine, rated at a criticality level of 1, operates with minimal privileges and does not store sensitive data. Analyzing Spoofing and Repudiation is deemed unnecessary, as the OpenSSL Engine is accessible to all users. Additionally, concerns related to the Elevation of Privilege are addressed by subsequent authentication layers. In contrast, the GP Client API holds a criticality level of 2 due to its role as the gateway to Kernel Space on the Host Processor. This component incorporates authentication mechanisms, effectively mitigating Spoofing and Elevation of Privilege threats. Moreover, the GP standard's Session feature tracks accessing processes, ensuring comprehensive logging and monitoring to prevent Repudiation. In the Kernel Space, the Device Driver (criticality level 5) effectively mitigates threats associated with Broken Access Control and Broken Authentication, thereby limiting Spoofing and Elevation of Privilege threats. The kernel log system manages Repudiation and Logging issues. Both the Kernel Space's Mailbox Drivers and the Secure Space share a common architectural framework. However, they differ in execution privileges (5 vs. 7) owing to their distinct Logging and Monitoring capabilities.

The Secure Firmware remains isolated and secure by default, with its primary vulnerability limited to potential tampering with corrupted mailbox data. Our data flow analysis specifically focuses on the potential leakage of sensitive data between the Device Driver and the Host Processor's mailbox, as such a compromise could pose a threat to the entire system architecture. It's crucial to emphasize that other data flows are fortified against Wrong Input Validation or Buffer Overflow, effectively preventing unauthorized access beyond buffer boundaries. Moreover, the system emphasizes the integrity and confidentiality of data exchanged through the mailbox by incorporating two key features: *i) Secure Boot* and *ii) Memory Protection*. Secure Boot allows OpenTitan to verify and authenticate all firmware and software components executed within CVA6 and OpenTitan itself, thereby ensuring that the code remains unaltered. Memory Protection restricts write accesses to the mailbox exclusively to the processor's

S-mode, an execution mode compliant with the RISC-V supervisor specification and activated only during the operation of internal Linux kernel functionalities.

7 Conclusion

In this paper we introduced a software stack designed to use OpenTitan as a Security Controller (SC) within a RISC-V based System-on-Chip (SoC). This software stack empowers applications operating in an unsafe environment to seamlessly interact with the OpenTitan Security Controller for the execution of cryptographic operations or the generation and storage of secrets. To achieve this integration, we developed a custom OpenSSL Engine (TitanSSL-SCE), a driver (TitanSSL-SCD), and a Security Controller firmware (TitanSSL-SCF). Consequently, applications can use the OpenSSL API to perform cryptographic operations within the SC.

To evaluate the effectiveness of TitanSSL-SCE and TitanSSL-SCD components, we conducted benchmarks on a Linux environment running on an FPGA prototype of a RISC-V-based CVA6 SoC. Additionally, we profiled the execution of TitanSSL-SCF using a Register Transfer Level (RTL) simulator to characterize the overhead introduced by the OpenTitan firmware and measure the execution cycles of accelerators. Despite introducing overhead, TitanSSL yielded performance speed-ups of 3.4x for SHA-256 and 2.4x for AES-256-CBC with payloads of 64 bytes and 256 bytes, respectively. Furthermore, for RSA operations, TitanSSL outperformed the default OpenSSL software implementation, achieving nearly a 1.8x speed-up for rsa-public and a 3.5x speed-up for rsa-private. Moreover, these performance enhancements are achieved while benefiting from other OpenTitan features, including isolation, secure boot, and root-of-trust capabilities.

In future work we will expose TitanSSL to real-world network workloads generated by communication protocols like HTTPS or TLS.

References

1. Andrade, G., Lee, D., Kohlbrenner, D., Asanovic, K., Song, D.: Software-Based Off-Chip Memory Protection for RISC-V Trusted Execution Environments. UC Berkeley (2020)
2. Bach-Nutman, M.: Understanding the top 10 owasp vulnerabilities. arXiv preprint arXiv:2012.09960 (2020)
3. Cheang, K., Rasmussen, C., Lee, D., Kohlbrenner, D.W., Asanović, K., Seshia, S.A.: Verifying risc-v physical memory protection (2022)
4. Ciani, M., et al.: Cyber security aboard micro aerial vehicles: an opentitan-based visual communication use case. In: 2023 IEEE International Symposium on Circuits and Systems (ISCAS), pp. 1–5 (2023). https://doi.org/10.1109/ISCAS46773.2023.10181732
5. lowRISC CIC. Opentitan official documentation (2019). https://opentitan.org/book/doc/introduction.html

6. Costan, V., Devadas, S.: Intel sgx explained. Cryptology ePrint Archive, Paper 2016/086 (2016). https://eprint.iacr.org/2016/086
7. Davide Schiavone, P., et al.: Slow and steady wins the race? a comparison of ultra-low-power RISC-V cores for internet-of-things applications. In: 2017 27th International Symposium on Power and Timing Modeling, Optimization and Simulation (PATMOS), pp. 1–8 (2017). https://doi.org/10.1109/PATMOS.2017.8106976
8. Enumeration, C.W.: 2022 CWE top 25 most dangerous software weaknesses (2022). https://cwe.mitre.org/top25/archive/2022/2022_cwe_top25.html
9. Fadiheh, M.R., Stoffel, D., Barrett, C., Mitra, S., Kunz, W.: Processor hardware security vulnerabilities and their detection by unique program execution checking. In: 2019 Design, Automation and Test in Europe Conference and Exhibition, 2019, pp. 994–999 (2019). https://doi.org/10.23919/DATE.2019.8715004
10. Foundation, O.S.: Source code for the openssl software (1998). https://github.com/openssl/openssl
11. Gautschi, M., et al.: Near-threshold RISC-V core with DSP extensions for scalable IOT endpoint devices. IEEE Trans. Very Large Scale Integr. Syst. **25**(10), 2700–2713 (2017). https://doi.org/10.1109/TVLSI.2017.2654506
12. Group, G.P.: Global platform client API documentation (2023). https://optee.readthedocs.io/en/stable/architecture/globalplatform_api.html#tee-client-api
13. Group, G.P.: Global platform official website (2023). https://globalplatform.org/
14. Joannou, A., et al.: Efficient tagged memory. In: 2017 IEEE International Conference on Computer Design (ICCD), pp. 641–648 (2017). https://doi.org/10.1109/ICCD.2017.112
15. Johnson, S., Rizzo, D., Ranganathan, P., McCune, J., Ho, R.: Titan: enabling a transparent silicon root of trust for cloud. In: Hot Chips: A Symposium on High Performance Chips, vol. 194 (2018)
16. Lee, D., Kohlbrenner, D., Shinde, S., Asanović, K., Song, D.: Keystone: an open framework for architecting trusted execution environments. In: Proceedings of the Fifteenth European Conference on Computer Systems (EuroSys 2020). Association for Computing Machinery, New York (2020).https://doi.org/10.1145/3342195.3387532
17. Lee, D., Kohlbrenner, D., Shinde, S., Song, D., Asanović, K.: Keystone: an open framework for architecting tees (2019)
18. Lu, T.: A survey on RISC-V security: hardware and architecture (2021)
19. Nasahl, P., Mangard, S.: Scramble-cfi: mitigating fault-induced control-flow attacks on opentitan (2023)
20. NXP. Edgelock secure enclave (2019). https://www.nxp.com/products/nxp-product-information/nxp-product-programs/edgelock-secure-enclave:EDGELOCK-SECURE-ENCLAVE
21. Parno, B., McCune, J.M., Perrig, A.: Roots of Trust, pp. 35–40. Springer, New York (2011). https://doi.org/10.1007/978-1-4614-1460-5_6
22. Pinto, S., Santos, N.: Demystifying arm trustzone: a comprehensive survey. ACM Comput. Surv. **51**(6) (2020). https://doi.org/10.1145/3291047
23. Potter, B.: Microsoft SDL threat modelling tool. Netw. Secur. **2009**(1), 15–18 (2009)
24. Schrammel, D., et al.: Donky: domain keys–efficient in-process isolation for RISC-V and x86. In: Proceedings of the 29th USENIX Conference on Security Symposium, pp. 1677–1694 (2020)

25. Weiser, S., Werner, M., Brasser, F., Malenko, M., Mangard, S., Sadeghi, A.R.: Timber-v: tag-isolated memory bringing fine-grained enclaves to risc-v
26. Zaruba, F., Benini, L.: The cost of application-class processing: energy and performance analysis of a Linux-ready 1.7-GHZ 64-bit RISC-V core in 22-nm FDSOI technology. IEEE Trans. Very Large Scale Integrat. Syst. **27**(11), 2629–2640 (2019). https://doi.org/10.1109/TVLSI.2019.2926114

A Lightweight and Responsive On-Line IDS Towards Intelligent Connected Vehicles System

Jia Liu[1,2,3], Wenjun Fan[2(✉)], Yifan Dai[3], Eng Gee Lim[2], and Alexei Lisitsa[1]

[1] Department of Computer Science, University of Liverpool, Liverpool L69 7ZF, UK
{J.Liu154,lisitsa}@liverpool.ac.uk
[2] School of Advanced Technology, Xi'an Jiaotong-Liverpool University, Suzhou 215123, Jiangsu, China
{Wenjun.Fan,Enggee.Lim}@xjtlu.edu.cn
[3] Suzhou Automotive Research Institute, Tsinghua University, Suzhou 215200, Jiangsu, China
Daiyifan@tsari.tsinghua.edu.cn

Abstract. The current intelligent connected vehicles (ICV) system often shares the detected intrusion event to the cloud for further collaborative investigation. The upstream channel leading to the Internet of Vehicles (IoV) cloud is typically vendor-proprietary and costly, and the congestion caused by false alarms even exacerbates the situation. Machine learning (ML) can improve intrusion detection performance by reducing the false alarm rate. However, as a computation-intensive approach, traditional ML is not appropriate for real-time detection. Therefore, this paper proposes a lightweight and responsive on-line intrusion detection approach aiming for the ICV system requiring real-time detection. More specifically, we design a model termed Machine Learning integrated with Blacklist Filter (ML-BF), which leverages the feature engineering and the Bloom filter techniques built on ML to enhance both detection and real-time performances. To evaluate the proposed solution, several experiments are conducted by using the Car-Hacking and CIC-IDS-2017 datasets. The experimental results show that our approach can detect intrusion at a microsecond level with a lower computational cost as well as a lower false positive rate than that in the state-of-the-art.

Keywords: Intelligent Connected Vehicles · Intrusion Detection · Responsive Detection · Bloom Filter · Machine Learning

1 Introduction

Over the past decades, the rapid development of the automobile industry has facilitated the evolution of vehicular networks through multiple generations, transitioning from vehicular ad-hoc network (VANET) to vehicle-to-everything (V2X) [4,19,29]. Also, leveraging computer technology and artificial intelligence, the traditional the vehicular system is gradually being replaced by ICV system, which integrates human, vehicle, road, and cloud [10]. Other than the areas of autonomous vehicles and IoV [16], ICV focuses on security issues in the context of vehicles since the security standardization in this field has progressed [24]. In particular, as the UNECE R155 regulation and ISO/SAE 21434 standard [3] were ratified, automotive manufacturers are requested to respond to security issues for the whole vehicle life cycle [8].

In this regard, an efficient intrusion detection system (IDS) towards ICV becomes a cutting-edge research area. Basically, an IDS of ICV consists of two components [17]: vehicle on-board intrusion detection and cloud situational awareness. At present, on the one hand, the rule-based on-board intrusion detection has a high false-alarm ratio due to the coarse granularity of the rules; on the other hand, the cloud situational awareness lacks applying historical information or threat intelligence. Such drawbacks result in an inefficient detection performance in the context of ICV. Machine learning (ML) methods can be leveraged to address the above drawbacks. However, it is inappropriate to take ML for granted, especially in such a resource-constrained ICV environment that demands real-time detection. This introduces a trade-off between achieving higher detection performance and utilizing lower computational resources in the IDS.

Therefore, this paper is motivated to propose a lightweight, responsive, and high-detection performance on-line IDS for the ICV system. To this end, several key technical problems should be addressed: 1) first, irrelevant features[1] like the redundant and little discriminatory ones should be removed from the datasets, i.e., Car-Hacking [22] and CIC-IDS-2017 [23] used in this work; 2) second, imbalanced datasets, where one class has significantly fewer samples, need to be adjusted to avoid biased modeling, with a feature selection method assisting by prioritizing features that contribute to distinguishing the minority class; 3) third, the presence of noise and outliers in the training data can result in overfitting, which can be addressed by using a feature selection method to reduce the complexity of the model; 4) fourth, given that model loading and data regularizing consume great computational resources and decrease response efficiency, it is rational to consider other computationally efficient methods, such as the Bloom filter [2], to detect attacks. Based on the above points, this paper applies ML algorithms, emphasizing the feature selection and Bloom filter, to enhance intrusion detection within the resource-constrained vehicular Controller Area Network (CAN) bus. The objective of this paper is to increase the intrusion detection performance and decrease the real-time latency as well as the computational cost. The contributions of this paper are summarized as follows:

- First, this paper optimizes the feature selection based on the empirical experiments over the datasets and so enhances the detection performance.

---

[1] In the context of machine learning, "features" refer to individual measurable properties or attributes of the data that are used to represent and characterize the input for a predictive model.

- Second, this paper reduces the false positive rate (FPR) to alleviate the issue of congested and expensive data transmission in the upstream channel.
- Third, compared with prior research, the proposed approach of combining ML with feature selection and Bloom filter greatly reduces the detection cost while maintaining accuracy, making it a breakthrough for lightweight and responsive intrusion detection for the ICV system.

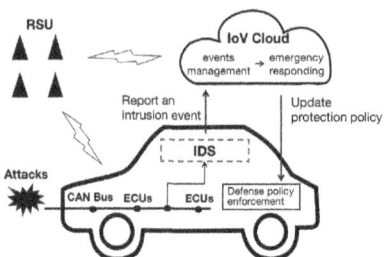

Fig. 1. A generic ICV system.

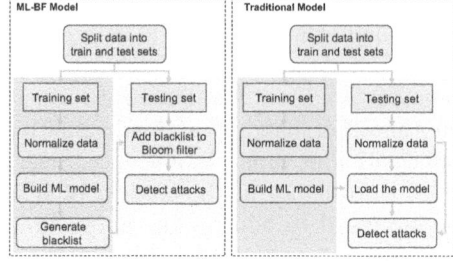

Fig. 2. ML-BF vs. Traditional ML.

The rest of the paper is organized as follows: Sect. 2 elucidates the proposed approach; Sect. 3 demonstrates the implementation; Sect. 4 presents the experimental results; Sect. 5 performs a comprehensive literature review; Sect. 6 concludes the paper and proposes avenues for future research.

2 Methodology

This section proposes the methodology, including an ICV system overview, the threat model, the ML-BF model, the feature engineering, and the blacklist filter.

2.1 System Overview

Figure 1 shows a general overview of the main functional components involved in the context of the ICV system, including the roadside units (RSU), IoV cloud, and the CAN bus. These three components can transmit data with each other. This work focuses on the IDS marked by the red dotted line in Fig. 1.

The on-board intrusion detection module mainly collects vehicle security status and data information and analyzes possible intrusion events from the CAN bus. The detected intrusion events are uploaded to the IoV cloud analyzer through the vehicle's network connection function. Cloud servers typically manage and present all vehicle-related events, including reported intrusion events. The IoV cloud server is also responsible for analyzing the intrusion events and updating the security protection policies on the vehicle's side. The defense policy enforcement can execute information security protection policies delivered by the IoV cloud server.

2.2 Threat Model

Built on the system overview, this subsection presents the threat model. First of all, it is assumed that the unauthorized user is able to access the target vehicular network, which includes the RSU, IoV cloud, and the CAN bus. With such an assumption, the potential threats and risks can be described as follows:

- **Unauthorized Access**: Threat actors attempt to gain unauthorized access to the vehicle's CAN bus or electronic control unit (ECU), compromising vehicle control and safety.
- **Data Tampering**: Intruders attempt to manipulate or spoof data transmitted via the CAN bus, leading to system malfunctions or false information.
- **Denial of Service (DoS)**: Malicious entities could launch DoS attacks against the vehicle's communication systems, disrupting normal operation and safety-critical functions.
- **Insider Threats**: Unauthorized or rogue elements within the vehicle's network exploit vulnerabilities to access or manipulate sensitive resources.

These potential threats form the basis of the in-vehicle threat model, emphasizing the need for robust and prompt intrusion detection, secure communications, and policy enforcement mechanisms to safeguard connected vehicles. Therefore, this paper is motivated to underscore the real-world implications of these threats and the on-line measures needed to mitigate them, amplifying the relevance and urgency of addressing these challenges in the context of connected vehicles.

2.3 ML-BF Model

This subsection introduces the proposed ML-BF model utilizing feature engineering and the Bloom filter, as depicted in Fig. 2, to address the lightweight and responsive intrusion detection issue. It is worth noting that this design is not limited to the ICV system, but should be applicable in the similar resource-constrained environment.

To this end, first, we use feature engineering for data normalization and then build the ML model. Second, we generate a blacklist, which is a rule base based on a trained model. Third, we leverage the Bloom filter to match the data with the blacklist to detect attacks. Here, we use the ML model for blacklist generation, which effectively increases the comprehensiveness and inclusiveness of the rule base. Compared with traditional ML, we maintain feature selection in the model training part and introduce the Bloom filter in the prediction part without normalizing data and loading the model, which eases system load. Since loading the model and processing the data increases computational costs during the detection when deploying IDS to a real-time system.

In this model, feature engineering reduces data dimensionality by eliminating irrelevant features, thereby reducing training consumption and the false alarm rate. The Bloom filter helps avoid resource-intensive steps, thereby lowering overall resource consumption and improving responsiveness. To illustrate, resource consumption is calculated based on training and testing, where the dark gray and

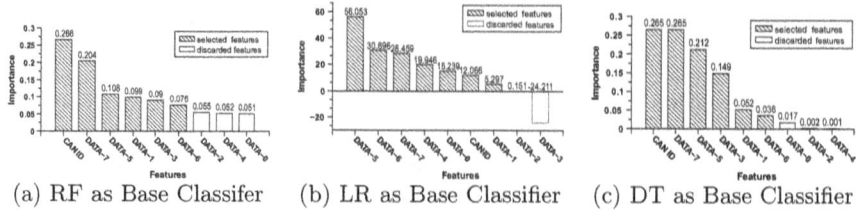

(a) RF as Base Classifer   (b) LR as Base Classifier   (c) DT as Base Classifier

Fig. 3. The feature importance of the Car-Hacking dataset.

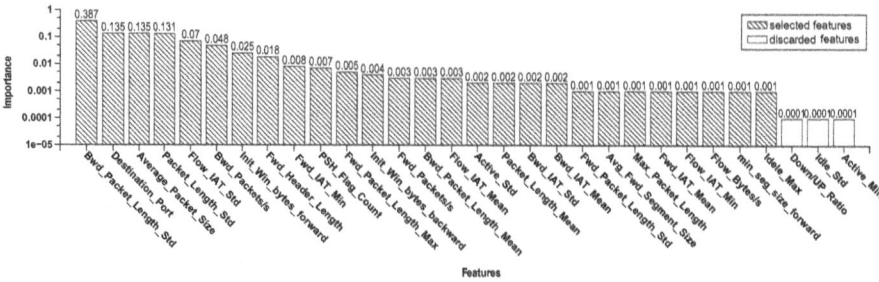

Fig. 4. The feature importance of the CIC-IDS-2017 dataset (which only shows 30 out of 80 features since the importance of the rest is negligible).

light gray areas in Fig. 2 represent these respective aspects. Feature engineering primarily focuses on the training phase by decreasing the model's execution time through data normalization and extracting important features to enhance performance and reduce false alarms. Conversely, the Bloom filter primarily focuses on the detection phase, accelerating the detection process through direct matching of data and rules. The ML-BF model enhances IDS performance through these two stages, ensuring a high detection rate and low false alarm rate while meeting lightweight and responsive requirements.

2.4 Feature Engineering

Intra-vehicle/inter-vehicle communications often generate a large amount of network traffic. When building a ML model, it's crucial to carefully select and engineer relevant features, as they directly impact the model's ability to learn patterns, make accurate detection. Using appropriate features leads to faster model training and inference times as well as higher detection performance. Also, finding a suitable method to reduce the feature dimension can reduce the risk of over-fitting. To this end, extracting features with high correlation is the focus of this part. Each feature has a contribution coefficient to the target variable, which is the feature importance.

Our study employs the recursive feature elimination (RFE) method [7], which wraps another learning base classifier to select features by repeatedly building multiple models, each time picking and removing the minor essential features according to the importance of coefficients and then repeating this process on the remaining features until all features have been traversed. Thus, the importance of the feature is returned according to the sum of the gains brought by the segmentation, and the feature set is narrowed to a range. Thereafter, RFE is used to remove features that are useless or have a negative impact on the model.

In detail, for the selection of the base classifier, we compare Random Forest (RF), Logistic Regression (LR), and Decision Tree (DT), obtaining the optimal number of features via 5-fold cross-validation. Then, we utilize the importance rankings of the features to select the specific features. After repeated tests on the Car-Hacking dataset, we determine that using the DT model as the base classifier and selecting 7 features, further introducing it to the IDS model training, results in less execution time and high accuracy of approximately 99.99% - the best performance amongst the aforementioned three methods. The result is illustrated in Fig. 3, which shows the importance of rankings and the feature selection process. On the CIC-IDS-2017 dataset, the base classifier DT maintains excellent performance, which selects 27 features, and the model's accuracy reaches 99.86%. Figure 4 indicates the feature selection process upon this dataset.

As different base classifiers utilize different methods to calculate feature importance, the range of the calculated feature importance is different. Specifically, RF and DT obtain the importance of features through the feature_importance_ attribute while LR uses the weight of the logistic regression formula as the feature importance. On the one hand, RF and DT first calculate the Gini impurity of the features across all nodes of the tree. Each node splitting calculates the impurity reduction of a feature. The split results of all nodes are averaged to obtain the importance of each feature, and then this value is normalized to obtain the final importance. On the other hand, regarding LR, we assume bigger coefficients have more contribution to the model and use coef_ attribute to show feature importance. LR formula can be expressed as $y = \frac{1}{1+e^{-z}}$, where z is the linear layer output of the LR model, $z = b + w_1 x_1 + w_2 x_2 + ... + w_n x_n$. In this case, coef_ stands for w_n, which is the weight. The value of the weight depends on the scale of features. It should be noted that the coefficient's value is unlimited and can even be negative, so the range of feature importance in the LR case extends across both the positive and negative sides of the y-axis.

2.5 Blacklist Filter

The proposed blacklist filter (BF) approach aims to overcome the problem of model loading and data normalization in deploying the traditional ML-based IDS model in a real-time vehicular detection system. It consists of the blacklist and Bloom filter. As mentioned earlier, the blacklist is generated by the trained model that contains multiple attack features to form an attack rules library. However, as the blacklist grows, it will consume significant memory and cause system load. Therefore, we insert the blacklist into the Bloom filter to achieve real-time detection.

Bloom filter is a memory-efficient, probabilistic data structure that can confirm whether an element exists in a set. It is based on multiple hash tables, which have significant advantages for processing large datasets that require a lot of time and memory. This characteristic is particularly useful for vehicle network data. Bloom filter consists of a bit list of length m, and when n items are added to the filter, k different hash functions are required. However, this method may produce false alarms (p), and the length of the filter will directly affect the FPR. The longer the filter, the smaller the FPR, but the computational time required will also increase. The number of hash functions also needs to be considered. The greater the number, the less efficient the filter will be, but if there are too few, the FPR will increase. Therefore, when designing a filter, it is necessary to select appropriate parameters based on the length of the rule's features and the number of rules, as shown in Eqs. (1) and (2). Based on the sample size, FPR, and time efficiency, we have determined that the optimal number of hash functions is 2, and the filter length should be set to 5000.

$$m = -\frac{n \ln p}{(\ln 2)^2} \tag{1}$$

$$k = \frac{m}{n} \ln 2 \tag{2}$$

3 Implementation

This section describes the implementation, including testbed setting, dataset selection, data pre-processing, ML approach adoption, and model training.

3.1 Testbed Setting

The specification of the testbed is detailed as follows. The host is the MacBook Pro having an Apple Silicon M1 Pro chip with RAM of 16GB. It installs the operating system (OS) macOS Big Sur (Version 14.2) and has the latest security patches and features. On the host, we use the terminal to run the code. The Python Interpreter used is Python 3.10. In addition, we utilize external libraries and frameworks to implement the models. Specifically, we utilized the scikit-learn library [21] to deploy the DT model.

3.2 Dataset Selection

Many public datasets in network security can be used to evaluate the performance of IDS. Imantha et al. [23] have studied 11 datasets since 1998, most datasets have been outdated and lack practicality now. In response to the above problems, the datasets with reliability and diversity of attack types under the premise of meeting the standards should be adopted. Regarding the IDS of the CAN bus, the Car-Hacking dataset [22] used by the Korean HCLR laboratory is applied. This dataset includes four categories: DoS attacks, obfuscation attacks, drive gear spoofing attacks, and Revolutions Per Minutes (RPM) meter spoofing attacks. It is constructed by recording CAN traffic using actual vehicle On-Board Diagnostics-II (OBD-II) ports while performing message injection attacks. Its attributes include Timestamp, CAN ID, data length code (DLC), data (CAN data field), and Flag (T or R, T represents intrusion information, R represents normal information).

Table 1. Normal Traffic and Attack Distribution in Car-Hacking.

Class label	Number of entries	Class label	Number of entries
Normal	14,037,293	Gear	597,252
DoS	587,521		
Fuzzy	491,847		
RPM	654,897		

Table 2. Normal Traffic and Attack Distribution in CIC-IDS-2017.

Class label	Number of entries	Class label	Number of entries
Normal	2,271,313	PortScan	158,804
WebAttack	2180	DoS	379,737
BruteForce	13,843	Bot	1956
Infiltration	36		

Also, the CIC-IDS-2017 dataset [23] developed by the Canadian Institute of Cyber Security contains common and latest attack types used since it provides network traffic data and was widely used by other typical related works as well. It contains 80 fields, and the data capture starts at 9:00 am and ends at 5:00 pm from Monday to Friday for a total of five days. It includes the network traffic analysis results using CICFlowMeter with labeled flows based on the time stamp, source, destination IPs, source and destination ports, protocols, and attack (CSV files). The implemented attacks include Brute Force File Transfer Protocol (FTP), Brute Force Secure Shell (SSH), DoS, Heartbleed, Web Attack, Infiltration, Botnet, and Distributed Denial of Service (DDoS).

3.3 Data Pre-processing

An initial processing is carried out to polish the datasets, like removing the redundant entries or omitting the missing values. The distribution of the two datasets after initial processing is shown in the Table 1 and Table 2. The curated CIC-IDS-2017 dataset contains one normal class and six intrusion classes, while the Car-Hacking dataset contains one normal class and four intrusion classes. The dataset is divided into a training set and a test set in an 8:2 ratio without requiring any data preprocessing for the BF module. The training set is processed first, followed by the test set during the prediction phase of the traditional model.

In order to further obtain a dataset suitable for the ML model, data pre-processing is performed as follows over the two datasets separately.

1. Numerical normalization: Since the characteristics of the dataset are distributed in different ranges, it is easy to cause deviations in the results. In order to remove the imbalance of large values in the dataset, all features are scaled, and the data is normalized to the range of 0 to 1. In this study, we use the MinMaxScaler method [21].

2. Label digitization: To facilitate the model's learning, the label needs to be converted into a numeric type. Normal behavior is assigned a value of 0, and all attack types are labeled from 1 to n.
3. Balanced datasets: The distribution of normal behaviors and attacks in two datasets is not even, and the gap is noticeable. It will affect the model's classification performance, so small attack types need to be expanded. This study uses the Synthetic Minority Over-Sampling Technique (SMOTE) [9] to synthesize data by combining over-sampling minority classes and under-sampling majority classes. The Car-Hacking dataset expands each attack to more than 1,000,000, and the CIC-IDS-2017 dataset expands a small number of attack classes to more than 10,000.
4. Feature selection: As mentioned before, this study uses the RFE-based method to perform the feature selection. In terms of our empirical experiments, we select 7 out of 9 features for the Car-Hacking dataset (see Fig. 3), and 27 out of 80 features for the CIC-IDS-2017 dataset (see Fig. 4).

3.4 Machine Learning Approaches Adoption

In this work, we focus on using the supervised ML algorithm, Decision Tree (DT) [13], to perform intrusion detection since it has been shown by the previous research [26] (which we compared with) that DT has a better detection performance than other ML algorithms in the CAN bus. DT is one of the non-linear supervised classification models. It is a dendrogram structure, the branch represents the predicted trend, and the node represents the predicted result [12]. Due to the large amount of data flow in network communication, the structure of DT is suitable for large datasets, and its high performance is also helpful for real-time detection. Its complexity depends on the dataset and tree shape and can be optimized by feature selection.

3.5 Model Training

Using the previously selected ML algorithm, a model is constructed from the pre-processed training set. The DT classifier calculates information gain based on information entropy, and the rest of the parameters are default values. The trained model is stored for testing purposes. In addition, a Bloom filter-enabled blacklist of attacking rules is generated based on the DT model.

4 Experimental Results

Building upon the implementation, we have undertaken a series of experiments to scrutinize the factors that impinge upon detection performance.

4.1 Evaluation Metrics

To evaluate the intrusion detection performance, we apply five metrics in terms of the confusion matrix, which consists of true positive (TP), true negative (TN), false positive (FP), and false negative (FN).

Accuracy $(=\frac{TP+TN}{TP+TN+FP+FN})$ indicates the proportion of normal activities and attacks that are correctly classified. **Precision** $(=\frac{TP}{TP+FP})$ indicates the proportion of the real attacks over the samples predicted as attacks. **Recall** $(=\frac{TP}{TP+FN})$ represents how many attacks are detected among all attack samples, also known as the detection rate. **FPR** $(=\frac{FP}{FP+TN})$ indicates the proportion of normal behaviors wrongly judged as attacks among all normal behaviors. To comprehensively evaluate IDS, considering the harmonic value of Precision and Recall, we also use **F1-Score** $(=\frac{2 \cdot \text{Precision} \cdot \text{Recall}}{\text{Precision} + \text{Recall}} = \frac{2TP}{2TP+FN+FP})$.

The above metrics are utilized to assess the accuracy and false alarm rate of the IDS model. Additionally, we introduce execution time and detection time to evaluate the consumption of computing resources in order to obtain a lightweight and responsive model.

Table 3. Compare results on the Car-Hacking dataset.

Model	FS Stacking [26]	DT [26]	RF [26]	RFE+DT	RFE+DT-BF
Accuracy (%)	99.99	99.99	99.99	**99.99**	**99.99**
Precision (%)	–	–	–	**99.99**	**99.99**
Recall (%)	99.99	99.99	99.99	**99.99**	**99.99**
FPR (%)	0.0006	0.006	0.0003	**0.0001**	**0.0002**
F1-Score (%)	99.90	99.99	99.99	**99.99**	**99.99**
Execution time (s)	325.6	328.0	506.8	**113.1**	**114.2**
Detection time per data (s)	–	–	–	**0.079**	**0.0000292**

Table 4. Compare results on the CIC-IDS-2017 dataset.

Model	FS Stacking [26]	DT [26]	RF [26]	RFE+DT	RFE+DT-BF
Accuracy (%)	99.82	99.72	99.37	**99.86**	**99.92**
Precision (%)	–	–	–	**99.87**	**99.89**
Recall (%)	99.75	99.30	99.29	**99.87**	**99.89**
FPR (%)	0.011	0.029	0.039	**0.009**	**0.008**
F1-Score (%)	99.70	99.80	98.30	**99.86**	**99.87**
Execution time (s)	2774.8	126.7	2421.6	**105.7**	**107.1**
Detection time per data (s)	–	–	–	**0.068**	**0.0000329**

4.2 Detection Performance

We build an efficient IDS by adjusting parameters to optimize the ML model. Based on the relevant indicators mentioned in the previous subsection, we analyze and evaluate the model's performance over the two datasets. We especially stress evaluating the FPR and detection time latency.

Table 3 presents the experimental results using the Car-Hacking dataset. We compare our approaches, which include both innovative and traditional models RFE+DT-BF and RFE+DT, with the related work [26], utilizing feature selection stacking (FS Stacking), DT, and RF. As [26] does not include precision testing, the corresponding precision rates are not included in the table. The RFE+DT-BF approach excels in performance on the Car-Hacking dataset, with all metrics achieving 99.99%, the FPR reduced to 0.0002%, and improved data detection time to 0.0000292 s. In comparison, the RF method in [26] achieves a lower FPR at 0.0003%, which is still 1.5 times higher than our FPR of 0.0002%, and has an execution time four times longer than ours. Compared to the traditional RFE+DT model, although the false alarm rate differs by only 0.0001%, the detection time has improved significantly by 2700 times, leading to a significant reduction in resource consumption. In general, the FS Stacking method in [26] outperforms other methods by selecting 4 features for learning and training with a threshold of 0.1. By contrast, our approach selects 7 features based on importance rankings. As a result, after integrating feature selection, all indicators have shown significant improvement, and the execution time has been notably reduced.

Table 4 shows the testing results using the CIC-IDS-2017 dataset. Once again, a number of values are absent from the table due to their omission in the related works. Amongst all the results, as the table shows, our RFE+DT-BF still works best. Importantly, the detection time for each data is as low as 0.0000329 s, which is 2000 times that of the traditional RFE+DT model. Whereas among the results from the previous works, the FS Stacking method is superior to the others (other than the execution time). However, even compared with FS Stacking [26], the accuracy rate, recall rate, and F1-Score are improved by 0.10%, 0.14%, and

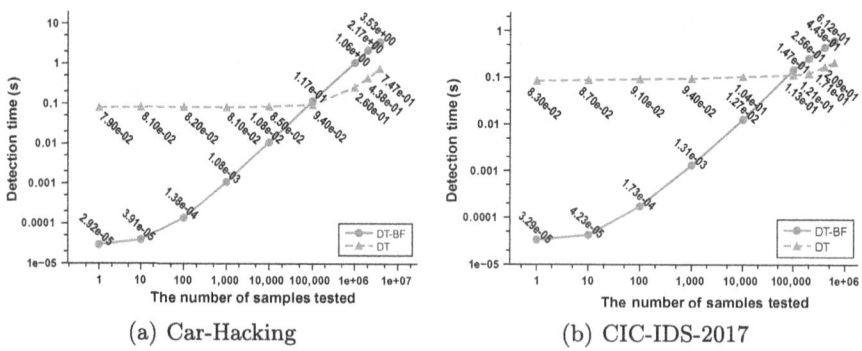

(a) Car-Hacking   (b) CIC-IDS-2017

Fig. 5. Detection time comparison of DT and DT-BF on datasets.

0.17%, respectively, and the FPR is decreased by 0.003%. From the perspective of execution and detection time, the proposed method is more efficient than those proposed by others, significantly reducing the time complexity.

On both datasets, upon analyzing the detection metrics, it is evident that though RFE+DT also performs well, its detection performance is still lower than that of RFE+DT-BF. Therefore, it shows that RFE can effectively reduce false alarms, while BF can greatly improve response efficiency. Furthermore, our execution time has significantly decreased compared to previous work, meeting lightweight needs.

4.3 Computational Consumption

Based on the DT method, we evaluate the detection time on the traditional ML model and ML-BF model to further analyze the computing resources. According to Fig. 5, the two datasets show similar trends in detection time. We analyze the changes in detection performance by detecting 1, 10, 100, ..., and 1000000 data until all data are processed. The traditional DT algorithm has advantages when processing large amounts of data in batches, but it still takes too much time when processing only 1 or even 100,000 data. In terms of single data processing speed, take Car-Hacking as an example, DT-BF can detect in microseconds (0.000029 s), while DT takes 0.079 s, a difference of nearly 2700 times. This result is particularly important for deploying IDS on the vehicle, as real-time data processing is required instead of batch processing. This research is constructive for systems that aim to achieve real-time response.

4.4 Analysis and Discussion

The model's performance, particularly the reduction in false alarm rate and computational consumption on both datasets, is satisfactory. The proposed RFE+DT-BF model achieves real-time and low-overhead IDS for the ICV system. Although Our experiments were conducted on a computer rather than a dedicated vehicular device, our RFE+DT-BF approach is feasible for deployment in a vehicle to execute high-performance detection.

In summary, leveraging ML to extract effective information and rules from large datasets, combined with feature engineering and BF, significantly improves detection efficiency. However, there are still some issues to address.

Balanced Datasets: ML requires balanced datasets with a sufficient number of normal and abnormal samples. In security applications, the dataset often contains a large number of normal samples and a small portion of abnormal samples, leading to accuracy issues in detection. Data augmentation methods can help mitigate this bias.

Unexpected Environmental Noises: Once ML model training is complete, each piece of data within the model has its own internal judgment. While most data aligns with known categories and receives accurate classification, the presence of unstable factors (such as system failures, routing delays, and network packet loss) in the real environment can lead to the generation of significant abnormal traffic. These abnormal characteristics may cause the data to be mistakenly identified as attacking traffic, resulting in numerous false alarms and undermining the accuracy of blacklist generation. To address this issue, a potential solution involves devising a combined mechanism that incorporates both rule-based detection and anomaly-based detection.

5 Related Work

This section reviews the related work of ML algorithms, lightweight and responsive IDS in ICV.

5.1 Machine Learning for Intrusion Detection in ICV

Numerous studies have utilized ML for CAN data intrusion detection. Yang et al. [26] proposed a stacking ensemble framework for IoV intrusion detection, and the accuracy on Car-Hacking reached 99.99% while the accuracy on CIC-IDS-2017 reached 99.82%. Later, Yang et al. introduced LCCDE [27] that combines multiple gradient-boosting algorithms, which results in lower execution time than the stacking ensemble framework while maintaining high detection performance. Panigrahi et al. [20] addressed dataset imbalance by proposing a model that combines the CTC algorithm and a detector, achieving 99.9% detection accuracy on NSL-KDD and CIC-IDS-2017, with notably low false negatives on Car-Hacking. The combined model exhibited longer running times than individual models, highlighting the trade-offs when applied in real-world environments.

However, methods that rely on supervised ML with labeled data are vulnerable to detecting unknown attacks. Enhancing detection accuracy can be achieved by utilizing multiple rounds of transfer learning to provide pseudo-labels for new attacks [15]. Wang et al. [25] extensively evaluated deep learning (DL) models using CSE-CIC-IDS-2018, including convolutional neural network (CNN), recurrent neural network (RNN), long short-term memory network (LSTM), and combined CNN-RNN models, all demonstrating strong performance with classification accuracy exceeding 98%. However, DL models often spend longer time for training and testing.

Though using ML-based approaches, our work differs from the prior research as we focus more on lightweight and responsive on-line intrusion detection.

5.2 Lightweight IDS in ICV

Due to the limited computational resources in the on-board device of the vehicle, the traditional ML-based IDS methods are not appropriate for the ICV context. During IDS data processing, irrelevant or redundant features can slow down model training. Gharaee et al. [5] proposed a genetic-based feature selection method with the SVM classifier, leading to significantly reduced training time with high accuracy and low FPR. Yang et al. [26] enhanced the stacking model with a combined feature selection method, halving execution time while maintaining 0.99 classification accuracy. Other researchers, including Hadeel et al. [1], have leveraged swarm intelligence algorithms to optimize model performance and improve running speed. Additionally, a feature selection method combining FGLCC and CFA algorithms demonstrated impressive detection, accuracy, and false alarm rates based on KDDCUP99 [18].

Our work uses the similar datasets, however, we apply the RFE method to select the most useful features to avoid overload, our work enhances the efficiency of building IDS model.

5.3 Responsive IDS in ICV

Real-time detection is a crucial characteristic of IDS for practical applications. Therefore, it is necessary to establish a high-response model to detect attacks promptly. To accelerate detection, [6] used the Bloom filter to solve the problem of insufficient memory when deploying IDS on the vehicle bus. Message processing time is reduced to the microsecond level, enabling effective detection of replay or modification attacks. Similarly, on the CAN dataset, [14] created an IDS whitelist and blacklist using Bloom filters to correct false positives and false negatives, respectively. Experimental results demonstrated that Bloom filters can effectively enhance IDS detection time. [28] established a unique fingerprint for each ECU based on clock skew and utilized the empirical rule and dynamic time warping to build IDS. It had high real-time performance, achieving millisecond-level detection for each instance. [11] used Logical Analysis of Data to extract rules, facilitating real-time intrusion detection.

Most previous work has relied on rule-based methods to enhance real-time detection efficiency. However, it will result in many false negatives. We develop an IDS based on the aforementioned ML. Our work applies model generation rules during actual deployment and utilizes the Bloom filter to identify attacks. It enhances IDS performance in terms of both detection accuracy and response.

6 Conclusion

The integration of electronics in modern vehicles has led to a surge in attacks targeting in-vehicle networks. This paper introduces a ML-BF model that utilizes the capabilities of RFE and the Bloom filter. Through this approach, intrusion detection performance shows significant improvement compared to previous related works. Specifically, our RFE+DT-BF approach not only increases detection performance but also reduces execution and detection time, a crucial aspect for real-time in-vehicle control systems. For future work, given that many attacks are unknown and supervised learning is unable to detect anomaly attacks, we plan to explore the use of semi-supervised methods such as Auto-Encoder for anomaly-based detection, with dynamic updates to detect new attacks timely.

Acknowledgment. This work is supported in part by Suzhou Science and Technology Plan - Key Technologies of Scenarios Generation and Application for Automated Driving Safety Evaluation (SYG202309), the XJTLU AI University Research Centre, Jiangsu Province Engineering Research Centre of Data Science and Cognitive Computation at XJTLU and SIP AI innovation platform (YZCXPT2022103), XJTLU Research Development Funding RDF-21-02-012, and XJTLU Teaching Development Funding TDF21/22-R24-177.

References

1. Alazzam, H., Sharieh, A., Sabri, K.E.: A feature selection algorithm for intrusion detection system based on pigeon inspired optimizer. Expert Syst. Appl. **148**, 113249 (2020)
2. Bloom, B.H.: Space/time trade-offs in hash coding with allowable errors. Commun. ACM **13**(7), 422–426 (1970)
3. Costantino, G., De Vincenzi, M., Matteucci, I.: In-depth exploration of ISO/SAE 21434 and its correlations with existing standards. IEEE Commun. Stand. Magaz. **6**(1), 84–92 (2022)
4. Gerla, M., Kleinrock, L.: Vehicular networks and the future of the mobile internet. Comput. Netw. **55**(2), 457–469 (2011)
5. Gharaee, H., Hosseinvand, H.: A new feature selection ids based on genetic algorithm and SVM. In: 2016 8th International Symposium on Telecommunications (IST), pp. 139–144 (2016)
6. Groza, B., Murvay, P.S.: Efficient intrusion detection with bloom filtering in controller area networks. IEEE Trans. Inf. Forens. Secur. **14**(4), 1037–1051 (2018)
7. Guyon, I.M., Weston, J., Barnhill, S.D., Vapnik, V.N.: Gene selection for cancer classification using support vector machines. Mach. Learn. **46**, 389–422 (2002)
8. Hoppe, T., Kiltz, S., Dittmann, J.: Applying intrusion detection to automotive it-early insights and remaining challenges. J. Inf. Assur. Secur. **4**(6), 226–235 (2009)
9. Jeatrakul, P., Wong, K.W., Fung, C.C.: Classification of imbalanced data by combining the complementary neural network and SMOTE algorithm. In: Wong, K.W., Mendis, B.S.U., Bouzerdoum, A. (eds.) ICONIP 2010. LNCS, vol. 6444, pp. 152–159. Springer, Heidelberg (2010). https://doi.org/10.1007/978-3-642-17534-3_19
10. Ji, B., Zhang, X., Mumtaz, S., Han, C., Li, C., Wen, H., Wang, D.: Survey on the internet of vehicles: network architectures and applications. IEEE Commun. Stand. Magaz. **4**(1), 34–41 (2020)
11. Kumar, A., Das, T.K.: Cavids: real time intrusion detection system for connected autonomous vehicles using logical analysis of data. Vehicul. Commun. **43**, 100652 (2023)
12. Kumar, M., Hanumanthappa, M., Kumar, T.V.S.: Intrusion detection system using decision tree algorithm. In: 2012 IEEE 14th International Conference on Communication Technology, pp. 629–634 (2012)
13. Kumar, M., Hanumanthappa, M., Kumar, T.S.: Intrusion detection system using decision tree algorithm. In: 2012 IEEE 14th International Conference on Communication Technology, pp. 629–634. IEEE (2012)
14. Lee, S., Kim, H., Cho, H., Jo, H.J.: Fids: Filtering-based intrusion detection system for in-vehicle can. Intell. Automat. Soft Comput. **37**(3), 2941–2954 (2023)
15. Li, X., Hu, Z., Xu, M., Wang, Y., Ma, J.: Transfer learning based intrusion detection scheme for internet of vehicles. Inf. Sci. **547**, 119–135 (2021)

16. Liu, Y., Fan, Y., Huang, D., Mi, B., Huang, L., et al.: Formal model and analysis for the random event in the intelligent car with stochastic petri nets and z. Secur. Commun. Netw. **2022**, 1–18 (2022)
17. Lokman, S.F., Othman, A.T., Abu-Bakar, M.H.: Intrusion detection system for automotive controller area network (can) bus system: a review. EURASIP J. Wirel. Commun. Netw. **2019**, 1–17 (2019)
18. Mohammadi, S., Mirvaziri, H., Ghazizadeh-Ahsaee, M., Karimipour, H.: Cyber intrusion detection by combined feature selection algorithm. J. Inf. Secur. Appl. **44**, 80–88 (2019)
19. Olariu, S., Weigle, M.C.: Vehicular Networks: From Theory to Practice. Chapman and Hall/CRC (2009)
20. Panigrahi, R., et al.: A consolidated decision tree-based intrusion detection system for binary and multiclass imbalanced datasets. Mathematics **9**(7), 751 (2021)
21. Pedregosa, F., et al.: Scikit-learn: machine learning in Python. J. Mach. Learn. Res. **12**, 2825–2830 (2011)
22. Seo, E., Song, H.M., Kim, H.K.: Gids: gan based intrusion detection system for in-vehicle network. In: 2018 16th Annual Conference on Privacy, Security and Trust (PST), pp. 1–6 (2018)
23. Sharafaldin, I., Habibi Lashkari, A., Ghorbani, A.: Toward Generating a New Intrusion Detection Dataset and Intrusion Traffic Characterization, pp. 108–116 (2018)
24. Sharma, S., Kaushik, B.: A survey on internet of vehicles: applications, security issues and solutions. Vehicul. Commun. **20**, 100182 (2019)
25. Wang, Y.C., Houng, Y.C., Chen, H.X., Tseng, S.M.: Network anomaly intrusion detection based on deep learning approach. Sensors **23**(4), 2171 (2023)
26. Yang, L., Moubayed, A., Hamieh, I., Shami, A.: Tree-based intelligent intrusion detection system in internet of vehicles. In: 2019 IEEE Global Communications Conference (GLOBECOM), pp. 1–6. IEEE (2019)
27. Yang, L., Shami, A., Stevens, G., De Rusett, S.: LCCDE: a decision-based ensemble framework for intrusion detection in the internet of vehicles. In: GLOBECOM 2022-2022 IEEE Global Communications Conference, pp. 3545–3550. IEEE (2022)
28. Zhao, Y., Xun, Y., Liu, J.: Clockids: a real-time vehicle intrusion detection system based on clock skew. IEEE Internet Things J. **9**(17), 15593–15606 (2022)
29. Zheng, K., Zheng, Q., Chatzimisios, P., Xiang, W., Zhou, Y.: Heterogeneous vehicular networking: a survey on architecture, challenges, and solutions. IEEE Commun. Surv. Tutor. **17**(4), 2377–2396 (2015)

Evaluating the Vulnerability Detection Efficacy of Smart Contracts Analysis Tools

Silvia Bonomi[1](✉), Stefano Cappai[1], and Emilio Coppa[2]

[1] Sapienza University of Rome, Rome, Italy
bonomi@diag.uniroma1.it, cappai.1844363@studenti.uniroma1.it
[2] LUISS University, Rome, Italy
ecoppa@luiss.it

Abstract. Smart contracts on modern blockchains pave the way to the development of novel application design paradigms, such as Distributed Applications (DApps). Interestingly, even some safety-critical systems are starting to adopt such a technology to devise new functionalities. However, being software, smart contracts are susceptible to flaws, posing a risk to the security of their users and thus making crucial the development of automatic tools able to spot such flaws.

In this paper, we examine 11 real-world DApps that participated in security auditing contests on the Code4rena platform. We first conduct a manual analysis of the vulnerabilities reported during the contests and then assess whether state-of-the-art analysis tools can identify them. Our findings suggest that current tools are unable to reason on business logic flaws. Additionally, for other root causes, the detectors in these tools may be ineffective in some cases due to a lack of generality or accuracy. Overall, there is a significant gap between auditors' findings and the results provided by these tools.

Keywords: Smart Contract · Vulnerability · Testing Tools · Blockchain

1 Introduction

Distributed Ledger Technologies (DLTs) and Decentralized Applications (DApps) [10,12], implemented through smart contracts, have become essential components in various modern solutions across diverse application domains, ranging from finance to agriculture. Among all the interested domains from this technological advent, there are also safety-critical systems [24] that are starting to adopt blockchains and smart contracts to support new functionalities. Secure information sharing of sensible data with integrity and confidentiality requirements [13], decentralized access control with privacy guarantee [1], continuous monitoring enabling certified removal of data in safety-critical databases [15] and (food) tracking along the supply chain [34] are just few examples of the increasing presence of DLTs and DApps in domains characterized by really high-quality dependability and security standards. For instance, the community is exploring the use of blockchains in the context of railway industry [22,25,26,32,35].

In safety-critical systems, more than in other domains, there is a huge attention to the system development life-cycle that should follow *security-by-design* and *continuous monitoring* principles to ensure that bugs and anomalies are timely detected and resolved. Indeed, the system development is usually supported by guidelines provided by certification standards and whose aim is to give recommendations to developers regarding all the development process activities with a particular emphasis on the *verification and validation* tasks and the maintenance aspects. Thus, it is a crucial requirement for a safety-critical system that its software components are deeply tested to discover the highest number of bugs and vulnerabilities *before* the system starts to be operational and that monitoring is in place when the system gets operational to ensure fast detection and patching at runtime.

This requirement becomes even more relevant if the system includes a software module developed by using smart contracts. Indeed, differently from traditional software, a smart contract is a piece of code deployed on top of a distributed ledger through the execution of a transaction. The direct consequence is that, once it is deployed, it is very hard to patch due to the immutability property of the underlying ledger. The direct consequence is that the adoption of DLTs- and DApps-based solutions in safety-critical systems requires increased attention to the verification and validation phase which translates into the need for testing tools able to detect and discover bugs and vulnerabilities with the highest accuracy possible[1].

In response to these problems, over the past decade, both the research community and the industry have initiated efforts to develop and explore software testing methods and security tools. The goal is to potentially identify bugs and vulnerabilities in smart contracts before deploying these software components on blockchains. Similar to traditional software, various techniques [3,17,20,29–31,33,40–42] can help developers detect the flaws.

The first aid to developers can come from *lightweight static analysis* tools [17,40], which can provide quick feedback about common security issues by performing a local pattern-driven analysis. Being quite efficient, the community is starting to integrate them even in continuous integration setups. However, they can be inaccurate, missing crucial security flaws and reporting a large number of invalid reports, i.e., false positives. To mitigate such problems, the community has explored different *heavyweight analysis* frameworks based, e.g., on symbolic execution [29–31] and fuzzing [20,33,42], aiming at better accuracy in exchange for worse scalability. Finally, formal approaches [3,41] have been investigated to bring stronger security guarantees but often require rethinking the blockchain technologies or imposing constraints on the development.

---

[1] https://ethereum.org/en/developers/docs/smart-contracts/formal-verification/.

As for traditional software, the community has launched several security auditing platforms, such as Code4rena and Immunefi, where DApp developers can ask experts to assess the quality and security of smart contracts in exchange for monetary rewards.

Considering the history of DLTs and smart contracts and the actual spreading of these technologies between different application domains, most of the auditing are performed on applications developed for the financial domain. However, many of the existing vulnerabilities are not context-dependent (i.e., they affect the Solidity programming language or the distributed environment hosting the blockchain) and thus the capability of a testing tool to discover it has a global interest beyond the financial domain (e.g., *Access Control Vulnerabilities* in Solidity Smart Contracts may lead to sensitive data exposure in the healthcare domain).

Our Contributions. In this practical experience paper[2], we examine 11 real-world DApps that participated in security auditing contests on the Code4Arena platform from March to August 2023. We first conduct a manual analysis of the vulnerabilities reported during the contests by the auditors to understand what are the most common security issues emerging in real-world smart contracts. These contracts are particularly interesting because, being part of well-known and relevant DApps, have been implemented by experienced developers. We then report the results of running three state-of-the-art tools on these DApps, assessing how their findings fare against auditors' discoveries. In particular, we considered two lightweight static analysis tools, namely *SmartCheck* and *Slither*, and one heavyweight analysis framework based on symbolic execution called *Mythril*. Our assessment suggests that current tools are unable to reason on business logic flaws, the most frequent root cause of vulnerabilities for the 11 DApps that we considered. Additionally, for other root causes, the detectors in these tools are ineffective in some cases due to a lack of generality or accuracy. Overall, unfortunately, there is still a significant gap between auditors' findings and the results provided by these tools suggesting the need for new developments in this context.

2 Background

Distributed Ledger Technologies and Blockchains. DLTs are an emerging class of decentralized distributed systems that allow the recording of transactions of assets in a digital ledger. Among them, blockchains gained huge popularity and are currently the most widespread DLT. A blockchain, as the name suggests, is a particular type of ledger where transactions are collected and stored in blocks and blocks are linked among them using pointers i.e., inserting in a block

---

[2] This paper is an extended version of a preliminary 2-page fast abstract presented at ISSRE 2023 [8] where the analysis was restricted to only 4 DApps and a single tool i.e., Mythril.

i the hash of the content of previous block $i-1$. Each block is constructed by selecting submitted transactions and validating their correctness and consistency with respect to the current ledger state. Once a block is created, it needs to be chained to the current ledger. This is done collaboratively by participants in the distributed systems that need to agree on the *next* block becoming the head of the chain (i.e., on the last ledger state). Blockchains are currently attracting a lot of attention due to their *immutability* property i.e., once a new block is created and attached to the chain, it can not be easily altered. This is achieved by combining and using together cryptographic primitives and robust consensus algorithms.

The Ethereum Blockchain. Currently, many different blockchain implementations exist but currently the most widespread can still be considered the Ethereum blockchain [39]. Firstly proposed by Vitalik Buterin in late 2013 the network became live in 2015 with its native cryptocurrency called *Ether* (ETH). Since then, it evolved adapting its internal mechanism and supporting additional features like smart contracts. The Ethereum cryptocurrency is used to incentivise participants who perform computations and validate transactions and it can also be used to pay for transaction fees and services on the Ethereum network. In Ethereum, every operation on the network requires a certain amount of computational resources that are measured in *gas units* i.e., the gas is the unit to measure the amount of computational effort required to execute operations or run smart contracts. Users must pay for gas in Ether when they execute transactions.

Smart Contracts. A smart contract is a self-executing contract (i.e., a piece of running software) with the terms of the agreement directly written into code. It is executed on top of a blockchain and it automatically runs its functions when predefined conditions are met by generating transactions on the blockchain. Smart contracts are written in programming languages specifically designed for blockchain platforms, such as *Solidity* for Ethereum. Smart contracts are deployed over all the blockchain nodes and are executed by all the participants. This decentralization ensures that the contract's execution is transparent, secure, and resistant to censorship or manipulation.

Ethereum supports the execution of smart contracts thanks to a Turing-complete virtual machine called the *Ethereum Virtual Machine* (EVM). Once deployed, smart contracts operate autonomously and independently of any human intervention. In addition, they are *immutable* i.e., once the code is deployed on the blockchain, it cannot be easily altered, patched or tampered with. While on one hand, such immutability ensures the integrity and reliability of the contract's execution, on the other hand, it raises potential significant dependability and security issues due to bugs and vulnerabilities.

3 Related Works

Smart contracts are currently used across many different domains ranging from DeFi where they are used to offer financial services[3] to *Supply Chain Management* where smart contracts are used to improve transparency and traceability of products and goods along the supply chain. This fast and large spreading is also pushing the scientific community to investigate security aspects connected to DApps. In particular, several smart contract analysis tools have been proposed by the community building on top of different techniques, such as *lightweight static analysis* [17,29,40], *symbolic execution* [30,31], *fuzzing* [20,33,42], and other *verification* techniques [3,41].

Different experimental studies have already tried to assess and compare the detection capabilities of such tools. In [43], authors analyzed several Code4rena contests using Oyente [29], a symbolic execution framework which is, however, now deprecated and unmaintained. They split findings into two groups: *Machine Auditable Bugs (MABs)* and *Machine Unauditable Bugs (MUBs)*. Their study aims to show the current state of automatic tools, and how those tools seem not able to find MUB vulnerabilities. Our contribution continues this investigation by enlarging the set of analyzed contests and taking an orthogonal perspective, trying to shed light on the strengths and weaknesses of different automatic tools when compared with human auditors. In [4], authors analyze thousands of smart contracts already deployed on the blockchain, i.e., they scan the bytecode of those contracts. They propose the *skeleton* concept: they found common patterns in the bytecode of smart contracts, grouping them accordingly. However, this approach could flatten the dataset excessively since most smart contracts are characterized by external calls to others, which often could be the actual source of vulnerabilities. Moreover, by just looking at the bytecode, it could be very hard to identify some types of vulnerabilities. In our study, we consider also the Solidity code of the DApps, including tools that can process both Solidity and the resulting bytecode.

4 Experimental Study Methodology

This section presents key aspects characterizing our experimental study: the source for our dataset (i.e., the auditing platform and the considered contests), the analysis tools, and our experimental setup.

Security Auditing Platform. *Code4rena (C4)* is a community-driven competition for smart contract audits. Any DApp developer can use it to launch a new contest involving a specific project and define what code is in scope and what is not. There are three main roles on this platform:

- *Wardens:* auditors in charge of reporting issues affecting a DApp. They are typically human experts who, exploiting both manual and automatic tools,

---

[3] Examples of DeFi DApps are Compound, Aave, Uniswap, and MakerDAO.

produce a detailed report. Wardens are not forced to disclose their analysis strategies;
- *Sponsors:* DApp developers sponsoring the contest with a *prize pool*. The higher the prize pool is, the higher the interest from advanced security auditors will be;
- *Judges:* experts rating the performance of wardens and deciding the severity and validity of their findings. Judges are chosen by the C4 community among the best wardens, according to their impartiality and accountability.

The prize pool is split into the following categories: *High* findings, *Medium* findings, *Quality Assurance (QA)* findings (composed of *Low* and *Non-Critical* findings), *Gas Optimization* findings, *Analysis*, and *Bot Races*. Usually, a *contest* lasts 1–3 weeks. High, Medium, and Quality Assurance findings are reported by wardens within the contest's period. Wardens report these issues establishing the severity. After the submission phase, the contest ends and judges analyze reports of wardens, checking for validity of issues and correctness in severity assignment. *Gas Optimization* findings are suggestions proposed by wardens to rewrite a better code that saves gas. *Analysis* reports are reports that analyze the DApp as a whole, identifying the architecture weaknesses. *Bot Races* are competitions among automatic tools developed by wardens. In the first hour of every contest, bot race participants can submit reports obtained by the execution of their bot. Only a few participants, who have already qualified during monthly bot qualifications, can report findings during the bot races.

Considered Contests. C4 contests can naturally provide a significant dataset for smart contracts. Indeed, they provide complex and real-world contracts that have reached a strong maturity. Moreover, the evaluation performed by the judges over the reported issues can naturally provide a *ground truth* that we can exploit in our experimental study. Overall, we analyzed 11 contests from March to August 2023. Table 1 summarizes the main characteristics of such contests, which involved 73 contracts, 15,562 SLOCs, and 130 vulnerabilities. Our selection of contests was aimed at exposing different *types* of DApps and different complexities (in terms of number of smart contracts and SLOCs). In the remainder of this section, we provide more details about these contests.

Ajna. The Ajna protocol [2] is a lending and borrowing protocol. There were 3 main contracts in the contest: `GrantFund`, which holds the treasury, i.e., an amount of tokens that are used for the governance; `PositionManager`, a position of a lender in a given pool; `RewardsManager`, which rewards a lender who decides to stake token.

Asymmetry. Asymmetry [5] aims at providing a solution to the centralization of the staked Ether market through *Liquid Staked Ethereum Derivatives*. The contest involved four smart contracts: `SafEth`, which allow a user to stake some funds; `Reth`, `WstEth`, and `SfrxEth`, that contain methods to acquire rETH, wstETH, and sfrxETH tokens.

Table 1. Smart Contracts Dataset Overview.

DApp	Date	Type	Contracts	SLOCs	Humans High	Humans Medium	Bots High	Bots Medium
Ajna	2023/05	Lending	3	1,391	11	14	0	2
Asymmetry	2023/03	Derivative	4	460	8	12	0	0
Caviar	2023/04	DEX	4	741	3	17	0	0
Eigenlayer	2023/04	Derivative	7	1,393	2	2	0	0
ENS	2023/04	Service	9	2,022	0	7	0	0
Frankencoin	2023/04	Stablecoin	8	949	6	15	0	0
Juicebox	2023/05	Service	1	160	0	3	0	0
Livepeer	2023/08	Social	4	1,605	2	3	0	0
Llama	2023/06	Service	11	2,047	2	3	0	1
Shell	2023/08	Service	1	460	1	0	0	0
Stader	2023/06	Derivative	21	4,334	1	14	0	1
Overall			73	15,562	36	90	0	4

Caviar. Caviar [11] is an on-chain, gas-efficient automated market maker (AMM) protocol for trading non-fungible tokens (NFTs), that allow users to deposit NFTs and associated assets inside *Liquidity Pools* (LP). The contest covered three implementation bits of the `Custom Pools: Factory`, which creates NFTs and holds protocol fees; `PrivatePool`, which allows a developer to set which operations (buy, sell, exchange) can be performed on the NTFs; `EthRouter`, used to perform actions across pools.

Eigenlayer. Eigenlayer [14] is stacking platform that allows users to (re-)stake their ETHs and ERC20 tokens based on custom strategies, which can then be accepted by validators. The contest involved 7 contracts: `StrategyManager`, which tracks stakers' deposits of tokens; `StrategyBase`, which represents a base strategy; `EigenPodManager`, which allows users to stake ETH; `EigenPod`, which allows users to stake ETHs on Ethereum and restake them on EigenLayer; `DelayedWithdrawalRouter`, which controls withdrawals of ETHs from `EigenPods`; `Pausable` and `PauserRegistry`, which can extend other contracts and makes them pausable, i.e., stops their tasks.

ENS. The Ethereum Name Service (ENS) [16] is a distributed domain naming system (DNS) based on the Ethereum blockchain. The most interesting contract proposed in the contest are: `DNSRegistrar`, which is in charge of the domain registration process exploiting a `registry` using a DNSSEC Oracle; `DNSClaimChecker`, which verifies DNS name claims; `OffchainDNSResolver`, which handles the domain name resolution.

Frankencoin. Frankencoin [19] is a collateralized stablecoin token, dubbed ZCHF, that tracks the value of the Swiss Franc. Out of 8 contracts included in the contest, four were extremely crucial: `Position`, which represents the position in tokens for a user; `MintingHub`, which creates positions; `Frankencoin`, which is the actual token; and `StablecoinBridge`, which allows users to swap Frankencoin with other stablecoins.

Juicebox. The Juicebox protocol [23] is a *programmable* treasury, allowing users to automatically mint NTFs when new funds are received in the context of a specific treasury. A single contract (`JBXBuybackDelegate`) was investigated in the contest: it is in charge of distributing tokens to a contributor according to its donation to the treasury.

Livepeer. Livepeer [27] aims at providing a distributed video live-streaming service built on top of Ethereum. The contest covered the handling of the protocol rewards for *broadcasters, transconders*, and *orchestrators*: `BondingManager`, which manages the protocol staking and rewards; `Treasury`, which holds funds of treasury and executes proposals; `LivepeerGovernor`, an OpenZeppelin Governor implementation; `BondingVotes`, which is tied to the transcoders selection process.

Llama. Llama [28] provides a governance framework to make life easier for DApp developers. The contest included the main contracts behind the Llama architecture: `LlamaCore`, which checks the execution of *actions*; `LlamaExeutor`, the actual executor of the actions; `LlamaPolicy`, an ERC721 contract that defines roles and permissions.

Shell Protocol. Shell [37] offers a platform, called Ocean, that can compose any type of DeFi primitive: AMMs, lending pools, algorithmic stablecoins, and NFT markets. The contest included only the contract `EvolvingProteus`, which can offer a primitive of the AMM implementation, i.e., a liquidity pool and its evolution over time.

Stader Labs. Stader [38] is a non-custodial staking platform for multiple Proof-of-Stake networks through the liquid staking token ETHx. The contest involved: `ETHx`, the actual ERC20 token; `PermissionedPool` and `PermissionlessPool` handle the deposit of ETHs; `StaderOracle` provides a source of exchange rates; `StaderStakePools Manager` allows users to stake ETHs, mint ETHxs, manage staking rewards.

Analysis Tools. As discussed in Sect. 2, there exists a large number of smart contract analysis tools. However, in this experimental study, we decided to focus on tools that are well-known in the Ethereum ecosystem and that have been used in different academic works as well as in security evaluations publicly disclosed by auditors. In particular, we considered one lightweight pattern-based static analysis tool, called *SmartCheck* [40], a more advanced yet still lightweight static analysis tool, called *Slither* [17], and then one advanced heavyweight analysis tool based on symbolic execution [7], called *Mythril* [31]. We did not consider dynamic tools, e.g., fuzzers, or tools using formal verification due to their non-trivial setup or lack of support for complex smart contracts.

SmartCheck. SmartCheck is a lightweight static analyzer that translates the Solidity code of a smart contract into an XML-based intermediate representation. Then, it checks the result with XPath patterns to detect four categories of issues: *security*, i.e., issues that may lead to vulnerable states; *functional* and *operational*, i.e., issues affecting the logic or leading to performance degradation; *developmental*, i.e., issues causing problems at deployment time. Overall, it has 43 issue detectors. We used SmartCheck 0.2.0.

Slither. Slither is Python framework for analyzing smart contracts. By translating solidity code into a custom intermediate representation, called *SlithIR*, it allows to implement detectors. It supports more than 80 detectors, covering different vulnerabilities categories. Moreover, it is quite efficient at performing its analysis and thus developers are starting to consider it within continuous integration setups. We used Slither 0.9.4.

Mythril. Mythril is a symbolic execution framework for EVM bytecode. It explores the program state space of a smart contract by executing its code on symbolic inputs, i.e., inputs whose value is not *apriori* fixed. In particular, an interpreter evaluates the EVM bytecode, constructing formulas to represent the data flows of the program computation. Whenever the program meets a decision point, the framework evaluates using an SMT solver whether the formula related to the decision is feasible, i.e., the smart contract can take that specific path for an assignment of the inputs. Moreover, the framework can use the solver to evaluate conditions related to well-known vulnerable patterns. Mythril integrates 13 vulnerability detectors. In our experiments, we used Mythril 0.23.24.

Smart Contract Vulnerabilities. In the literature, smart contract vulnerabilities have been classified according to different criteria [6,21,36,43] and, in this paper, we group them in four categories [6,43]:

- **Business Logic**: any issue due to the erroneous application of the *business logic* model. These flaws are strongly tied to the nature of a DApp;
- **Solidity**: any flaw due to the erroneous use of the Solidity language or unexpected nuances of its design, e.g., rounding arithmetic rules;
- **EVM**: issues due to the EVM behavior, e.g. the gas block limit;
- **Blockchain**: problems resulting from the nature of the blockchain, e.g. untrustworthy oracles or dependence on transaction order.

Table 2. Overview of the security issues identified by humans, bot, and analysis tools, on the 11 contests, dividing them based on their vulnerability nature. Legend: ☐ : high severity issue. ○ : medium severity issue. ■○ : the issue was found only by humans. ☐◐ : the issue was found only by bots. ☐● : the issue was found only by analysis tools. ■● : the issue was found both by humans and tools. ■● : the issue was found both by bots and tools.

		Ajna	Asymmetry	Caviar	EigenLayer	ENS	Frankencoin	Juicebox	Livepeer	Llama	Shell	Stader	
Blockchain	Frontrunning	○ ○	☐ ○	○			☐ ☐ ○ ○ ○		○			● ○	13
	Erroneous Assumption						○						1
Business Logic	Erroneous Logic Model	○	☐ ○ ○	○	○		☐ ☐ ○ ○ ○ ○					○ ○ ○	15
	Freezing Active Position	☐ ☐ ☐	○				☐ ○ ○	○					7
	Erroneous Accounting	☐ ○ ○	☐ ☐ ○	○ ○ ○	☐	○ ○ ○	○ ○ ○	○	○	☐		○ ○ ○	22
	Missing Logic Checks	☐ ☐ ☐ ☐ ○ ○ ○	☐ ○ ○	○	☐	○	○		☐	☐	☐	○ ○ ○	20
	Wrong Implementation	○		○		○ ○		○	○ ○	○		○ ○	10
EVM	Gas Limit DoS	■							○				2
Solidity	Reentrancy			☐ ○ ○ ○									4
	Precision Loss	● ●	☐ ●				○						5
	Integer Overflow	☐ ○	☐	○ ○			☐		☐				7
	Arbitrarily Code Injection	☐		☐ ○								■	4
	Outdated Compiler Bug	●											1
	Unsafe Casting	○		☐		○							3
	Use of transferFrom()	○											1
	Use of _mint()	○							○				2
	Dangerous DelegateCall												0
	Relying on External Source		■ ● ●	○ ○ ○	●		○					○	9
	Division by Zero DoS		○										1
	Temporal DoS		○										1
	Erroneous ERC721 Impl.			○									1
	Erroneous Pausable Impl.											○	1
	Overall	27	20	20	4	7	21	3	5	6	1	16	130

Experimental Setup. We now describe the setup adopted during our study.

We executed our experiments on a server equipped with two Intel Xeon E5-4610v2 CPUs and 256 GB of RAM, running on Ubuntu 22.04. The tools were executed under Docker: for SmartCheck, we exploited the container image from the SmartBugs framework [18], while, for Slither, we installed it on a vanilla Python 3 container image, and finally, for Mythril, we used its official container image.

While SmartCheck and Slither are expected to terminate their analysis in just a few minutes, Mythril, as most symbolic execution frameworks, can easily run for several hours or even days. Hence, we decided to run the first two tools for up to two hours, while leaving Mythril running up to 7 days. Tools were executed in independent experiments and did not share their budget. SmartCheck and Slither do not have critical parameters that need tuning. On the other hand, Mythril can work with five *exploration strategies* (*DFS*, *BFS*, *naive-random*, *weighted-random*, and *pending*), which may affect the generation of program states, possibly bringing different results. Hence, we performed different experiments for each strategy and then considered the best result. Furthermore, Mythril defines a *Max Depth* parameter, which limits the maximum depth for an execution path, thus bounding the analysis over a path, avoiding to waste the entire time budget over a single path. We kept this parameter at its default value but then increased it when investigating the results related to some specific contests.

5 Experimental Study Results

We now report the results of our experimental study involving the 11 Code4rena contests presented in Sect. 4. To evaluate the effectiveness of the three state-of-the-art analysis tools, we relied on the reports submitted by humans, i.e., the wardens, and bots to Code4rena, which have been validated by expert judges. Hence, these reports constitute our *ground truth* about the critical issues affecting the smart contracts under auditing. Our investigation has been targeted around the following research questions:

RQ1: Do the contests experience similar security issues?
RQ2: Are the analysis tools considered in this study effective at detecting such issues?
RQ3: Do the results of the tools align with their claimed detection capabilities?
RQ4: Is a more complex analysis able to bring significantly better results compared to a lighter analysis?

5.1 RQ1: Contests Versus Vulnerabilities

We start by analyzing the distribution of security issues across the contests. Table 2 provides a visual summary of the security issues that were reported either by a human or by a bot (notice that an issue reported by a bot cannot be reported also by a human). The shape of the symbols in the table represents the severity of the flaw: square for high severity and circle for medium severity. Underlined symbols represent issues originally identified by bots, while not underlined symbols represent problems originally reported by humans. To help perceive the nature of vulnerabilities, we manually classified them according to the groups presented in Sect. 4.

From a quick look at the table, from top to bottom, we can see that most vulnerabilities, i.e., 74 out of 130, are related to the business logic of a DApp. Then, the second most frequent group, with 40 out of 130 flaws, is the one involving Solidity-specific aspects. Finally, 14 and 2 security problems derive from blockchain and EVM nuances, respectively. When instead we look at the table from left to right, we can observe that different contests were affected by different numbers and types of vulnerabilities. Nonetheless, we can quickly notice that the vast majority of flaws have been found by humans, with only 4 issues reported by bots. In particular, bots mostly identified uses of unsafe functions and their findings were always evaluated as medium-severity issues.

An interesting question is whether the DApp type is somehow connected to the number of discovered flaws. Table 1 reports the type of each DApp. Lending DApps, DEXs, and Stablecoin projects have reported significantly more vulnerabilities than Derivative, Service, and Social DApps. In particular, the first set of DApps has seen more than 19 issues every 1k of SLOC, while the second set has seen less than 7 issues every 1k of SLOC. We hypothesize that the first group contains DApps for which the community has seen several hacks in the past and thus is quite careful with some operations and more trained to recognize some vulnerable patterns. Strangely, Derivative DApps should be part of the first group but instead fall into the second one. Differently, Social DApps are a new thing in the blockchain environment, while Service DApps have a unique behavior and are strongly dependent on the context. Hence, it could be harder for auditors to spot uncommon issues in these types of DApps. Nonetheless, these considerations must be taken lightly due to our limited dataset size.

Table 3. Comparison between humans and bots findings versus tools findings. Legend: □ : high severity issue. ○ : medium severity issue. ■◐ : the issue found at least by two tools. ■● : the issue found only by one tool.

		Ajna	Asymmetry	Caviar	EigenLayer	ENS	Frankencoin	Juicebox	Livepeer	Llama	Shell	Stader	
Contest	Humans	25	20	20	4	7	21	3	5	5	1	15	126
Results	Bots	2							1			1	4
Tools Results	SmartCheck	■	○									□	3
	Slither	●● ●	■● ●○		●							□	9
	Mythril											□●	2

Due to the lack of space, we cannot present in detail the vulnerabilities affecting the contests. Hence, we refer to our technical report [9] for a more in-depth discussion.

5.2 RQ2: Tools Versus Vulnerabilities

We now include in the scope of our discussion the three analysis tools that we considered in our study. Table 2 visually depicts which security issues, originally reported either by a bot or by a human, have been also detected by at least one of the tools during our experiments: shapes with an empty right side have not been detected by any tool, while full black shapes have been detected also by at least one of the tools under consideration[4].

With a quick look, we can see that tools identified mostly flaws related to Solidity aspects. This makes sense as this group of vulnerabilities is quite well-known and has been targeted for a long time by state-of-the-art smart contracts analysis frameworks. Tools were able to spot a few issues tied to nuances of the blockchain or the EVM. Differently, they were unable to find any security flaw related to the business logic group, which, however, was also the one with more reported vulnerabilities.

A more clear view of the tool efficacy is given by Table 3, where we report the count of findings for each tool. Tools discovered 14 security flaws, among which 11 are unique (see full black shapes in Table 2). In particular, Slither has found most of these flaws (9 out of 14), followed by SmartCheck (3 out of 14) and Mythril (2 out of 14).

An important observation is that, since we are considering publicly available tools, DApp developers may have already run such tools on their projects. Hence, we believe that these tools may have identified additional flaws but they were fixed before the contest. We attempted to recover such information by looking for tool-specific configuration files in the DApp repositories. Interestingly, 7 out of 11 DApps show evidence of Slither, only one DApp was likely tested with Mythril, and none appear to have exploited SmartCheck. However, 5 DApps used Solhint, a common alternative to SmartCheck.

Overall, unfortunately, the efficacy of the tools appears to be quite limited compared to the auditors' findings. The next research questions attempt to investigate these results.

5.3 RQ3: Tools in Theory Versus Tools in Practice

Given the poor efficacy of the tools at finding the security issues, we investigated whether such tools were equipped with adequate *vulnerability*

---

[4] Since we based our ground truth on the validated findings from bots and humans, there is no case where a security issue has been identified only by a tool. Nonetheless, we manually investigated any additional issue reported only by a tool: we could not demonstrate the validity of such (unknown) flaws, i.e., we believe that these findings could be classified as false positives.

detectors that could allow them to report the problems that emerged in the contests. SmartCheck integrates 43 detectors, Slither has 83 detectors, and Mythril includes 13 detectors.

Based on our analysis [9], among the 22 vulnerability types listed in Table 2, around 12 of them are targeted by at least one tool detector. Tools mostly consider issues due to Solidity, EVM, and Blockchain, but struggle at devising detectors targeting the business logic[5]. Unfortunately, even when detectors are available for a vulnerability type, they were not always able to correctly find the related flaws for our contests.

One possible reason behind detector failures could be due to our time and memory budget limits. However, both SmartCheck and Slither did not experience any timeout or out-of-memory error. Differently, Mythril experienced several timeouts in the Ajna contest, which may explain some detection failures. For instance, we believe Mythril should be able to find the *Integer Overflow* in Ajna given a sufficient time budget. Estimating such budget is quite hard since this tool explores the program state space, which highly depends on the smart contract complexity: we observed a good correlation between the analysis time and the number of SLOC[6], requiring on average 42 hours for each contest. Given our already large time budget (one week), increasing this limit could be not a practical workaround for most developers. Another reason behind the failures in Mythril could be the impact of the *Max Depth* parameter, which aims at limiting state explosion. To investigate such hypothesis, we repeated the experiments setting the parameter to 1000, which generated additional timeouts without improving the results.

Investigating why the detectors may fail, after ruling out scalability issues, is not trivial as it requires a manual and issue-specific investigation. Due to the lack of space, we only report two case studies. The first one is related to *precision loss*, with five known instances but only three discovered by Slither and SmartCheck, which ship with a detector for this vulnerability type. One missed issue is in Asymmetry, where precision loss arises due to rounding rules in the case of a division after multiplication. However, the tools only consider the inverse pattern, i.e., division before multiplication. Similarly, the tools fail to detect an issue in Frankencoin because the division before multiplication is split across different portions of code, making static analysis harder. Mythril, thanks to its state space analysis, given the proper detector, should identify these flaws. Another interesting case study is related to *integer overflow*, with 7 known instances but zero detections from Mythril, which ships with a detector for this vulnerability type. In Ajna, Mythril fails to detect an overflow because its detector does not cope with OpenZeppelin's SafeMath library, which can

---

[5] The only exception is related to *Missing Logic Checks* where Mythril exploits assertions embedded in the code to check specific logic conditions.

[6] Table 1 does not report the SLOC of the imported libraries, which, however, are evaluated by Mythril. For instance, Ajna requires to reason on more than 15,000 SLOC.

handle overflows at execution time but does not avoid them. In particular, when SafeMath is used, Mythril incorrectly skips the overflow analysis.

Overall, tools may have the potential to improve their detection capabilities, even when considering only non-business logic issues. Mythril has great potential due to its powerful state space analysis but suffers from the lack of detectors and scalability issues, while the other tools may benefit from more general pattern definitions.

5.4 RQ4: Analysis Complexity Versus Tool Efficacy

One result emerging from our experiments is that SmartCheck and Slither found more security flaws than Mythril. Moreover, Mythril has not discovered any high-severity flaw. This is unexpected because Mythril is a state-of-the-art symbolic execution framework that may in principle outperform simpler analysis frameworks, such as SmartCheck and Slither, by exchanging scalability for efficacy. However, as already pointed out in RQ3, Mythril does not ship with the same broad set of detectors as the other tools. Nonetheless, looking at the detected flaws is only one side of the story when considering automatic analyses. Indeed, a prominent concern is often the false positive rate. In our experiments, Mythril produced 2 true positives (TP) and 61 false positives (FP), thus with a precision of 3.28%. Slither performed significantly worse, with 9 TPs and 1061 FPs, thus with a precision of 0.46%. SmartCheck positioned between the two, with a precision of 1.46%, 3 TPs, and 202 FPs. Overall, FPs are still a big problem for these three tools, likely hurting their adoption. The more in-depth analysis from Mythril has likely contributed to limiting the number of FPs but there is still room for improvement. For instance, SmartCheck reported more than 30 *Gas Limit DoS* invalid alerts, which can be ruled out when exploiting Mythril analysis. Additionally, we believe that Mythril analysis can likely support the implementation of more powerful detectors, integrating more general patterns. For instance, Mythril discovered a *frontrunning* vulnerability, whose detection is likely out of reach for the two other tools. Similarly, the *precision loss* issues split across two portions of code discussed in RQ3 is likely a good example where a more global analysis can favor the implementation of more robust detectors.

6 Conclusions

In this paper, we report our practical experience when running three state-of-the-art tools on smart contracts from 11 Code4rena contests. We have manually analyzed all the security issues reported during the contests by humans and bots, classifying them in common vulnerability types. Then, we observed that tools are unable to find business logic security flaws, which were the most common vulnerable cause. Our observation is consistent with another recent study [43]. We believe that a key idea to mitigate such limitation would be to ask developers to formalize logic conditions through, e.g., assertions or invariant checkers, allowing tools such as Mythril to exploit them during the analysis. Tools performed better on flaws related to Solidity, EVM, and the blockchain, but their vulnerability detectors come with several limitations in terms of accuracy and generality, which we believe could be mitigated by exploiting powerful analyses such as the one used by Mythril. However, Mythril suffers from scalability issues due to its underlying analysis, suggesting that it is not only a matter of engineering effort. It is interesting to note that, among all analyzed DApps, some vulnerable contracts implement common functions used as building blocks for safety-critical systems (e.g., the tracking coin logic implemented in Frankecoin is not different from the logic implemented to track contaminated food) and thus the same performance of such tools extends to other domains. There seems to exist still a big gap between the expectations of security levels required from safety-critical systems and the capability of smart contract testing tools.

Acknowledgements. This papers is partially supported by the Sapienza research project DYNASTY (protocol number RM123188F791C0F7).

References

1. Abid, A., Cheikhrouhou, S., Kallel, S., Tari, Z., Jmaiel, M.: A smart contract-based access control framework for smart healthcare systems. Comput. J. **67**(2), 407–422 (2022). https://doi.org/10.1093/comjnl/bxac183
2. Ajna: Paper. https://www.ajna.finance/pdf/Ajna_Protocol_Whitepaper_10-19-2023.pdf
3. Almakhour, M., Sliman, L., Samhat, A.E., Mellouk, A.: Verification of smart contracts: a survey. Perv. Mob. Comput. **67**, 101227 (2020). https://doi.org/10.1016/j.pmcj.2020.101227
4. di Angelo, M., Durieux, T., Ferreira, J.F., Salzer, G.: Evolution of automated weakness detection in ethereum bytecode: a comprehensive study. Empir. Softw. Eng. (2023). https://doi.org/10.1007/s10664-023-10414-8
5. Asymmetry: Whitepaper (2023). https://www.asymmetry.finance/whitepaper
6. Atzei, N., Bartoletti, M., Cimoli, T.: A survey of attacks on ethereum smart contracts (sok). In: Principles of Security and Trust (2017). https://doi.org/10.1007/978-3-662-54455-6_8
7. Baldoni, R., Coppa, E., D'Elia, D.C., Demetrescu, C., Finocchi, I.: A survey of symbolic execution techniques. ACM Comput. Surv. (2018). https://doi.org/10.1145/3182657
8. Bonomi, S., Cappai, S., Coppa, E.: On the efficacy of smart contract analysis tools. In: 2023 IEEE 34th International Symposium on Software Reliability Engineering Workshops (ISSREW). IEEE Computer Society (2023). https://doi.org/10.1109/ISSREW60843.2023.00041
9. Bonomi, S., Cappai, S., Coppa, E.: Extended version. Technical report (2024). https://github.com/niser93/SmartContractToolAnalysis/blob/master/TechReport.pdf
10. Casino, F., Dasaklis, T.K., Patsakis, C.: A systematic literature review of blockchain-based applications: current status, classification and open issues. Telematics Inf. **36**, 55–81 (2019). https://doi.org/10.1016/j.tele.2018.11.006
11. Caviar: Caviar docs (2023). https://docs.caviar.sh/

12. Chowdhury, M.J.M., et al.: A comparative analysis of distributed ledger technology platforms. IEEE Access **7**, 167930–167943 (2019). https://doi.org/10.1109/ACCESS.2019.2953729
13. Díaz, M., Soler, E., Llopis, L., Trillo, J.: Integrating blockchain in safety-critical systems: an application to the nuclear industry. IEEE Access **8**, 190605–190619 (2020). https://doi.org/10.1109/ACCESS.2020.3032322
14. EigenLayer: Whitepaper (2023). https://docs.eigenlayer.xyz/assets/files/EigenLayer_WhitePaper-88c47923ca0319870c611decd6e562ad.pdf
15. Elia, N., et al.: Smart contracts for certified and sustainable safety-critical continuous monitoring applications. In: Chiusano, S., Cerquitelli, T., Wrembel, R. (eds.) ADBIS 2022. LNCS, vol. 13389, pp. 377–391. Springer, Cham (2022). https://doi.org/10.1007/978-3-031-15740-0_27
16. ENS: Documentation (2021). https://docs.ens.domains/
17. Feist, J., Greico, G., Groce, A.: Slither: a static analysis framework for smart contracts. In: WETSEB 2019 (2019). https://doi.org/10.1109/WETSEB.2019.00008
18. Ferreira, J.a.F., Cruz, P., Durieux, T., Abreu, R.: Smartbugs: a framework to analyze solidity smart contracts. In: ASE 2020 (2021). https://doi.org/10.1145/3324884.3415298
19. Frankencoin: Documentation (2023). https://docs.frankencoin.com/
20. Grieco, G., Song, W., Cygan, A., Feist, J., Groce, A.: Echidna: effective, usable, and fast fuzzing for smart contracts. In: ISSTA 2020 (2020). https://doi.org/10.1145/3395363.3404366
21. Gupta, B.C., Kumar, N., Handa, A., Shukla, S.K.: An insecurity study of ethereum smart contracts. In: Security, Privacy, and Applied Cryptography Engineering (2020)
22. Hua, G., Zhu, L., Wu, J., Shen, C., Zhou, L., Lin, Q.: Blockchain-based federated learning for intelligent control in heavy haul railway. IEEE Access **8**, 176830–176839 (2020). https://doi.org/10.1109/ACCESS.2020.3021253
23. Juicebox: Documentation (2023). https://docs.juicebox.money/dev/
24. Knight, J.C.: Safety critical systems: challenges and directions. In: Proceedings of the 24th International Conference on Software Engineering, ICSE 2002 (2002). https://doi.org/10.1145/581339.581406
25. Kuperberg, M., Kindler, D., Jeschke, S.: Are smart contracts and blockchains suitable for decentralized railway control? Ledger **5** (2020). https://doi.org/10.5195/LEDGER.2020.158
26. Liang, H., Zhang, Y., Xiong, H.: A blockchain-based model sharing and calculation method for urban rail intelligent driving systems. In: 2020 IEEE 23rd International Conference on Intelligent Transportation Systems (ITSC), pp. 1–5 (2020). https://doi.org/10.1109/ITSC45102.2020.9294263
27. Livepeer: Whitepaper (2017). https://github.com/livepeer/wiki/blob/master/WHITEPAPER.md
28. Llama: Documentation (2023). https://docs.llama.xyz/
29. Luu, L., Chu, D.H., Olickel, H., Saxena, P., Hobor, A.: Making smart contracts smarter. In: CCS 2016 (2016). https://doi.org/10.1145/2976749.2978309
30. Mossberg, M., et al.: Manticore: a user-friendly symbolic execution framework for binaries and smart contracts. In: ASE 2019 (2019). https://doi.org/10.1109/ASE.2019.00133
31. Mueller, B.: Smashing ethereum smart contracts for fun and real profit (2018). https://github.com/muellerberndt/smashing-smart-contracts/blob/master/smashing-smart-contracts-1of1.pdf

32. Naser, F.: Review: the potential use of blockchain technology in railway applications. In: 2018 IEEE International Conference on Big Data (Big Data) (2018)
33. Nguyen, T.D., Pham, L.H., Sun, J., Lin, Y., Minh, Q.T.: sfuzz: an efficient adaptive fuzzer for solidity smart contracts. In: ICSE 2020 (2020). https://doi.org/10.1145/3377811.3380334
34. Oriekhoe, O.I., Ilugbusi, B.S., Adisa, O.: Ensuring global food safety: integrating blockchain technology into food supply chains. Eng. Sci. Technol. J. **5**(3), 811–820 (2024). https://doi.org/10.51594/estj.v5i3.905
35. Preece, J., Easton, J.: A review of prospective applications of blockchain technology in the railway industry. Technical report (2018). https://doi.org/10.13140/RG.2.2.15751.75681
36. Ruggiero, C., Mazzini, P., Coppa, E., Lenti, S., Bonomi, S.: Sok: a unified data model for smart contract vulnerability taxonomies. In: Proceedings of the 19th International Conference on Availability, Reliability and Security, ARES 2024 (2024)
37. Shell: The ocean (2022). https://github.com/Shell-Protocol/Shell-Protocol/blob/main/
38. Stader: Documentation (2023). https://www.staderlabs.com/docs-v1/intro
39. defillama team: Total value locked all chains. https://defillama.com/chains
40. Tikhomirov, S., Voskresenskaya, E., Ivanitskiy, I., Takhaviev, R., Marchenko, E., Alexandrov, Y.: Smartcheck: static analysis of ethereum smart contracts. In: WETSEB 2018 (2018)
41. Tolmach, P., Li, Y., Lin, S.W., Liu, Y., Li, Z.: A survey of smart contract formal specification and verification. ACM Comput. Surv. (2021). https://doi.org/10.1145/3464421
42. Torres, C.F., Iannillo, A.K., Gervais, A., State, R.: Confuzzius: a data dependency-aware hybrid fuzzer for smart contracts. In: EuroS&P 2021 (2021)
43. Zhang, Z., Zhang, B., Xu, W., Lin, Z.: Demystifying exploitable bugs in smart contracts. In: ICSE 2023 (2023). https://doi.org/10.1109/ICSE48619.2023.00061

Safety-Security Analysis via Attack-Fault-Defense Trees: Semantics and Cut Set Metrics

Reza Soltani[1]()✉, Milan Lopuhaä-Zwakenberg[1](), and Mariëlle Stoelinga[1,2]()

[1] University of Twente, Enschede, The Netherlands
{r.soltani,m.a.lopuhaa,m.i.a.stoelinga}@utwente.nl
[2] Radboud University, Nijmegen, The Netherlands

Abstract. Cyber-physical systems such as the advanced smart grid have a dynamic interaction between security, safety, and defense. Therefore, we need risk management strategies that take all three into account. This paper introduces a novel framework that seamlessly combines attack trees, fault trees, and defense mechanisms: Attack-Fault-Defense Trees (AFDTs). This model creates a common language using an easily understood visual aid designed for experts from various backgrounds, thereby accelerating multidisciplinary collaboration. We define the semantics and cut set metrics for AFDTs and explore how qualitative analysis can be done through cut set analysis. Furthermore, we provide a case study that revolves around a Gridshield lab, which is a set of remotely connected charging stations at the University of Twente campus, to demonstrate the practical implementation of AFDT analysis.

Keywords: Attack-Fault Tree · Attack-Defense Tree · Smartgrid · Safety · Security

1 Introduction

In the cyber-physical domain, safety and security have a complex relationship that entails trade-offs [6,22]. Specifically, taking steps to improve safety may also weaken security, and vice versa [21]. Security and safety are frequently looked into separately in different studies. As cyber-physical systems grow increasingly complex, so do the risks they face to their continued functioning. An example of such a complex system is a smart grid, an evolution of the power grid that integrates control and digitization advancements with the current power grid architecture [3]. Smart grids are an important tool for the electrification that is necessary for the green energy transition, such as the use of electric vehicles for commuting and industrial thermal pumps for heating. As a critical infrastructure, a smart grid needs to be resilient not only to the accidental failure that can come from its many high-tech components, but also from (cyber)attacks by malicious actors. To create effective defenses against such safety and security risks,

---

This work has been funded by the ERC Consolidator Grant 864075 (*CAESAR*).

© The Author(s), under exclusive license to Springer Nature Switzerland AG 2024
A. Ceccarelli et al. (Eds.): SAFECOMP 2024, LNCS 14988, pp. 218–232, 2024.
https://doi.org/10.1007/978-3-031-68606-1_14

experts from several fields (most notably electrical engineering and software design) must collaborate on their designs and implementation. Thus a common framework for assessing the safety and security impact of countermeasures, that can be understood by experts from many different domains, is paramount to create resilient smart grid architectures, and resilient cyber-physical systems in general.

Tree-based risk models, such as Fault Trees (FTs) for safety and Attack Trees (ATs) for security, are an often-used framework for allowing communication across disciplines. These are hierarchical diagrams consisting of a Top Level Event (TLE) denoting system disruption, and leaves denoting atomic events (component failures and basic attacker actions, respectively). In between are intermediate events, whose activation depends the activation of their children and their type (AND/OR). Such models are easily understandable by people from different disciplines, and each expert can contribute to the part of the tree that falls under their domain.

Since FTs and ATs cover only safety and security, respectively, more elaborate models are needed to cover the multitude of risks and their countermeasured faced by complex systems. Two such models are Attack-Fault Trees (AFTs) [12], that model joint safety-security risks, and Attack-Defense Trees (ADTs) [1,9], that model security risks and countermeasures to mitigate them. However, none of these models have the expressive power to model the interaction between safety, security and countermeasures that are inherent to smart grids.

Contributions. Our main contribution is the introduction of Attack-Fault-Defense Trees (AFDTs), a novel framework for capturing safety, securty and countermeasures in a single model. This combines elements from AFTs and ADTs, but is considerably more involved than a simple combination: First, existing ADT definitions [1,9] presume a dichotomy between events that augment the system's functioning, and events that weaken it. However, it has been recognized that interactions between safety and security can be more complicated, and countermeasures that increase safety may harm security and vice versa [11]. We add such *antagonism* into our model by allowing events to inhibit not only events of the opposite 'type', but any other event.

Second, we introduce the qualitative and quantitative analysis based on minimal cut sets (MCSs), which represent minimal combinations of adverse events that together cause system disruption. MCS analysis has been used extensively in FT/AT analysis [14,17]. However, defenses make this more complicated, as what constitute an MCS now depends on the system owner's chosen countermeasures. Turning this around, a countermeasure's efficacy can be measured by its impact on the MCSs. We develop methods to assess this impact both qualitatively, based on MCS number and size, and quantitatively, by attaching failure probability and attacker cost values to each MCS and assessing how these are impacted by defenses.

We showcase our methods by applying them to *Gridshield*, a protection mechanism for smart grids that reduces charging power at charging stations in the event of a smart gird asset overload [2,19]. In particular, we consider Gridshield

Lab, a set of remotely connected charging stations at the University of Twente campus. While Gridshield is designed to increase the grid's reliability, it's impact on the grid's overall architecture must be carefully assessed to ensure it does not introduce additional vulnerabilities. We show how AFDTs can be used for a system-wide analysis. Summarizing, our contributions are:

- The introduction of AFDTs as a modeling mechanism for the interplay of safety, security, and defenses;
- Methods to assess the effect of defenses on safety and security via their effect on cut sets and cut set metrics;
- A demonstration of our approach, using the Gridshield lab as a case study.

Paper Organization. The paper is organized as follows. We summarized the related works in Sect. 2. Section 3 describes the Gridshield case study. Section 4 provides background for AFDT. We define the AFDT model in Sect. 5, and then we elucidate its qualitative analysis in Sect. 6. Section 7 is about safety and security dependencies and their visualization. Finally, Sect. 8 concludes the paper and provides suggestions for future work.

2 Related Work

Tree extensions make up the majority of formalisms that capture connections between security and safety [15]. Fault trees [5,13] were designed to address systems failures, whereas attack trees [20] address system attacks. In a survey on models for safety-security co-analysis, Nicoletti et al. found that there is no model that specifically models safety-security interactions [15]. Rather, constructs are taken from frameworks that focus solely on security or safety and then merged using various techniques. Metrics are no different; none are specially designed with safety/security interactions in mind. Furthermore, existing formalisms model dependencies in small- and medium-size case studies, and There is still a lack of large-scale case studies.

Due to the differences in how they are used, fault trees and attack trees are extended either with additional gates and system recovery [16,24] or defenses [9,10]. According to Kordy et al. [9], attack-defense trees are attack trees with countermeasures put in place as defenses. Furthermore, the connection between attack-defense trees and game theory is established by Kordy et al. in [8], who also examine the expressiveness of this relationship. Also, they prove that attack-defense trees are equivalent to a two-player binary, zero-sum game.

The authors in [4] investigate the best sets of countermeasures in ADTs. In attack graphs, an alternative risk model for security, Khouzani et al.'s methods for choosing the best countermeasures in [7] depend on the actions of each defender, influencing the likelihood that an attack will succeed. The authors in [18] proposed a Failure-Attack-Countermeasure Graph framework for safety and security alignment at early development phases in cyber-physical systems. The graph incorporates safety, security, and countermeasures only in the early development phases. The paper lacks a semantic and qualitative analysis of the graph.

Table 1 indicates a summary of the comparison of related works with the proposed approach. As can be seen, AFDT is novel in combining both attacks, failures and defenses. Furthermore, we include both qualitative and quantitative analysis and a case study.

Table 1. Related works comparison to the proposed approach

Model		Attack	Failure	Defense/Countermeasure	Qualitative/Quantitative analysis	Case study
FT	[24]				✓	
ADT	[8]	✓		✓		
	[9]	✓		✓	✓	
	[16]	✓		✓	✓	✓
	[4]	✓		✓	✓	✓
AT	[7]	✓		✓	✓	✓
AFDT		✓	✓	✓	✓	✓

3 Case Study: Gridshield

The expansion of electric vehicles (EVs) within the low-voltage distribution grid raises the electrical load, accelerating the risk of overloading and damaging grid assets. Energy management systems are being developed for smart EV charging, aimed at averting congestion. Nevertheless, such systems have an intrinsic sensitivity to failure, which underscores the necessity for backup mechanisms [19]. GridShield, serving as a local (on-site) congestion management control system, fulfills this critical role [2,19].

Gridshield is a safety mechanism that reduces charging power at charging stations in the event of a smart grid asset overload, such as transformers. Moreover, Gridshield is proposed as a last-resort defense mechanism that must prevent power grid service disruptions. That is, it is an extra line of defense in addition to the current digital technologies and physical safety measures (such as fuses). The general notion behind this control is as follows. The load on this asset is regularly monitored by a sender module. In the event that this sender detects an overload on this asset, then it will broadcast a message to all nearby charging stations instructing them to lower their power consumption. The charging stations are equipped with receiver modules that regulate the charging power according to the received message. After a shift in the asset's load is observed, a fresh message can be sent out to either further cut the charging power or bring it back to normal. More information about how Gridshield operates will be provided at the end of Sect. 5 after introducing Attack-Fault-Defense Trees.

4 Background

When it comes to safety, Fault Trees (FT) analysis [13] is frequently used to identify potential problems, such as component failures that propagate throughout a system. From a security perspective, we use Attack Trees (AT) [20] to understand how attackers might break things down into smaller steps and elements to achieve their goals. Combining these approaches, Attack-Fault Trees (AFT) [12] are utilized for a thorough study that considers elements related to safety and security, providing a comprehensive view of possible system vulnerabilities. The critical system under consideration constitutes the main element within the context of AFTs, namely the Top Level Event (TLE), representing an outcome we aim to prevent. Numerous factors, whether intentional attacks or component failure, can trigger the TLE. The TLE is systematically broken down into distinct sub-goals, categorized under security or safety considerations. These sub-goals, referred to as intermediate events, are connected by AND/OR gates determining how basic events lead to the TLE. The activation of intermediate events relies on the activation of their respective children, creating a recursive refinement process until further breakdown is not feasible. At this stage, the tree's leaves emerge, representing individual entities: basic component failures (BCFs), which refer to specific failures within the system components, and basic attack steps (BASs), which denote individual actions taken by an attacker.

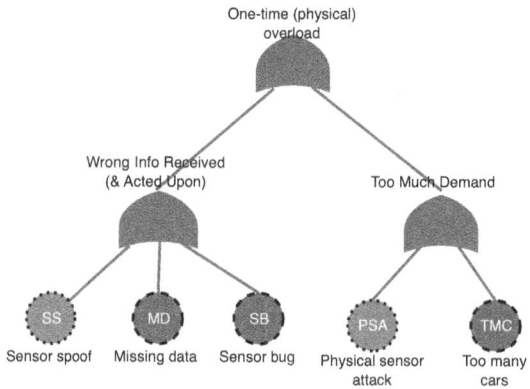

Fig. 1. An example of AFT for a smart grid

An example of AFT for a smart grid with three BCFs and two BASs is shown in Fig. 1. BCFs and BASs are represented in this AFT by blue dashed-circular and red dotted-circular nodes, respectively. All gates are shown in purple. TLE's (one-time overload) failure is due to either receiving wrong information or high demand. When the attacker does Sensor spoof, or any of the two components, missing data or sensor bug, fails, the TLE fails through receiving wrong info.

Physical sensor attack or an excessive number of cars (too many cars) can also lead to TLE failure through too much demand.

Although AFTs consider elements related to safety and security, Attack-Defense Trees (ADTs) are a robust modeling framework for comprehensive cybersecurity analysis. ADTs add defenses and inhibitors to ATs, enhancing the ability to model potential threats. Unlike AFTs, which mainly depict potential failure scenarios, attack-defense trees offer a more comprehensive approach by combining defensive strategies alongside offensive approaches. By analyzing the ADT and evaluating the effect of each attack and defense disruption, we can determine the minimal possible attacks that propagate to a successful TLE attack. In the following section, we will explain in detail how defenses and inhibitors influence possible attacks that can lead to a successful TLE attack.

5 Attack-Fault-Defense Trees

Table 2. Summary of acronyms and symbols

Acronyms, Symbols	Meaning	Acronyms, Symbols	Meaning
FT	Fault Trees	CS	Cut Set
AT	Attack Trees	MCS	Minimal Cut Set
AFT	Attack-Fault Trees	τ	Trigger
ADT	Attack-Defense Trees	A_T	set of all Basic Attack Steps
AFDT	Attack-Fault-Defense Trees	$\mathscr{A}_T$	set of all attacks
TLE	Top Level Event	a	element of the $\mathscr{A}_T$ set
BCF	Basic Component Failure	F_T	set of all Basic Component Failures
BAS	Basic Attack Step	$\mathscr{F}_T$	set of all failures
BDS	Basic Defense Step	f	element of the $\mathscr{F}_T$ set
INH	Inhibition (gate)	D_T	set of all Basic Defense Steps
ch(v)	Children of node	$\mathscr{D}_T$	set of all defenses
		d	element of the $\mathscr{D}_T$ set

Table 2 summarizes all acronyms and symbols used in this paper. In this section, We propose AFDT, which integrates ADTs with AFTs into a single framework. This allows for a detailed, comprehensive examination of faults, attacks, and defensive tactics in one package. Compared to AFTs and ADTs, AFDT features two additional elements, shown in Fig. 2. *Basic Defense Step (BDS)* is a leave that models the actions of the defender and is depicted as green pentagon node. The *Inhibition (INH)* gate, shown as a hexagon in Fig. 2b, has two inputs, w_1 and w_2, where w_2 is called the *inhibitor*. The INH gate is deactivated in the presence of an active inhibitor. On the contrary, its output equals its input when the inhibitor is not activated. The behavior of the INH gate can be written mathematically as $INH(w_1|w_2) = w_1 \wedge \neg w_2$.

The color of the INH gate stems from its positive or negative effect, where green represents a positive effect that contributes to preventing the failure of

TLE, and purple signifies a negative effect that contributes to the failure of TLE. In general, such a coloring might not be possible because an event can have both positive and negative consequences. In the *INH* gate, negation is possible for both positive and negative elements. This is demonstrated in Fig. 3 by *BCF* inhibitor (sanity check failure), which has the ability to deactivate *BDS* Event (sanity check implementation), or *BDS* inhibitor (Gridshield on) activation to safeguard the system. The *Inhibition (INH)* gate proposed in this paper is the generalized version of the ADT proposed by Kordy et al. [9], that the possibility of a negative element can even inhibit another negative element, which is not feasible within the context of [9].

The attacker (depicted in red) can gain access to the root (cause the TLE to fail) via sensor spoof or physical sensor attack. The defender, shown in a green pentagonal node, can intervene to stop this possible breach through the sanity check implementation and Gridshield on. Also, three *BCF*s are identified as potential contributors to the TLE failure. Nevertheless, defenses play a role in preventing the propagation of these *BCF*s. The Gridshield on defense system, for example, can prevent too many cars from increasing demand. On the other hand, sensor bug and missing data can't cause false information if we use the sanity check implementation defense mechanism. Nevertheless, in this specific case, sanity check implementation's efficacy could be compromised in the case that sanity check fails.

(a) BDS

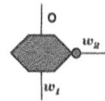

(b) Inhibition gate

Fig. 2. The additional leaf and gate for AFDT.

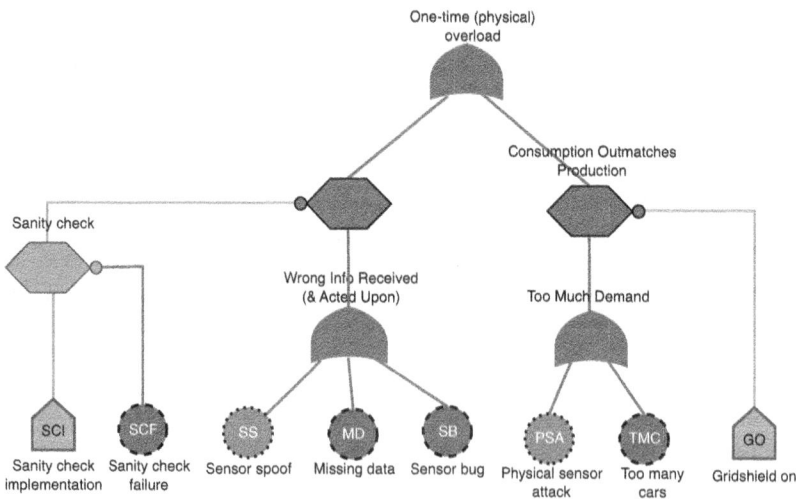

Fig. 3. An exemplary smart grid AFDT. The scenario is in line with Fig. 1.

5.1 Formal Definition of AFDT

A formal definition of AFDT provides a structured framework for explaining cut set analysis and cut set metrics. Formally, AFDTs are defined as follows:

Definition 1 (AFDT). *An Attack-Fault-Defense Tree is a quadruple $T = \langle V, E, \gamma, \tau \rangle$, such that:*

- *E is the set of all edges where $\langle V, E \rangle$ is a rooted directed acyclic graph.*
- *γ is a function $\gamma : V \to \{AND, OR, INH, BAS, BCF, BDS\}$.*
- *$\gamma(v) \in \{BAS, BCF, BDS\}$ if and only if v is a leaf of $\langle V, E \rangle$.*
- *If $\gamma(v) = INH$ then v has exactly two children.*
- *τ is a function $\tau : \{v \in V | \gamma(v) = INH\} \to V$ s.t. $\langle V, \tau(v) \rangle \in E$.*

The root of T is denoted as Root_T, and the set of children of a node v as $ch(v) = \{w \in N \mid (v, w) \in E\}$. In the above definition, τ indicates the trigger node for the *INH* gate, i.e. $v = INH(w_1|w_2)$, $ch(v) = \{w_1, w_2\}$, and $\tau(v) = w_2$.

The set of Basic Attack Steps (*BASs*) on T is denoted $A_T = \{\alpha_1, \ldots, \alpha_n\}$. An attack is a set of *BASs*. The set of all attacks on T is denoted $\mathscr{A}_T$, which indicates the family of subsets of A_T identified as binary vectors; $\boldsymbol{a}$ denotes an element of $\mathscr{A}_T$. For example, if $A_T = \{\alpha_1, \ldots, \alpha_5\}$, the vector $\boldsymbol{a} = (0, 1, 0, 1, 0)$ represents the set $\{\alpha_2, \alpha_4\}$. In other terms, $\boldsymbol{a} \in \mathscr{A}_T = \mathbb{B}^\kappa$ signifies the attacks resulting from activating all α_i for which $a_i = 1$.

Similarly, the set of all Basic Component Failures (*BCFs*) on T is denoted $F_T = \{\beta_1, \ldots, \beta_m\}$. A failure is a set of *BCFs*. The set of all failures on T is denoted $\mathscr{F}_T$, which indicates the family of subsets of F_T identified as binary vectors; $\boldsymbol{f}$ denotes an element of $\mathscr{F}_T$. For example, if $F_T = \{\beta_1, \ldots, \beta_m\}$, the vector $\boldsymbol{f} = (0, 1, 0, 1, 0)$ represents the set $\{\beta_2, \beta_4\}$.

the set of Basic Defense Steps (*BDSs*) on T is denoted $D_T = \{\delta_1, \ldots, \delta_p\}$. A defense is a set of *BDSs*. The set of all defenses on T is denoted $\mathscr{D}_T$, which indicates the family of subsets of D_T; $\boldsymbol{d}$ shows an element of the $\mathscr{D}_T$ set, while $\boldsymbol{d} \in \mathscr{D}_T = \mathbb{B}^p$ represents the defenses induced by all δ_i for which $d_i = 1$.

The extent to which attacks, failures, and defenses together cause or prevent TLE disruption is captured by the structure function. The structure function of AFDT acts as a formal representation of the relationships and dependencies among the attack, fault, and defense nodes, providing a systematic framework for analyzing the potential propagation of attacks and faults and the effectiveness of defense mechanisms within the system.

Definition 2 (Structure function). *Let T be an AFDT with $A_T = \{\alpha_1, \ldots, \alpha_n\}, F_T = \{\beta_1, \ldots, \beta_m\}$, and $D_T = \{\delta_1, \ldots, \delta_p\}$. The structure function of T is the function $S_T : \mathscr{A}_T \times \mathscr{F}_T \times \mathscr{D}_T \times V \to \mathbb{B}$ defined as:*

$$S_T(\boldsymbol{a},\boldsymbol{f},\boldsymbol{d},v) = \begin{cases} a_i, & \text{if } v = \alpha_i; \\ f_i, & \text{if } v = \beta_i; \\ d_i, & \text{if } v = \delta_i; \\ \bigwedge_{w \in ch(v)} S_T(\boldsymbol{a},\boldsymbol{f},\boldsymbol{d},w), & \text{if } \gamma(v) = AND; \\ \bigvee_{w \in ch(v)} S_T(\boldsymbol{a},\boldsymbol{f},\boldsymbol{d},w), & \text{if } \gamma(v) = OR; \\ S_T(\boldsymbol{a},\boldsymbol{f},\boldsymbol{d},w_1) \wedge \neg S_T(\boldsymbol{a},\boldsymbol{f},\boldsymbol{d},w_2), & \text{if } \gamma(v) = INH(w_1|w_2). \end{cases} \quad (1)$$

Definition 3 (Successful disruption). *An attack $\boldsymbol{a} \in \mathscr{A}_T = \mathbb{B}^n$ and a failure $\boldsymbol{f} \in \mathscr{F}_T = \mathbb{B}^m$ are called successful if for a given defense $\boldsymbol{d} \in \mathscr{D}_T = \mathbb{B}^p$ the $S_T(\boldsymbol{a},\boldsymbol{f},\boldsymbol{d},\mathrm{Root}_T) = 1$.*

5.2 Gridshield AFDT

Figure 4 is the AFDT of Gridshield that shows the comprehensive overview of the overall interactions in Gridshield. Figures 1 and 3 depict a small portion of Gridshield, while Fig. 4 shows more details about the interaction of attack, fault, and defense nodes in Gridshield. The left-hand side of the AFDT models the risks attached to the existing Dynamic Load Management Systems (DLMS). These systems can prevent connection overloading during times of simultaneous charging and/or react to external energy market incentives for economic benefit. However, failure of DLMS is possible. DLMS may respond to wrong sensor readings or incomplete data (MD). Furthermore, many DLMS use cloud services that may malfunction (e.g., network disconnection), which results in a lack of control. It is expected that basic sanity checks w.r.t. sensor readings are implemented in DLMS (*SCI* BDS).

The right-hand side of the AFDT (Fig. 4 part 2) models the implementation of Gridshield and its effect on consumption. Gridshield safeguards against physical overload by monitoring and controlling consumption to prevent it from exceeding production. It consistently manages demand, production, and dynamic load to ensure optimal operation. If the sender detects an overload, it will broadcast a message to all nearby charging stations (shown as Receiver 1 to Receiver N) instructing them to lower their power consumption. If a certain number of these charging stations (K out of N) do not receive the message, the Gridshield will be disrupted. This is indicated by the K/N gate, also known as a voting gate, which is disrupted if at least K out of N inputs are disrupted. Gridshield can also wrongfully interfere, represented by the purple outgoing edge of GO, which is interrupted by the green outgoing edge of *Consumption Outmatches Production*.

6 Qualitative Analysis of AFDT

Determining a Minimal Cut Set (MCS) is one of the most used methods for qualitative risk analysis in AFT. MCS is the smallest combination of component

Safety-Security Analysis via Attack-Fault-Defense Trees 227

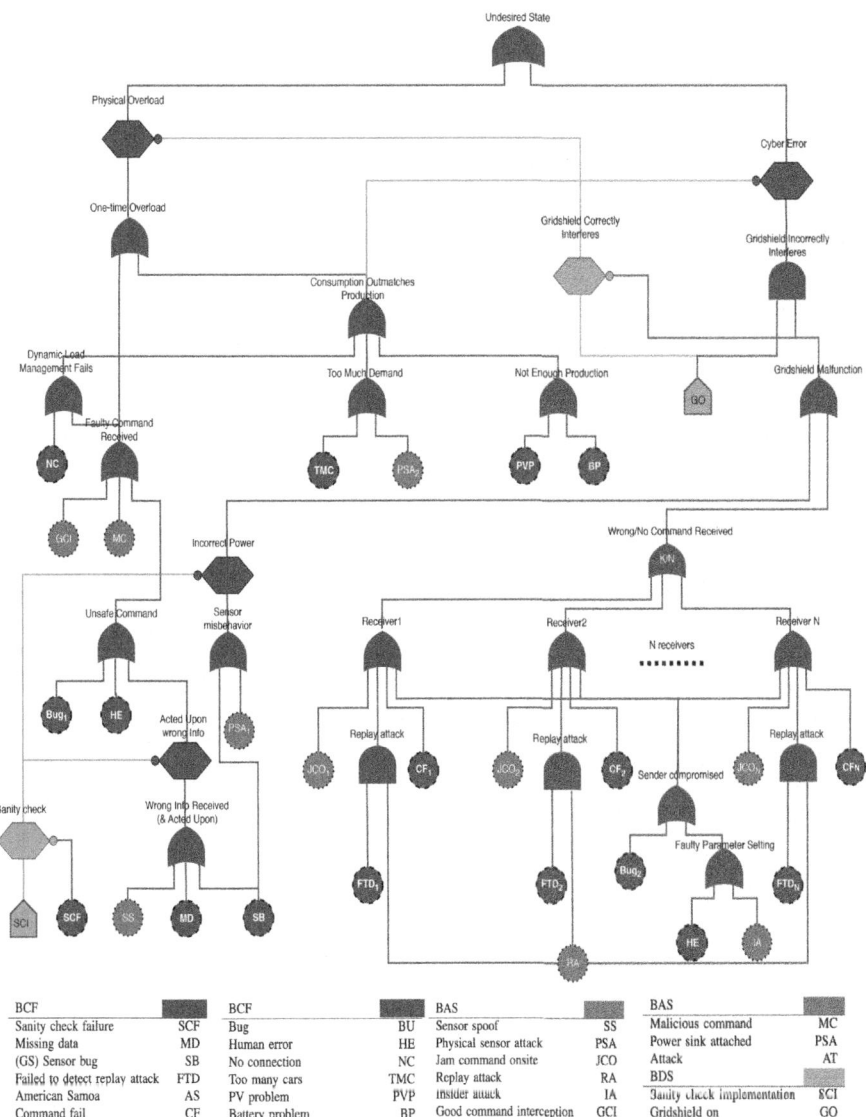

Fig. 4. Gridshield AFDT

(safety/security) disruptions that can cause TLE to fail. MCSs provide details on the system's vulnerability and outline the most concise path to the TLE's occurrence. Each MCS contains a certain number of elements, and the dificulty of propagating TLE depends on the number of these elements. Analyzing MCS provides important information about the system's reliability since it has the potential to be made more reliable overall by reducing the likelihood of an attack or failure with these cut sets. This section defines the terms Cut Set and MCS for AFDT and illustrates how defense affects system reliability.

Definition 4 (Cut Set). *Let d be a defense, i.e., an element of the $\mathscr{D}_T$ set. A Cut Set (CS) is a pair (a, f) w.r.t. a given defense d, such that:*

- $a \in \mathscr{A}_T$
- $f \in \mathscr{F}_T$
- $S_T(a, f, d, \text{Root}_T) = 1$

Definition 5 (Minimal Cut Set). *Minimal Cut Set (MCS) is a CS of which no proper subset is a CS.*

In the AFT of the Fig. 1, there are five MCSs identified: {sensor spoof}, {missing data}, {sensor bug}, {too many cars}, and {physical sensor attack}. AFT and AFDT differ in the way of identifying MCS since AFDT has defenses that can stop an attack from spreading. We can see the effects of each defense by deploying them and pinpointing MCSs. MCS in AFDT significantly depends on the activation of defenses. If there are p defenses, there are at least 2^p collections of MCSs. Each collection contains a set of MCSs. This complexity results from the dynamic interaction between possible failure scenarios and defense activation. The collection can be an empty set {∅} as well, i.e., there is no MCS for a given set of activated defenses. We have four collections of MCSs since there are two BDSs in Fig. 3. An overview of the different MCSs corresponding to each defense's activation status in Fig. 3 is given in Table 3. Table 3A shows the MCSs w.r.t. each set of defenses, and how they change with the change of defense set. Table 3B shows Table 3A from another perspective. It shows the effective defense set against each MCS. Our methodology can also be applied to ADT by simply excluding failures from CS and MCS definitions.

Table 3. The interaction between MCSs and set of active defenses.

	Defense set			
	No defense	{SCI}	{GO}	{SCI, GO}
MCSs	{SS}	{SCF, SS}	{SS}	{SCF, SS}
	{MD}	{SCF, MD}	{MD}	{SCF, MD}
	{SB}	{SCF, SB}	{SB}	{SCF, SB}
	{TMC}	{TMC}	-	-
	{PSA}	{PSA}	-	-

MCSs	Effective defense set
{SS}	{SCI}
{MD}	{SCI}
{SB}	{SCI}
{TMC}	{GO}
{PSA}	{GO}

In general, defenses may have two beneficial results: eliminate MCS and increase the size of MCS. As indicated in Table 3, two MCSs are eliminated when the Gridshield defense mechanism is included, demonstrating its efficacy. The sanity check implementation impacts three MCSs and adds one more element to them and increase their size. System reliability is increased when two defenses are activated simultaneously, creating a three two-element MCSs. An overview of the effects of each defense and how they relate to system dependability is provided by MCS analysis in AFDT. Finding a balance between activating defenses and system reliability, defense cost and failure likelihood, etc., is made easier by taking MCSs into account. Note that, in principle, defenses can also have negative effects, such as introducing new MCS or decreasing cut set size.

7 Safety and Security Dependencies via MCS

As highlighted in the preceding section, we have a 2^p collection of MCSs for p number of defenses. Every MCS has components pertaining to security and safety, and each defense specifically have the potential to affect safety, security, or both. The domains of safety and security are distinct. There are various possible metrics for security analysis, such as cost, attacker's skill, probability of successful attack, attacker's budget, etc. In this paper, security is evaluated in terms of cost, and safety is evaluated in terms of probabilities. Therefore, probability and cost are two domains that can exist in every MCS. For each MCS with BCFs and BASs, we have:

- The cost $C : A_T \to \mathbb{R}_{\geq 0}$,
- The probability $P : F_T \to [0,1]$.

Therefore, for $m = (\vec{a}, \vec{f}) \in MCS$, we can calculate the total cost $TC : \mathbb{B}^n \to \mathbb{R}_{\geq 0}$ of attack, and the total probability $TP : \mathbb{B}^o \to [0,1]$ of failure such that:

$$TC(\vec{a}) = \sum_{i=1}^{n} C(\alpha_i) a_i,$$
$$TP(\vec{f}) = \prod_{i=1}^{o} P(\beta_i)^{f_i},$$
(2)

The formula 2 calculates the total cost and total probability, which means if the attacker pays $TC(u)$, the probability of system disruption via m is $TP(f)$. The set of all cost-probability pairs indicates system reliability.

It is important to note that the total cost TC or total probability of failure TF is calculated consistently based on given defenses $\vec{d}$. Thus, the impact of defenses can be quantified by their impact on TC and TF. Figure 5 shows the graphical representation of Table 3, which indicates the MCSs of the AFDT illustrated in Fig. 3. This graph indicates the five MCSs considering the probability of failure and cost of attacks, alongside defenses' effect on MCSs. It is worth noting that the figure is exemplary, without any numerical attack costs or probabilities of failure. Furthermore, attacks are positioned at the rightmost side of the graph, signifying a probability of one with the condition that the attacker successfully performs specific steps. The effect of each defense is differentiated through variants of green-colored arrows with distinct arrowheads. For example, in this figure, consider sensor spoof (SS) attack. Activating Gridshield does not affect the SS attack, while the Sanity Check Implementation (SCI) reduces the probability of the attack, as now the attack is dependent on saity check failure (SCF). In this figure, dashed arrows denote cases where a defense results in the elimination of an MCS. This visual illustration allows a thorough overview of the intricate give-and-take between attacks, faults, and defenses within the AFDT framework.

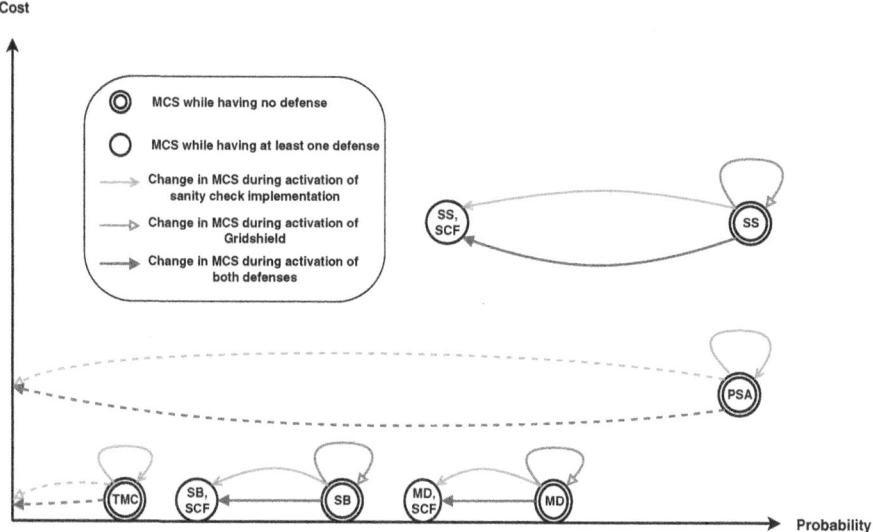

Fig. 5. The graphical representation of Table 3. It indicates the MCSs alongside the defenses' effect on each MCS

8 Conclusion and Future Work

In this paper, we bring forward the Attack-Fault-Defense Tree (AFDT) framework that combines Attack-Fault Trees with defense mechanisms. We demonstrate our approach by using Gridshield lab, a set of remotely connected charging stations in the University of Twente, as a case study. We present the mathematical definition of AFDT and its structure function. Furthermore, we perform a qualitative analysis via minimal cut sets, demonstrating the interdependencies between safety and security.

Scalability is an issue in FT, AT, AFT, and ADT since the large-scale systems increase models' size. As the number of attacks, failures, and defenses grows, the trees become exponentially larger. AFDT also inherited this problem. In future work, we will analyze the scalability of AFDT. Another interesting prospect for future work is to further refine the cut set analysis using game theory. In earlier work, we have modeled spare management in FTs as a $1\frac{1}{2}$-player game [23]. One could use an analogous approach to model a AFDT as a $2\frac{1}{2}$-player game, with the attacker and defender taking the role of the two players. Such a time-game automaton could then be subject to comprehensive quantitative safety-security-defense analysis using model checkers. A hurdle to this approach is that it can be hard to obtain realistic parameter settings, especially for security.

References

1. Arias, J., Budde, C.E., Penczek, W., Petrucci, L., Sidoruk, T., Stoelinga, M.: Hackers vs. security: attack-defence trees as asynchronous multi-agent systems. In: Lin, SW., Hou, Z., Mahony, B. (eds.) Formal Methods and Software Engineering. ICFEM 2020. Lecture Notes in Computer Science(), vol. 12531. Springer, Cham (2020). https://doi.org/10.1007/978-3-030-63406-3_1
2. ElaadNL: Smoothems met gridshield, elaad.nl/projecten/smoothems-met-gridshield/
3. Fang, X., Misra, S., Xue, G., Yang, D.: Smart grid - the new and improved power grid: a survey. IEEE Commun. Surv. Tutorials **14**(4), 944–980 (2012). https://doi.org/10.1109/SURV.2011.101911.00087
4. Fila, B., Wideł, W.: Exploiting attack-defense trees to find an optimal set of countermeasures. In: 2020 IEEE 33rd Computer Security Foundations Symposium (CSF), pp. 395–410 (2020). https://doi.org/10.1109/CSF49147.2020.00035
5. IEC, I.: 61025: Fault tree analysis (fta). Tech. rep., Technical Report (2006)
6. Jepsen, S.C., et al.: A research setup demonstrating flexible industry 4.0 production. In: 2021 International Symposium ELMAR, pp. 143–150 (2021).https://doi.org/10.1109/ELMAR52657.2021.9550961
7. Khouzani, M., Liu, Z., Malacaria, P.: Scalable min-max multi-objective cyber-security optimisation over probabilistic attack graphs. Eur. J. Oper. Res. **278**(3), 894–903 (2019). https://doi.org/10.1016/j.ejor.2019.04.035
8. Kordy, B., Mauw, S., Melissen, M., Schweitzer, P.: Attack-defense trees and two-player binary zero-sum extensive form games are equivalent. In: Alpcan, T., Buttyán, L., Baras, J.S. (eds.) Decision and Game Theory for Security, pp. 245–256. Springer, Berlin Heidelberg, Berlin, Heidelberg (2010)
9. Kordy, B., Mauw, S., Radomirović, S., Schweitzer, P.: Attack-defense trees. J. Log. Comput. **24**(1), 55–87 (2014)
10. Kordy, B., Piètre-Cambacédès, L., Schweitzer, P.: Dag-based attack and defense modeling: don't miss the forest for the attack trees. Comput. sci. rev. **13**, 1–38 (2014)
11. Kriaa, S., Pietre-Cambacedes, L., Bouissou, M., Halgand, Y.: A survey of approaches combining safety and security for industrial control systems. Reliab. Eng. Syst. Saf. **139**, 156–178 (2015)
12. Kumar, R., Stoelinga, M.: Quantitative security and safety analysis with attack-fault trees. In: 2017 IEEE 18th International Symposium on High Assurance Systems Engineering (HASE), pp. 25–32. IEEE (2017)
13. Lee, W.S., Grosh, D.L., Tillman, F.A., Lie, C.H.: Fault tree analysis, methods, and applications a review. IEEE Trans. Reliab. **34**(3), 194–203 (1985)

14. Lopuhaä-Zwakenberg, M., Budde, C.E., Stoelinga, M.: Efficient and generic algorithms for quantitative attack tree analysis. IEEE Trans. Dependable Secure Comput. 20(5), 4169–4187 (2022)
15. Nicoletti, S.M., Peppelman, M., Kolb, C., Stoelinga, M.: Model-based joint analysis of safety and security: survey and identification of gaps. Comput. Sci. Rev. **50**, 100597 (2023). https://doi.org/10.1016/j.cosrev.2023.100597
16. Roy, A., Kim, D.S., Trivedi, K.S.: Attack countermeasure trees (ACT): towards unifying the constructs of attack and defense trees. Secur. Commun. Netw. **5**(8), 929–943 (2012). https://doi.org/10.1002/sec.299
17. Ruijters, E., Stoelinga, M.: Fault tree analysis: a survey of the state-of-the-art in modeling, analysis and tools. Comput. sci. rev. **15**, 29–62 (2015)
18. Sabaliauskaite, G., Mathur, A.P.: Aligning cyber-physical system safety and security. In: Cardin, M.A., Krob, D., Lui, P.C., Tan, Y.H., Wood, K. (eds.) Complex Systems Design & Management Asia, pp. 41–53. Springer International Publishing, Cham (2015)
19. van Sambeek, H.L., Zweistra, M., Hoogsteen, G., Varenhorst, I.A.M., Janssen, S.: GridShield-optimizing the use of grid capacity during increased EV adoption. World Electr. Veh. J. **14**(3) (2023).https://doi.org/10.3390/wevj14030068, https://www.mdpi.com/2032-6653/14/3/68
20. Schneier, B.: Modeling security threats. Dr. Dobb's J. **24**(12) (1999)
21. Soltani, R., Kang, E.Y., Mena, J.E.H.: Towards energy-aware cyber-physical systems verification and optimization. In: Position and Communication Papers of the 16th Conference on Computer Science and Intelligence Systems. Annals of Computer Science and Information Systems, vol. 26, pp. 205–210. PTI (2021)https://doi.org/10.15439/2021F125, http://dx.doi.org/10.15439/2021F125
22. Soltani, R., Kang, E.Y., Mena, J.E.H.: Verification and optimization of cyber-physical systems: Preprint for FedCSIS. arXiv preprint arXiv:2109.01574 (2021)
23. Soltani, R., Volk, M., Diamonte, L., Lopuhaä-Zwakenberg, M., Stoelinga, M.: Optimal spare management via statistical model checking: a case study in research reactors. In: Cimatti, A., Titolo, L. (eds.) Formal Methods for Industrial Critical Systems, pp. 205–223. Springer Nature Switzerland, Cham (2023)
24. Čepin, M., Mavko, B.: A dynamic fault tree. Reliab. Eng. Syst. Saf. **75**(1), 83–91 (2002). https://doi.org/10.1016/S0951-8320(01)00121-1

Safety Verification

Coyan: Fault Tree Analysis – Exact and Scalable

Nazareno Garagiola[1](✉)[iD], Holger Hermanns[1][iD], and Pedro R. D'Argenio[2][iD]

[1] Saarland University, 66123 Saarbrücken, Germany
{garagiola,hermanns}@depend.uni-saarland.de
[2] Universidad Nacional de Córdoba - CONICET, 5000 Córdoba, Argentina
pedro.dargenio@unc.edu.ar

Abstract. We revisit one of the oldest problems in probabilistic safety analysis and present a breakthrough in scalability and precision. This paper develops a novel, *scalable and exact* method to compute the unreliability of a system modelled by a *Fault Tree* (FT). This unreliability corresponds to the *top event probability* (TEP) of the tree. Our method harvests recent advances in *Weighted Model Counting* (WMC), together with tailored encodings mapping each FT to an equisatisfiable Boolean formula and tailored weight assignments. We then resort to state-of-the-art WMC solvers to obtain an exact value of the TEP. This approach delivers precise results in high reliability scenarios and it scales to very large and complex trees, where so far one needed to resort to approximate analysis (e.g., based on minimal cut sets). We provide empirical evidence regarding superiority of our method by comparing to state-of-the-art tools for Fault Tree Analysis across a large set of Fault Tree benchmarks.

1 Introduction

Fault Trees (FT) are powerful tools in the field of reliability engineering and risk analysis, offering a systematic approach to model and analyze a failure in a complex system. A FT embodies a Boolean function, illustrating the interdependence of overall system failures on the malfunction of fundamental system components, this feature gives them utility across diverse domains, including aerospace [33], nuclear power [37], chemical processes [4], and others. By offering a comprehensive visualization of failure pathways, FTs enable engineers and analysts to enhance system resilience, optimize maintenance strategies, and make informed decisions to ensure the robustness of intricate systems in the face of potential failures.

Fault Tree Analysis (FTA) consists in the computation of relevant metrics on the modeled system, where one of the most important metrics is the unreliability of the FT, corresponding to the TEP, the *top event probability* of the tree. Common approaches for computing the TEP proceed by translating the tree into a Binary Decision Diagram (BDD) [9] then computing directly the TEP [25] or by using Minimal Cut Sets (MCS). The first method can in principle provide an exact value of the TEP, but there are limits to the size of the models for which the methods succeeds, partly rooted in its susceptibility to the variable ordering of the BDD representation used. The second method utilizes MCS to compute an upper bound of the TEP, which results in a conservative approximation of the unreliability of the FT, therefore, trading scalability for precision.

The contribution of this paper lies in the development of an approach for computing the exact value of the TEP, taking advantage of the recent improvements in the area of *Model Counting*, which is about counting all the interpretations that satisfy a boolean formula. More concretely, we translate the FT into an equisatisfiable formula in *Conjunctive Normal Form* that can then be submitted to a *Weighted Model Counter* with the probabilities of the basic events appearing as weights for some of the literals and indifferent weights for the literals created by the translation. We prove that the weighted model count computed in this way agrees with the TEP of the original tree.

The approach is implemented in a software prototype named COYAN, written in *Rust*. We show empirically that COYAN can outperform state-of-the-art tools for the calculation of the TEP across a large collection of benchmarks harvested from existing sources as well run time and in memory usage.

Organization of the Paper: Section 2 presents *Static Fault Trees* and *Weighted Model Counting*. Section 3 introduces the problem we study together with providing an overview of other works in the area. Section 4 develops the theoretical underpinning, thereby proving the correctness of our approach. In Sect. 5 we discuss details of our implementation. Section 6 presents an elaborate empirical evaluation and discusses the results in the context of industrial-size problems. Section 7 concludes the paper.

2 Preliminaries

A *Fault Tree* (FT) is a directed acyclic graph that models how the failure of atomic components can propagate, eventually resulting in the failure of the entire system. A FT is composed of two types of nodes, the *basic events* and the *gates*. The basic events are the leafs of the tree and represent failures of individual pieces of equipment, external events or operator errors, as atomic failures. The gates are the intermediate nodes that connect the failures of basic events, propagating their effect along the tree. The top level event of the tree represents the failure of the entire system. Often, basic events are assigned an occurrence probability.

We consider traditional static Fault Frees, however, there are various extensions of Fault Trees, such as *Dynamic Fault Trees* (DFT) [14] or *SD Fault Trees* [20] which combines both static and dynamic features. A review analysis of the different types of Fault Trees can be found on [30].

Definition 1. *A Fault Tree is a tuple* $\mathcal{F} = (V, \sigma, T, top, \Gamma)$ *where*

 – V *is a set of nodes, where* $top \in V$ *is the top event of* $\mathcal{F}$,
 – $\sigma \colon V \to \mathcal{P}(V)$ *maps each node to its set of children,*

- $T\colon V \to Types$ maps each node to its type, where $Types = \{BE, AND, OR\} \cup \{VOT(k, N) \mid 0 < k \le N\}$, and
- $\Gamma\colon V_{BE} \to [0, 1]$ assigns a probability to each basic event in $V_{BE} = \{v \in V \mid T(v) \in BE\}$.

In addition, we require that $\mathcal{F}$ is a directed acyclic graph, i.e., $v \notin \sigma^*(v)$ for all $v \in V$, every leaf is a basic event and vice-versa, i.e., $\sigma(v) = \emptyset \iff T(v) = BE$, and top is the root, i.e., $\sigma^*(top) = V$.

The function T associates a type to each node, which – apart from basic events at the leaves of the tree – can be one of three variants of gates, each of which propagates the failures of its children differently:

- An AND gate fails if all of its children fail;
- An OR gate propagates the failure if any of its children fails;
- A $VOT(k, N)$ (voting) gate fails if at least k of its N children fail.

The function Γ which assigns probabilities to basic events has its origin in either a discrete distribution or a continuous distribution of the time to failure. In the discrete case, the values present directly correspond to the respective failure probability per basic event. In the other case one fixes a certain time bound t for the analysis. Then, given a FT $\mathcal{F}$ and a basic event b_e representing the event e, the probability that event e occurs within time bound t is $P_t(e) := \mathbf{P}(b_e < t)$ and this is the value of $\Gamma(b_e)$.

The unreliability of fault tree $\mathcal{F}$ is defined as $P_t(\mathcal{F}) := \mathbf{P}(\mathcal{F}_{top} < t)$ for the same t, i.e., the probability that the top event happens within the time bound t. This is what is referred to as top event probability (TEP).

We observe that any VOT gate can be rewritten into a proper combination of AND and OR gates. More logical constructs like NOT or XOR can be used to add expressiveness. Finally, the logical structure of a fault tree $\mathcal{F}$ can be represented as a boolean formula $\Phi_{\mathcal{F}}$ where each propositional variable corresponds to a basic event. We call this the *implicit boolean formula*.

Weighted Model Counting (WMC) is an extension of the traditional problem of *Model Counting*, usually known as #SAT. While the aim of #SAT is to count all the interpretations that satisfy the original boolean formula, WMC calculates the sum of weights associated with each of these interpretations. The method has risen as an effective way to solve different problems like probabilistic inference [7], Bayesian inference [31] [10], probabilistic programming problems [17] and others. Mediated by the *Model Counting Competition* [8], there has been an increase in the number of WMC solvers in recent years, each solver sporting different types of methods utilized, including Algebraic Decision Diagrams [12], Dynamic Programming [13], Tree Decompositions [19], Knowledge Compilation [21] and even approximate solvers that work via sampling [32], among others. We refer to the *solvers* section in [8] for an exhaustive list of available solvers.

Before the formal definition of WMC, we need some notation. Given a propositional formula Φ, let $Var(\Phi)$ be the set of its propositional variables and let $Lit(\Phi) = Var(\Phi) \cup \{\neg x \mid x \in Var(\Phi)\}$ be the set of its possible literals. A function

$I \colon \mathit{Var}(\Phi) \to \{0,1\}$ is an *interpretation* for Φ and it is extended to any formula recursively as usual (e.g. $I(\Phi \wedge \Psi) = 1$ iff $I(\Phi) = I(\Psi) = 1$, etc.). We say that I *satisfies* Φ (or I is a *model* for Φ), denoted by $I \models \Phi$, whenever $I(\phi) = 1$. Let $\mathit{Mod}(\Phi) = \{I \mid I \models \Phi\}$ be the set of all models of Φ.

The problem of model counting of Φ is to determine the size of $\mathit{Mod}(\Phi)$. The problem of *weighted* model counting is to find the sum of *weights* of all models in $\mathit{Mod}(\Phi)$. The weight of each model is determined using a *literal-weight function* $w \colon \mathit{Lit}(\Phi) \to \mathbb{R}$ that provides the weight of the propositional variable either in its positive or negative form. (It is also common to restrict the weight function to the interval $[0,1]$.) Formally WMC is defined as follows.

Definition 2. *The* weighted model count *of Φ w.r.t. to w is given by*

$$\mathsf{WMC}_w(\Phi) = \sum_{I \in \mathit{Mod}(\Phi)} W_w(I)$$

where $W_w(I) = \prod_{I(x)=1} w(x) \cdot \prod_{I(x)=0} w(\neg x)$ is the weight of model I.

Valiant showed in his seminal work [36] that when the function Φ is in *Conjunctive Normal Form* (CNF) computing the literal-weighted model count is #P-Complete, this is why almost all of the tools developed to solve this problem takes as input a CNF formula, usually in *DIMACS* [11] format.

3 Computing Unreliability Values

Fault trees have been studied and revisited along the years, and one of the most accepted methods to process FTs is the use of Binary Decision Diagrams (BDD), first introduced in [25]. A BDD is a graph-based data structure that provides a compact representation of Boolean functions, where each inner node represents a variable, with two outgoing edges that encode an `if-then-else` regarding the variable valuation. The nodes commonly adhere to a specific ordering regarding the variables they represent, the leafs represent the constants 0 and 1. Different variable orderings induce different BDD representations, and good orderings are often based on heuristics [26] [28]. There are various approaches to simplify or exploit specifics of the basic BDD concept [18].

Another broadly accepted method for studying FTs uses Minimal Cuts Sets (MCS) [30]. A Cut Set is a set of basic events that, if occurring together, lead to the occurrence of the top event. Minimal Cut Sets are the smallest such sets. The collection of all Minimal Cut Sets encompasses all the possible ways in which the top event can event occur. MCS-based reasoning can be combined with BDD-based analysis [25].

With the objective of computing the TEP of a FT, each method provides an algorithmic way to compute it. In the BDD approach the value can be computed without approximation by using Shannon Decomposition:

$$P_t(\Phi_{\mathcal{F}}) = P_t(x) \times P_t(\Phi_{\mathcal{F}}|_{x=1}) + (1 - P_t(x)) \times P_t(\Phi_{\mathcal{F}}|_{x=0})$$

where $\Phi_{\mathcal{F}}$ is the implicit boolean formula of the FT, and $x \in \mathit{Var}(\phi)$.

In practice, the issue of facing a non-optimal variable ordering can induce the calculation of the exact TEP to become inefficient. The MCS-based method is therefore considered the method of choice when dealing with large or very large FTs. From the set of MCS of a FT, one can easily derive a conservative approximation of the TEP, meaning that the true probability of the TEP is at most the one computed via MCS. This permits the processing of considerably larger trees, but at the cost of a precision loss regarding the value.

Our approach to the problem is instead to take the implicit Boolean formula of a given FT, and to transform that formula into a CNF formula that can then be submitted to a WMC solver. We keep the probabilities for the basic events, and set indifferent weights for newly introduced literals, thereby ensuring that the value computed by the WMC solver will be the TEP we are looking for.

This method gives us an exact unreliability value of the FT while enabling the exploitation of recent advances made by the WMC solver community. We harvest insights from the *MCCompetition* [8] and make use of state-of-the-art solvers presented within the WMC section of the competition. It is worth noting that some the tools further restrict the *literal weight-functions* to have *normal* weights functions (adopting terminology used in [5,31]), this is $\forall x \in Var(\Phi)\colon w(x) = 1 - w(\neg x)$, but this is undesirable in our case, because we want to give to specific literals *indifferent* weights, i.e. $x \in Var(\Phi)\colon w(x) = w(\neg x) = 1$. We particularly considered the solvers ADDMC [12] and GPMC [15], since these are the ones that allowed us to set the weights in the way we required.

4 Unreliability Through WMC of Tseitin Transformation

To analyse a FT using WMC solvers, we consider its implicit boolean formula. Since WMC solvers expect its input to be boolean formulas in CNF format, we could proceed by translating the formula to its equivalent CNF and run the solver on the latter. An appropriate definition of the weight function and some straightforward calculations guarantee this to be correct (see Theorem 2 below). Nevertheless, the equivalent CNF can grow exponentially large with respect to the original formula, thus making the solution not scalable. Instead, the *Tseitin transformation* [35] delivers a linearly growing equisatisfiable CNF that comes in handy. However the models of the original formula and those of the Tseitin transformation are not the same since they run on a different set of propositional variables. This may have some impact on the weighted counting (contrarily to what happens in SAT solving). In this section we introduce the formal framework and theorems that provide the mathematics foundations of the tool.

The Tseitin transformation of a boolean formula Φ provides a formula Ψ in CNF so that Φ and Ψ are equisatisfiable, that is, Φ has a model if and only if so does Ψ, with the property that Ψ is only linearly bigger than Φ. In the following, and for simplicity, we consider binary operators but the Tseitin transformation can be generalized to conjunctions and disjunctions of arbitrary arities.

For a propositional formula Φ, let $Sub(\Phi)$ be the set of all subformulas of Φ. Notice that, in particular, $\Phi \in Sub(\Phi)$ and $Var(\Phi) \subseteq Sub(\Phi)$. Let $Sub^-(\Phi) =$

$Sub(\Phi) \setminus Var(\Phi)$ be the set of all subformulas that are not propositional variables. For each subformula $\Psi \in Sub^-(\Phi)$ that is not a variable, we introduce a fresh new variable $x_\Psi \notin Var(\Phi)$ ensuring also that these new variables are different from each other. If instead $\Psi = y$ is a propositional variable, we let $x_\Psi = y$.

Now, for every $\Psi \in Sub^-(\Phi)$ define the encoding $\mathsf{Enc}(\Psi)$ by

$$\mathsf{Enc}(\Psi) := \begin{cases} (\neg x_\Psi \vee x_{\Psi_1}) \wedge (\neg x_\Psi \vee x_{\Psi_2}) \wedge (\neg x_{\Psi_1} \vee \neg x_{\Psi_2} \vee x_\Psi) & \text{if } \Psi \equiv \Psi_1 \wedge \Psi_2 \\ (\neg x_\Psi \vee x_{\Psi_1} \vee x_{\Psi_2}) \wedge (\neg x_{\Psi_1} \vee x_\Psi) \wedge (\neg x_{\Psi_2} \vee x_\Psi) & \text{if } \Psi \equiv \Psi_1 \vee \Psi_2 \\ (\neg x_\Psi \vee \neg x_{\Psi'}) \wedge (x_{\Psi'} \vee x_\Psi) & \text{if } \Psi \equiv \neg \Psi' \end{cases}$$

The *Tseitin transformation* of Φ, denoted by $\mathsf{T}(\Phi)$, is defined by

$$\mathsf{T}(\Phi) := x_\Phi \wedge \bigwedge_{\Psi \in Sub^-(\Phi)} \mathsf{Enc}(\Psi)$$

Notice that $\mathsf{T}(\Phi)$ is a formula in CNF and that it only grows linearly with respect to the size of Φ. In addition, we have the following propositions.

Proposition 1. *1.* $\mathsf{T}(\Phi) \leftrightarrow \left(x_\Phi \wedge \bigwedge_{\Psi \in Sub^-(\Phi)} x_\Psi \leftrightarrow \Psi \right)$.
2. Φ *and* $\mathsf{T}(\Phi)$ *are equisatisfiable, that is,* $\exists I : I \models \Phi$ *iff* $\exists I' : I' \models \mathsf{T}(\Phi)$.

The first item follows by induction on the size of the subformulas and the observation that, if $\Psi \equiv \Psi_1 \wedge \Psi_2$ then $\mathsf{Enc}(\Phi)$ is equivalent to $x_\Psi \leftrightarrow (x_{\Psi_1} \wedge x_{\Psi_2})$ and similarly for disjunction and negation. The second item is a well known fact [35] which, anyway, is also a consequence of Lemmas 1 and 2 below.

In the following, we show how to calculate the unreliability value of a fault tree represented in a boolean formula Φ through an appropriate weighted model count of the Tseitin transformation $\mathsf{T}(\Phi)$. In particular, Lemmas 1 and 2 help to understand which are the models of $\mathsf{T}(\Phi)$ in terms of the models of Φ. Later, Theorems 1 and 2 prove that unreliability can be calculated using WMC.

Let $I : Var(\Phi) \to \{0, 1\}$ be an interpretation for Φ. We define function f that extends the interpretation I to an interpretation for $\mathsf{T}(\Phi)$ in such a way that guarantees equisatisfiability. Thus, $f(I) : Var(\mathsf{T}(\Phi)) \to \{0, 1\}$. Notice that $Var(\mathsf{T}(\Phi)) = Var(\Phi) \cup \{x_\Psi \mid \Psi \in Sub^-(\Phi)\}$. So, formally, f is defined by

$$f(I)(x) = \begin{cases} I(x) & \text{if } x \in Var(\Phi) \\ I(\Psi) & \text{if } x = x_\Psi \text{ and } \Psi \in Sub^-(\Phi). \end{cases} \qquad (1)$$

The next two lemmas together state that a propositional formula Φ is satisfied in the interpretation I if and only if $\mathsf{T}(\Phi)$ is satisfied in $f(I)$.

Lemma 1. *Let Φ be a propositional formula and let* $I : Var(\Phi) \to \{0, 1\}$ *be an interpretation for Φ. Then,* $I \models \Phi$ *if and only if* $f(I) \models \mathsf{T}(\Phi)$.

The previous lemma is not sufficient to guarantee equisatisfiability since there might be some interpretation I' for $\mathsf{T}(\Phi)$ which does not have the form of some $f(I)$ but yet $I' \models \mathsf{T}(\Phi)$. The next lemma states that this is not possible.

Lemma 2. *Let I': $Var(\mathsf{T}(\Phi)) \to \{0,1\}$ be an interpretation for $\mathsf{T}(\Phi)$. Then, if $I' \notin Img(f)$ (i.e., I' is not in the image of f), $I' \not\models \mathsf{T}(\Phi)$.*

As discussed before, the models for $\mathsf{T}(\Phi)$ are not the same as those for Φ because many new propositional variables are introduced. Therefore the relation between the WMC of a formula and the one of its Tseitin transformation is not direct. Nonetheless, one still can make this connection by properly manipulating the weight function as shown in the next theorem.

Theorem 1. *Let Φ be a boolean formula. Then $\mathsf{WMC}_w(\Phi) = \mathsf{WMC}_{w'}(\mathsf{T}(\Phi))$ provided*

$$w'(\ell) = \begin{cases} w(\ell) & \text{if } \ell \in \{x, \neg x\} \text{ and } x \in Var(\Phi) \\ 1 & \text{otherwise.} \end{cases} \quad (2)$$

Proof. First notice that for any interpretation $I'\colon Var(\mathsf{T}(\Phi)) \to \{0,1\}$ of $\mathsf{T}(\Phi)$

$$\begin{aligned} W_{w'}(I') &= \prod_{I'(x)=1} w'(x) \cdot \prod_{I'(x)=0} w'(\neg x) \\ &= \prod_{I'|_{Var(\Phi)}(x)=1} w(x) \cdot \prod_{I'|_{Var(\Phi)}(x)=0} w(\neg x) = W_w(I'|_{Var(\Phi)}) \end{aligned} \quad (3)$$

The first and last equality are by definition (see Def. 2) while the middle equality follows from the definition of w' in (2). Now, we can calculate

$$\begin{aligned} \mathsf{WMC}_{w'}(\mathsf{T}(\Phi)) &= \sum_{I' \in Mod(\mathsf{T}(\Phi))} W_{w'}(I') && \text{(by Def. 2)} \\ &= \sum_{I'} I'(\mathsf{T}(\Phi)) \cdot W_{w'}(I') && \text{(since } I'(\mathsf{T}(\Phi)) = 1 \text{ iff } I' \models \mathsf{T}(\Phi)) \\ &= \sum_{I' \in Img(f)} I'(\mathsf{T}(\Phi)) \cdot W_{w'}(I') && \text{(by Lemma 2)} \\ &= \sum_{I' \in Img(f)} I'|_{Var(\Phi)}(\Phi) \cdot W_w(I'|_{Var(\Phi)}) && \\ & && \text{(by Lemma 1 and observation (3))} \\ &= \sum_I I(\Phi) \cdot W_w(I) && \text{(since } f \text{ is injective)} \\ &= \mathsf{WMC}_w(\Phi) && \text{(since } I(\Phi) = 1 \text{ iff } I \models \Phi, \text{ and Def 2)} \end{aligned}$$

This concludes the proof. □

So, w' is just like w, but with the weights of fresh variables and their negations both set to 1, while for each original variable $x \in Var(\Phi)$ it holds that $1 = w(x) + w(\neg x)$. In the literature of weighted model counting [5,31], the former is known as *indifferent* weights and the latter as *normal* weights.

It remains to make the connection between unreliability and WMC. For this, we use the implicit boolean formula $\Phi_{\mathcal{F}}$ and define the weighted function w as follows. For each basic event e define $w_t(x_e) = P_t(e)$ and $w_t(\neg x_e) = 1 - P_t(e)$, where x_e is the propositional variable in $\Phi_{\mathcal{F}}$ associated to the basic event e in $\mathcal{F}$. Now we can state the following theorem.

Theorem 2. *Let $\Phi_{\mathcal{F}}$ be the implicit boolean formula of a FT $\mathcal{F}$. Then $P_t(\mathcal{F}) = \mathsf{WMC}_{w_t}(\Phi_{\mathcal{F}})$.*

Proof. The following calculations prove the theorem.

$$P_t(\mathcal{F}) = P_t(\{I \mid I \models \Phi_\mathcal{F}\}) = \sum_{I \in Mod(\Phi_\mathcal{F})} P_t(I) \tag{4}$$

$$= \sum_{I \in Mod(\Phi_\mathcal{F})} \prod_{I(x)=1} P_t(x) \cdot \prod_{I(x)=0} 1 - P_t(x) \tag{5}$$

$$= \sum_{I \in Mod(\Phi_\mathcal{F})} \prod_{I(x)=1} w_t(x) \cdot \prod_{I(x)=0} w_t(\neg x) = \mathsf{WMC}_{w_t}(\Phi_\mathcal{F}) \tag{6}$$

The first equality in (4) follows from the fact that the top event in $\mathcal{F}$ becomes true if and only if $\Phi_\mathcal{F}$ becomes true, and this happens in each interpretation I that satisfies $\Phi_\mathcal{F}$. The second equality in (4) follows from the fact that, in a discrete domain, the probability of a set equals the sum of the probability of each element (since they are independent). (5) follows by the independence of occurrence of each basic event. The first equality in (6) follows by the definition of w_t while the second one follows by Def. 2. □

Now, Theorems 1 and 2 immediately yield $P_t(\mathcal{F}) = \mathsf{WMC}_{w'_t}(\mathsf{T}(\Phi_\mathcal{F}))$ which is the fundamental result on which our tool is standing.

5 Implementation

This section gives an overview of the implementation details of our tool COYAN. We discuss the specifics of the implementation of the theory developed, and provide details regarding the implementation of a random Fault Trees generator that was developed by us to diversify our benchmarks.

COYAN is a tool entirely developed in *Rust* [22]. The tool takes as input a static FT in standard GALILEO [34] format, together with a time bound t. Different from the theoretical exposition, the format is not restricted to binary logical operators. After parsing the input, two pre-processing steps are performed, one of them is optional and consists of discarding intermediate gates with only one child, thereby reducing the final amount of gates and the final amount of CNF clauses. The other pre-processing step decomposes *VOT* gates into a equivalent representation made up with *AND* and *OR* gates. Subsequently, the Tseitin transformation is applied to each gate while keeping track of the newly added clauses and literals. Afterwards the weights are set as per Theorem 1, newly added literals obtain weights of 1, while other literals (corresponding to basic events) obtain the weight as $P_t(e)$, as stipulated by function w_t, described prior to Theorem 2, for each basic event e of the FT.

After having collected all clauses and weights, COYAN executes one of the supported WMC solvers and post-processes the result obtained. The tool provides direct support for the solvers ADDMC and GPMC, more can be added easily.

As part of our work we also implemented a flexible approach to create random Fault Trees for the purpose of extending the variety of the already existing benchmark sets. This is rooted in the observation that, putting human-made case studies aside, the available tool [24] for the generation of random FTs notoriously produce trees in which a small set of basic events have an overly significant influence in the top level event. This appears not to be representative for fault trees corresponding to real cases.

With this observation in mind, and inspired by the work of [1], we developed a small module that allows the creation of random FTs. The implemented algorithm can be configured with respect to the number of nodes, the proportion for each type of nodes, a starting gate from which nodes are added as children and a modifier for the probability of the basic event.

6 Experimentation

We now turn to our large scale empirical evaluation of the COYAN approach. We evaluate tool performance with respect to two metrics, runtime and memory consumption. In the experiments, we use COYAN with the WMC solver GPMC, because it proved to have a better score on our metrics than other WMC solvers, including ADDMC, and because it also supports more precision digits. Notably, while the theoretical underpinning can deal with exact arithmetics, the WMC solvers at hand are numerical in nature.

We compared our tool against the *state-of-the-art* publicly available software tools STORM-DFT [2] and XFTA [27].

- STORM [16] is a tool for the analysis of systems involving probabilistic phenomena. The tool supports different types of inputs, among them, DFTs in GALILEO format. Therefore, it can process static Fault Trees in their module STORM-DFT by employing the method of BDDs [2]. We experimented with version 1.8.1 of STORM-DFT and for our calculations executed it with the flags `-tm, -dft, --bdd` .
- XFTA is a calculation engine for FTs and related models. It supports different methods for processing trees, which are all well-documented in their textbook [27]. While XFTA does not provide native support for GALILEO, its user-friendly language for defining models provides enough flexibility to translate the FTs so it can be executed with XFTA. This tool provides different quantification methods to compute the TEP of a FT, among them an exact calculation, based on BBDs and Shannon Decomposition. But it also provides three distinct methods for approximating the TEP via MCS, *Rare Event Approximation*, *Mincut Upper Bound* and *Pivotal Upper Bound*. For our experiments we use version 2.0.1 of XFTA, and we experimented with the exact calculation as well as the *Pivotal Upper Bound*, which empirically performed better than the other two approximation alternatives.

6.1 Benchmarks

For our empirical evaluation, we used benchmarks from four sources.

- The benchmark collection appearing in [2]. The count of basic events/nodes ranges from 32/52 to 1567/3354 in size, and are further subdivided into the sets **Aralia, Industry, Railway,** and **Sprinkler** [23].

- The benchmark set presented in [29] ranging from 4/10 to 44/69 in basic events/nodes (**FFORT**).
- 35 Fault Trees were created using a script in the Scram tool [24] repository[1], five examples for seven different sizes each, in the spectrum from 40/103 up to 4000/10142 basic event/node counts, without VOT gates (**ScramGen**).
- Similarly, a set of FTs was generated using the method explained in Sect. 5, with essentially the same characteristics as for the item above (**CoGen**).

According to our knowledge this is the most diverse and complete set of benchmarks ever experimented with. All this was done to diversify the benchmarks with new large examples according to parameters considered meaningful in the generation. Table 1 summarizes relevant information regarding each of the sets.

Table 1. Characteristics of benchmark sets: Number of FTs, Basic Event count (Min–Max), Gate count (Min–Max) and CNF clauses resulting (Min–Max)

	Aralia	Industry	Railway	Sprinkler	FFORT	ScramGen	CoGen
#FTs	40	3	8	3	20	35	35
#BE	32 – 1567	36 – 184	27 – 54	32	6 – 44	40 – 4000	40 – 4000
#Gates	20 – 1787	21 – 67	69 – 259	35	4 – 25	63 – 6142	60 – 6000
#CNF clauses	93 – 6916	80 – 318	210 – 882	111		14 – 522	255 – 24636	322 – 34044

6.2 Results

We run all 144 models using the three tools STORM-DFT, XFTA and COYAN (with GPMC model counter) on a Intel®Core^{TM} i9 of 13th generation. We used 14 physical cores, 20 virtual cores and a maximum RAM size of 32GB. The time limit was set to 5 min for all experiments. In this context, *OOR* means *Out of Resources* and makes reference that the model was not processed in the time limit or that the model consumed all the available memory. Whenever two tools reported results, the results numerically agreed up machine precision (COYAN vs. STORM) or up to the fourth digit (XFTA). All the results can be found on the artifact[2].

Runtime. The left column of Fig. 1 shows the time needed to compute the TEP of each model, using logarithmic scales. Looking at the lower right triangle in each plot, a point below the dashed (--) line means an at least 10-fold speed-up, while below the dash-dotted line (-·) there is a speed-up by a factor of 25 at least. Analogously, on the upper left triangle, the dashed line represents a decrease by a factor of 10 and the dash-dotted line of 25.

---

[1] https://github.com/rakhimov/scram/tree/master.
[2] https://zenodo.org/records/11191730.

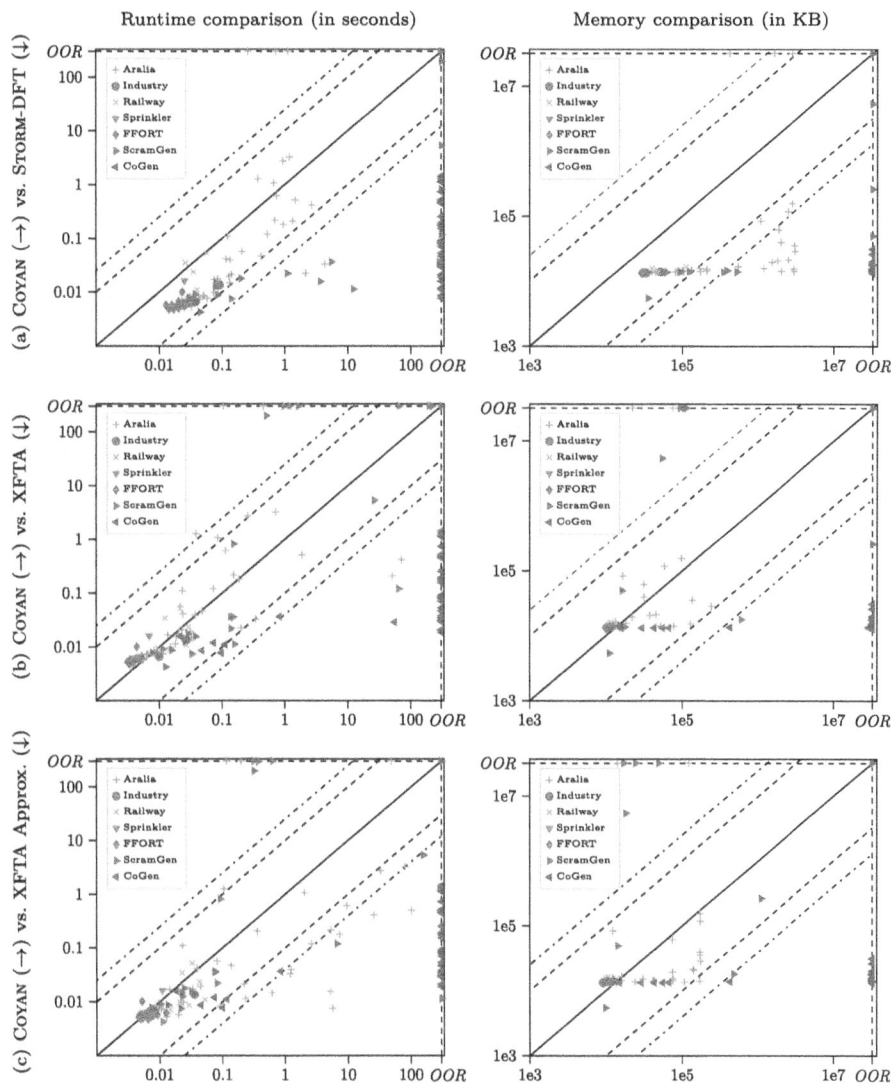

Fig. 1. Overview of experimental results. Vertical axes (↑) correspond to the outcomes of COYAN while horizontal axes (→) refer to the outcome of the tool compared with.

The first Fig. 1a presents the comparison of our tool against STORM-DFT. It shows that COYAN has a better performance in nearly all of the tested models, and a considerable amount of them by a factor close to 10. Most of the larger random models cannot be processed by STORM-DFT without running out of resources. Figures 1b and 1c compares our tool against XFTA, both in their exact computation of the TEP and in the approximate calculation. It shows that there is more parity in the obtained results in the not so large models from

FFORT, SPRINKLER, RAILWAY and INDUSTRY. On the ARALIA set, there are some models on which COYAN fails to process as fast as the exact version of XFTA, but obtains considerably better results against the approximate calculation of the TEP. On large random FTs, our tool outperforms the competitor on almost all the models that both tools can process, and there is a substantial amount of examples that cannot be analyzed by XFTA.

Memory Usage. The right column of Fig. 1 compares the memory consumption of the respective tools on each model. As before, the off-diagonal lines represent an increase/decrease in memory footprint by a factor of 10 (dashed line --), respectively of 25 (dash-dotted line -·).

Figure 1a compares our tool against STORM-DFT. The data shows that COYAN exhibits a near-constant memory consumption across most of the models, while for STORM-DFT the memory footprint increases with increasing FT size. Figures 1b and 1c compares our tool against XFTA, regarding the precise and the approximate computation of the TEP. The results shows that COYAN performs better in memory usage on the large models from the ARALIA set, while on the smaller models, the memory use is similar. It is worth to note that the larger FTs randomly created by COGEN make XFTA run out of memory.

6.3 Industrial Benchmarks

Notably, we experimented with many more randomly generated FTs, but without gaining additional insights worth reporting. Furthermore, we got in touch with the developers of the RISKSPECTRUM software, one of the leading commercial tools for probabilistic safety analysis, with our ambition being to showcase the effectiveness of COYAN on indisputably real, industrial-size cases, where no precise analysis is possible to date. RISKSPECTRUM provides professional Reliability & Availability Assessment and Probabilistic Safety Assessment, supporting among others the modelling with Fault Trees and Event Trees. It comes with an analysis tool that represents models as Minimal Cuts Sets whenever possible and can thereby efficiently derive different metrics, including the top event probability, sensitivity, time dependency and uncertainty.

We are grateful for having been granted access to an exemplary RISKSPECTRUM model that shares properties with real life models, and can thus be used to extract representative benchmarks for evaluating algorithm performance on nuclear PSA models. Much to our surprise however, there was a principled and a practical obstacle in the way that together turned out unsurmountable, except for relatively simple models. The principled problem was that *all* examples we extracted from RiskSpectrum contained negations, for reasons rooted in the combination of fault trees and event trees, discussed in [3]. Negations are outside the scope of GALILEO and no part of basic fault trees. They are however readily supported in our approach, as already hinted at in Sect. 4. The practical problem was the presence of very large disjunctions (of very large disjunctions) in many examples extracted, posing problems for the WMC instances COYAN needed to ask GPMC to solve. As a result of these two intertwined phenomena, we managed to extract only very few truly large cases from the RISKSPECTRUM models amenable to our tool chain.

One of them had 1784/4087 basic events/nodes, contained 12 negations and one large disjunction of disjunctions. The generated CNF has 8192 clauses and the precise TEP value of 0.00000003722287451304710 is computed by COYAN in 1.6 s, which is consistent with the very tight MCS approximation provided by RISKSPECTRUM of 3.73×10^{-8}. XFTA and STORM-DFT ran out of time (with a time limit of 20 min). Notably, RISKSPECTRUM can provide conservative approximations across *all* models extracted rapidly. In this respect, more work will be needed to understand the interplay of negations and large disjunctions. As a starter, we increased the average number of disjuncts in disjunctions appearing in random generated models, and found that COYAN is able to handle these models far better than both XFTA and STORM-DFT.

7 Concluding Remarks

We introduced a new method to compute the unreliability of a Fault Tree. Our method was put into practice in a new tool named COYAN which exploits advances in *Weighted Model Counting* and thus can efficiently compute the precise value for the top event probability of a Fault Tree. We empirically compared our tool against other state-of-the-art tools on the largest set of benchmarks we are aware of. The results are encouraging.

Our method is still fresh and can be further improved, for example by incorporating other WMC solvers to further enrich the capabilities of the tool. Even model counters that, instead of calculating an exact value, provide probably approximately correct (PAC) values of the model count [6], such as GANAK [32], can potentially be made use of, assuming this approach extends to weighted models.

We have also explored the use of *Projected WMC* (PWMC) which is supported by solvers such as GPMC [15]. In PWMC only a designated set of variables is considered for calculating the weight of each interpretation. This is in line with our framework since auxiliary variables in the Tseitin transformation are to be ignored. Our experiments with this alternative have however indicated that it is outperformed by our main approach of using WMC with tailored weights.

COYAN is open source and available to the public via https://github.com/NazaGara/Coyan under MIT License. All our new benchmarks will be donated to FFORT [29] in order to extend the benchmarking possibilities for Static Fault Trees. Furthermore, the CNFs generated by COYAN can serve as challenging benchmarks for the WMC community to further inspire advances in WMC tool development.

Acknowledgements. We are very grateful to Pavel Krcál and Ola Bäckström (both with RiskSpectrum AB) for enabling us to experiment with the RISKSPECTRUM software, sharing industrial examples with us, and giving us valuable insights into the tool functioning. Special thanks to Supratik Chakraborty (IIT Bombay) for sharing with us his insights regarding Model Counting, rooted in conversations at Dagstuhl Seminar 23241. This work is partially funded by DFG grant 389792660 as part of TRR 248 – CPEC (see https://perspicuous-computing.science), by the EU under MSCA grant agreement 101008233 – MISSION (see https://mission-project.eu/) and as part of STORM_SAFE (see https://www.interregnorthsea.eu/stormsafe), an Interreg project supported by the North Sea Programme of the ERDF, by the Agencia I+D+i project PICT 2022-09-00580 – CoSMoSS, and SeCyT-UNC project 33620230100384CB – MECANO, and by VolkswagenStiftung as part of grant AZ 98514 – EIS (see https://explainable-intelligent.systems).

References

1. Barrère, M., Hankin, C.: Fault tree analysis: identifying maximum probability minimal cut sets with maxsat. In: DSN 2020, pp. 53–54. IEEE (2020). https://doi.org/10.1109/DSN-S50200.2020.00029
2. Basgöze, D., Volk, M., Katoen, J., Khan, S., Stoelinga, M.: BDDs strike back - efficient analysis of static and dynamic fault trees. In: NASA Formal Methods Symposium, pp. 713–732 (2022). https://doi.org/10.1007/978-3-031-06773-0_38
3. Bäckström, O., Krćal, P.: A treatment of not logic in fault tree and event tree analysis. In: PSAM 11 (2012)
4. CCPS: Guidelines for Hazard Evaluation Procedures, 3rd Edition. Wiley-AIChE (2008), Center for Chemical Process Safety
5. Chakraborty, S., Fried, D., Meel, K.S., Vardi, M.Y.: From weighted to unweighted model counting. In: IJCAI 2015, pp. 689–695. AAAI Press (2015)
6. Chakraborty, S., Meel, K.S., Vardi, M.Y.: Approximate model counting. In: Handbook of Satisfiability - Second Edition, vol. 336, pp. 1015–1045. IOS Press (2021). https://doi.org/10.3233/FAIA201010
7. Chavira, M., Darwiche, A.: On probabilistic inference by weighted model counting. Artif. Intell. (2008). https://doi.org/10.1016/j.artint.2007.11.002
8. Model Counting Competition: https://mccompetition.org/solvers/solvers
9. Coudert, O., Madre, J.: Fault tree analysis: 10^{20} prime implicants and beyond. In: RAMS 1993, pp. 240–245 (1993). https://doi.org/10.1109/RAMS.1993.296849
10. Dilkas, P., Belle, V.: Weighted model counting with conditional weights for Bayesian networks. In: UAI 2021, pp. 386–396 (2021). https://proceedings.mlr.press/v161/dilkas21a.html
11. DIMACS: Satisfiability suggested format. https://www21.in.tum.de/~lammich/2015_SS_Seminar_SAT/resources/dimacs-cnf.pdf. Center for Discrete Mathematics and Theoretical Computer Science
12. Dudek, J.M., Phan, V., Vardi, M.Y.: ADDMC: weighted model counting with algebraic decision diagrams. In: AAAI-20, pp. 1468–1476 (Apr 2020). https://doi.org/10.1609/AAAI.V34I02.5505
13. Dudek, J.M., Phan, V.H.N., Vardi, M.Y.: DPMC: weighted model counting by dynamic programming on project-join trees. CoRR **abs/2008.08748** (2020). https://arxiv.org/abs/2008.08748
14. Dugan, J., Bavuso, S., Boyd, M.: Fault trees and sequence dependencies. In: RAMS 1990, pp. 286–293 (1990). https://doi.org/10.1109/ARMS.1990.67971

15. Hashimoto, K.: https://git.trs.css.i.nagoya-u.ac.jp/k-hasimt/GPMC. GPMC is an exact model counter for CNF formulas
16. Hensel, C., Junges, S., Katoen, J., Quatmann, T., Volk, M.: The probabilistic model checker storm. Int. J. Softw. Tools Technol. Transf. **24**(4), 589–610 (2022). https://doi.org/10.1007/S10009-021-00633-Z
17. Holtzen, S., den Broeck, G.V., Millstein, T.D.: Scaling exact inference for discrete probabilistic programs. Proc. ACM Program. Lang. **4**(OOPSLA), 1–31 (2020). https://doi.org/10.1145/3428208
18. Husung, N., Dubslaff, C., Hermanns, H., Köhl, M.A.: OxiDD: a safe, concurrent, modular, and performant decision diagram framework in Rust. In: TACAS 24 (2024). https://doi.org/10.1007/978-3-031-57256-2_13
19. Korhonen, T., Järvisalo, M.: SharpSAT-TD in Model Counting Competitions 2021-2023 (2023). https://doi.org/10.48550/arXiv.2308.15819
20. Krcál, J., Krcál, P.: Scalable analysis of fault trees with dynamic features. In: 2015 45th Annual IEEE/IFIP International Conference on Dependable Systems and Networks, pp. 89–100 (2015). https://doi.org/10.1109/DSN.2015.29
21. Lagniez, J.M., Marquis, P.: An Improved Decision-DNNF Compiler. In: IJCAI-17, pp. 667–673 (2017). https://doi.org/10.24963/ijcai.2017/93
22. Matsakis, N.D., Klock II, F.S.: The Rust language. ACM SIGAda Ada Lett. **34**, 103–104 (2014)
23. Moinuddin, K.A., Innocent, J., Keshavarz, K.: Reliability of sprinkler system in Australian shopping centres -a fault tree analysis. Fire Saf. J. **105**, 204–215 (2019). https://doi.org/10.1016/j.firesaf.2019.03.006
24. Olzhas, R.:https://doi.org/10.5281/zenodo.1146337. Scram PRA Tool
25. Rauzy, A.: New algorithms for fault trees analysis. Reliab. Eng. Syst. Saf. **40**(3), 203–211 (1993). https://doi.org/10.1016/0951-8320(93)90060-C
26. Rauzy, A.: Some disturbing facts about depth-first left-most variable ordering heuristics for binary decision diagrams. J. Risk Reliab. **222** (4), 573–582 (2008). https://doi.org/10.1243/1748006XJRR174
27. Rauzy, A.: Probabilistic Safety Analysis with XFTA. ALTARICA ASSOCIATION (2020)
28. Reay, K.A., Andrews, J.D.: A fault tree analysis strategy using binary decision diagrams. Reliab. Eng. Syst. Saf. **78**(1), 45–56 (2002). https://doi.org/10.1016/S0951-8320(02)00107-2
29. Ruijters, E., Budde, C., Chenariyan Nakhaee, M., Stoelinga, M., Bucur, D., Hiemstra, D., Schivo, S.: FFORT: a benchmark suite for fault tree analysis. In: ESREL 2019, pp. 878–885 (2019).https://doi.org/10.3850/978-981-11-2724-3_0641-cd
30. Ruijters, E., Stoelinga, M.: Fault tree analysis: a survey of the state-of-the-art in modeling, analysis and tools. Comput. Sci. Rev. **15-16**, 29–62 (2015). https://doi.org/10.1016/j.cosrev.2015.03.001
31. Sang, T., Beame, P., Kautz, H.A.: Performing Bayesian Inference by Weighted Model Counting. In: AAAI-05, pp 475–482 (2005)
32. Sharma, S., Roy, S., Soos, M., Meel, K.S.: GANAK: A scalable probabilistic exact model counter. In: Kraus, S. (ed.) IJCAI-19f, pp. 1169–1176. ijcai.org (2019). https://doi.org/10.24963/IJCAI.2019/163
33. Stamatelatos, M., Vesely, W., Dugan, J., Fragola, J., III, J.M., Railsback, J.: Fault Tree Handbook with Aerospace Applications. NASA (2002)
34. Sullivan, K., Dugan, J.B.: Galileo user's manual & design overview. https://www.cse.msu.edu/~cse870/Materials/FaultTolerant/manual-galileo.htm. v2.11 - University of Virginia

35. Tseitin, G.: On the Complexity of Derivation in Propositional Calculus, pp. 466–483. Springer (1983). https://doi.org/10.1007/978-3-642-81955-1_28
36. Valiant, L.G.: The complexity of enumeration and reliability problems. SIAM J. Comput. **8**(3), 410–421 (1979). https://doi.org/10.1137/0208032
37. Vesely, W.E., Goldberg, F.F., Roberts, N.H., Haasl, D.F.: Fault Tree Handbook. U.S, Nuclear Regulatory Commission (1981)

Safety Argumentation for Machinery Assembly Control Software

Julieth Patricia Castellanos-Ardila[✉], Sasikumar Punnekkat, Hans Hansson, and Peter Backeman

Mälardalen University, 721 23 Västerås, Sweden
{julieth.castellanos,sasikumar.punnekkat,hans.hansson,
peter.backeman}@mdu.se

Abstract. Assemblies of machinery commonly require control systems whose functionality is based on application software. In Europe, such software requires high safety integrity levels in accordance with the Machinery Directive (MD). However, identifying the essential regulatory requirements for the safety approval is not an easy task. To facilitate this job, this paper presents a process for *Safety Argumentation for Machinery Assembly Control Software* (SAMACS). We are inspired by patterns provided in the Goal Structuring Notation (GSN) and the use of contracts in safety argumentation. SAMACS contribution is aligning those methods with the MD by adopting EN ISO 13849. In particular, we define safety goals based on expected software contribution to control system safety and the standard guidance. Software safety goals are detailed into software safety requirements and expressed further as contracts, which shall be verified with prescribed techniques. We apply SAMACS to a case study from a European mining company and discuss the findings. This work aims at helping practitioners compose the safety case argumentation necessary to support machinery integration approval in Europe.

Keywords: Software Safety Case · GSN · Control systems · EN ISO 13849

1 Introduction

Machinery in Europe has to be CE-marked to be approved for operations. The CE (*"Conformité Européenne"*) is granted to machinery assemblies (i.e., machinery integrated to function as a whole [15]) if machinery and their protective functions conform to the health and safety requirements of the Machinery Directive (MD) [26]. EN ISO 13849:2023 [16] is a newly released version of the standard for safety-related parts of control systems (SRP/CS), which provides guidance that can be used to show conformance with the MD. In particular, protective measures based on control software require high safety integrity, which can be

---

This Research is supported by Vinnova via the project ESCAPE-CD, Ref: 2021-03662.

© The Author(s), under exclusive license to Springer Nature Switzerland AG 2024
A. Ceccarelli et al. (Eds.): SAFECOMP 2024, LNCS 14988, pp. 251–266, 2024.
https://doi.org/10.1007/978-3-031-68606-1_16

represented with PLs (Performance Levels). PLs limit the probability of dangerous failures of the safety functions and are defined in EN ISO 13849:2023. This standard is expected to be considered by the European Commission in the harmonization process of the Machinery Regulation [8] (to be enforced in 2027).

In general, the site integrator can demonstrate safety-related confidence levels for machinery assemblies by using safety assurance cases, i.e., an argument structure showing that the system is acceptably safe in a specific context [18]. For software, in particular, the assurance case can be constructed by arguing that software failure modes, i.e., the failures that can give rise to, or contribute to, hazards at the system level, are mitigated [20]. However, composing the software safety arguments for SRP/CS can be challenging. In particular, it is difficult to identify the mandatory aspects that must be strengthened at design time since creating convincing arguments requires skills and experience [7]. Thus, standard recommendations and mandatory requirements can be considered as a starting point to facilitate the definition of such arguments, which can be populated later with more refined arguments as the development life cycle evolves.

Following the previous reasoning, this paper presents a process for *Safety Argumentation for Machinery Assembly Control Software* (SAMACS). We took inspiration from the structural reasoning capabilities provided by GSN (Goal Structuring Notation) [25]) and the use of contract-based desing [21] in safety argumentation (e.g., [2,11]). SAMACS's particular contribution is aligning those methods with the MD requirements by adopting the guidance provided in the standard EN ISO 13849:2023. As a result, we present two levels of arguments. At the top level, we consider the definition of an argumentation structure based on expected software contribution to SRP/CS safety, i.e., providing and protecting the intended functionality. The contributions are mapped to the standard requirements to provide safety goals. At the lower level, we present supporting arguments. In particular, safety goals are detailed into software safety requirements and expressed as contracts. For safety, contracts provide a description made up of assumptions (pre-conditions) and guarantees (post-conditions) that must be ensured by executing the safety function [24]. Safety contracts facilitate verification techniques as prescribed by the standard's performance levels (PL). We also present a case study from a European mining company and discuss our findings. This work aims at helping practitioners compose the safety case argumentation necessary to support machinery integration approval in Europe.

This paper is structured as follows. Section 2 presents the essential background information required in this paper. Section 3 presents a detailed description of SAMACS. Section 4 presents an application of SAMACS to a case study. Section 5 presents a discussion of the findings. Section 6 presents related work. Finally, Sect. 7 presents conclusions and future work.

2 Background

2.1 EN ISO 13849:2023

EN ISO 13849:2023 [16] provides requirements for designing and integrating safety-related parts of control systems (SRP/CS) for machinery. It uses performance levels (PL), i.e., a level between a to e, with e being the most stringent, to specify the ability of a system to perform a safety function. In clause 7, the standard includes requirements for embedded and application software, allocated to the activities included in a V-like lifecycle model. All relevant activities, i.e., those defined as essential during the specific development, must be documented, including traceability links between those activities.

From the risk assessment at the system level, the designer decides the contribution of the SRP/CS to the risk reduction, i.e., the safety function specification. Based on this input, the software specification, which shall be expressed using the criteria in Table 1, is created and used during the software development. Verification activities are chosen according to the assigned PL. In particular, functional testing and reviews are required regardless of the PL, while extended functional testing is prescribed for PL c to d. Semi-formal methods must be used to describe data and control flow in software with PL c to e.

Table 1. Safety-related Software Specification Criteria

No.	Criteria
1	Safety Functions with required PL and associated operating modes
2	Performance criteria, e.g., reaction times.
3	Communication interfaces
4	Detection and control of hardware failure to achieve the required DC and fault reaction

2.2 Assurance Cases

Assurance is the ground for justified validity of a claim [14] (i.e., a true-false statement about the limitations on the values of a property). Assurance information is commonly collected in an assurance case, a document that presents arguments with specified confidence levels supporting the claims. Assurance cases are used in safety, where the argumentation structure is done to demonstrate that the system under consideration is acceptably safe [18]. The validity of the arguments is bound to the context in which the system will be operating, as well as specified assumptions and justifications regarding such operations. In particular, to demonstrate safety, the arguments have to be in accordance to the risk reduction expected by the system. For software, the arguments are commonly oriented to justify that its functionality does not contribute to system-level hazards [28].

There are different notations to document assurance cases. We focus on the Goal structuring Notation (GSN), which uses graphical elements (see Fig. 1) [25]. GSN argumentation starts with a top-level goal supported by a strategy that connects the goal with subgoals and solutions. Goals and strategies require assumptions and justifications in a particular context to explain why the claim in a goal is acceptable. Those elements (see Fig. 1a) are connected with two types of relationships: SupportedBy (link claims with strategies/solutions) and InContextOf (link claims/strategies with contextual information). GSN provides decorators (see Fig. 1b). For example, the hollow diamond, added to a goal, represents an undeveloped goal, i.e., a goal to which the line of argument still needs to be developed. GSN structures can also be partitioned into separate packages (see Fig. 1c), e.g., an away goal represents a claim presented in another module.

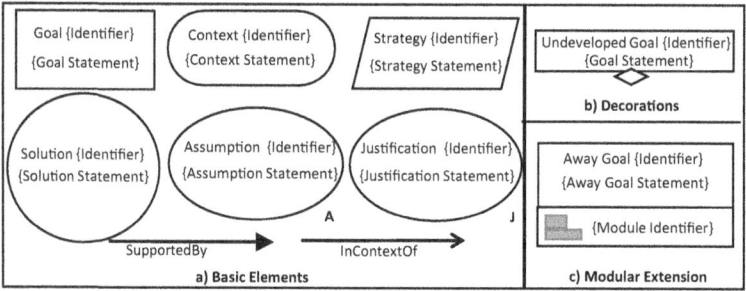

Fig. 1. GSN Elements.

2.3 Contract-Based Design

Contract-based design is an approach where correctness requirements are expressed as a contract between a method and its callers [21]. Contracts are made up of pre-conditions (that must be true before the operation call) and post-conditions (that must be ensured by the execution of the call given the pre-conditions satisfaction). In this way, contracts can ensure that the behavior of such interactions occurs as expected. Contracts have been used in safety assurance [2,11,24]. In such a context, they are called safety contracts (C). They explicitly handle a pair of properties representing the assumptions (A) on the environment (pre-conditions) and the guarantees (G) of the system under these assumptions (post-conditions). Assumptions and guarantees in a contract, which are represented as the pair $C = (A;G)$, can be used in the safety case to illustrate the agreed relationships in an argumentative way. Safety contracts support the generation of different types of evidence required to support confidence levels. In particular, each assumed safety requirement is satisfied by at least one safety contract, and each safety contract can have supporting evidence regarding consistency, completeness, and correctness regarding the represented requirements.

For example, evidence that supports contract correctness can be a report with analysis results used to derive the contract.

3 SAMACS: Safety Argumentation for Machinery Assembly Control Software

In this section, we present SAMACS, a practitioner-centric context-specific safety argumentation process targeting software in SRP/CS, which is used in the control system of machinery assemblies. In particular, software in SRP/CS needs to comply with the requirements included in the standard EN ISO 13849:2023 (see Sect. 2.1) to be in line with European regulatory frameworks, i.e., the Machinery Directive (MD). SAMACS is aimed at helping practitioners build safety assurance cases in GSN (see Sect. 2.2) supported with the definition of safety contracts (see Sect. 2.3), which facilitate the argumentation structure required for safety conformance. Figure 2 depicts an overview and the methodological steps of SAMACS, which include the elements previously mentioned to identify the information required to build a top-level and supporting arguments to demonstrate that the software in an SRP/CS is sufficiently safe.

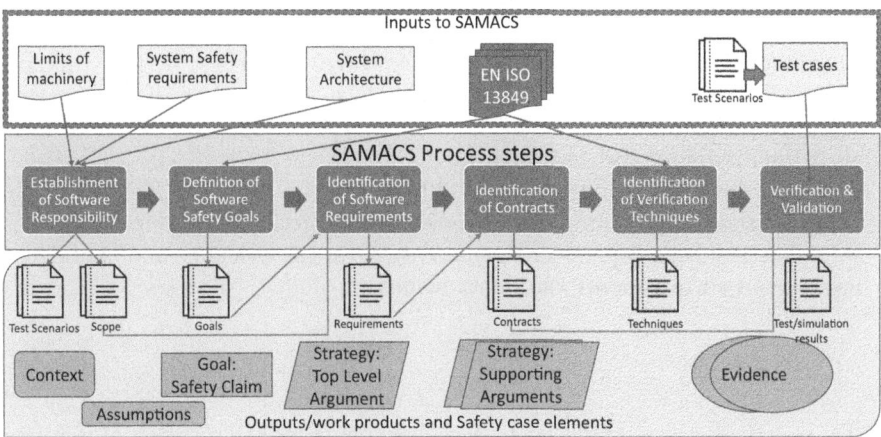

Fig. 2. SAMACS process overview

Task-1: Establishment of software responsibility

Protective measures identified at the system level are allocated to the control software via a safety specification (see Sect. 2.1). This specification, which contains the system safety requirements, the system architecture, and the limits of the machinery, is used in this task to frame the context of the top-level argument by providing the *scope* and the *test scenarios*.

Task-2: Definition of software safety goals
Software safety goals must ensure that the software does not contribute to system-level hazards. For this, the software in the SRP/CS needs to provide the intended (protective) safety functionality corresponding to the protective measurement established in the system-level risk assessment. As such functionality is essential (i.e., it shall not fail), it has to be protected from malfunctions and malpractices. Such contributions are mapped to the information requirements of EN ISO 13849:2023 (see Table 1). This information is used as the first argumentation strategy in the top-level structure.

Task-3: Identification of software requirements
Every software safety goal from Task-2 is then developed in supporting arguments. For this, each goal is detailed in terms of software safety requirements by considering the software scope resulting from Task 1.

Task-4: Definition of contracts
The software safety requirements resulting from Task-3 are expressed as contracts and are used to provide arguments regarding fulfilling such requirements by considering assumptions on the system (or the environment) and guarantees (the expected properties/functionality). As safety contracts are based on assumptions and guarantees (see Sect. 2.3), they facilitate input/output validation and the creation of error-handling specifications.

Task-5: Identification of verification techniques
Before proceeding with verification and validation, it is important to identify the right verification techniques as expected/prescribed by the standard (e.g., according to the specified PL). This is to ensure that the verification results/reports produced are in compliance with the applicable standards

Task-6: Verification and validation
Scenarios obtained in Task-1 are used to create the test cases, which are the basis for the actual validation and verification activities performed in this step. The resulting reports of this activity form the evidence and confidence levels required to support the safety claims.

4 Case Study

In this case study, we provide safety case arguments for the software of a Safety Control System (SCS) for traffic operations in an underground mine. The operations are mixed, i.e., autonomous haulers are used to transport the extracted ore, while manned-driven vehicles are used to transport personnel and materials.

4.1 Establishment of Software Responsibility

First, we collect system-level information. As depicted in Fig. 3a), the tunnel has an Autonomous Operating Zone (AOZ) (area in blue color) with entrance/exit areas where autonomous machines (shown in yellow) and manned vehicles (shown in orange) operate. Both types of vehicles have buffer areas (in gray) for waiting their turn to enter the AOZ. Manned vehicles have specific in/out areas (in orange). Meeting areas (red rectangles) and prospected drilling areas (a side tunnel ending in a dead-end room) are alongside the tunnel. We assume that the AOZ does not have human operators on foot.

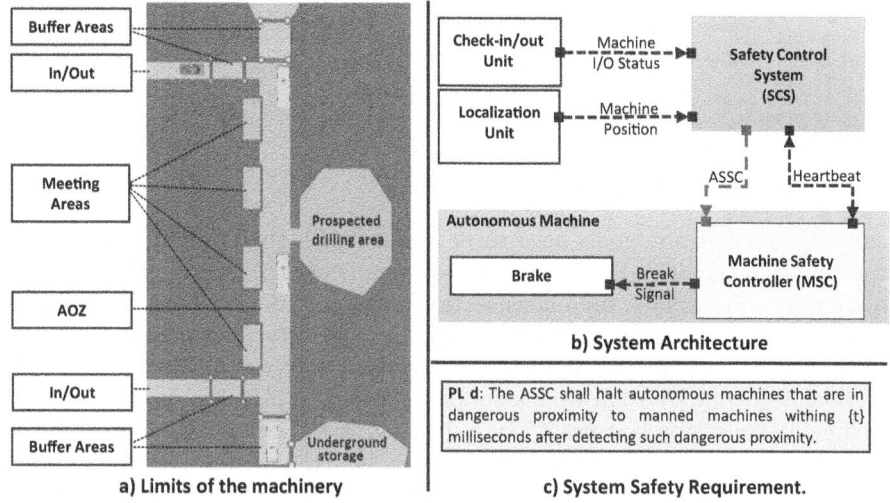

Fig. 3. Scope of the Control System.

The architecture (see Fig. 3b)) contains the system under consideration, i.e., a **Safety Control System (SCS)**, which is in charge of providing an automated safety stop command (ASSC) to meet the system safety requirement (see Fig. 3c)). The SCS receives inputs from the **localization unit**, i.e., machines position, and the **check-in/out unit**, i.e., a value indicating whether the machines enter and leave the AOZ. The SCS communicates the ASSC to the **machine safety controller (MSC)**, which converts it to a brake signal that is further sent to the **machine brakes**. The SCS and the MSC maintain a bidirectional heartbeat signal for communicating their operational availability.

Second, we determine the scope of the software function and the test scenarios. The scope is given by the system safety requirement (with PL d), which shall be allocated to the software. The limits of the machinery are used to select the test scenarios for verifying the software functionality. In our case, the control system is designed to halt autonomous machines that operate in an underground mine. So, the test scenarios can be derived systematically using the taxonomy describing the Operational Design Domain for Underground Mines (ODD-UM) provided in [5]. An excerpt of the ODD-UM related to the scenery with the case study specifications (colored in green) is depicted in Fig. 4. It says that the functionality of the SCS has to be tested in specific areas, i.e., moving and meeting, with one lane operating in both directions, i.e., up to down/down to up

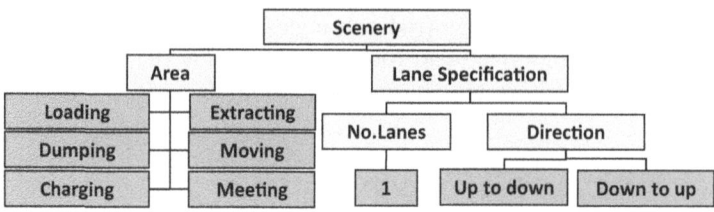

Fig. 4. ODD-UM-Scenery [5].

4.2 Definition of Software Safety Goals

Safety-related software goals have to be aligned with the mitigation strategies expected from the SCS. In general, the software is required to provide a control function (i.e., the ASSC). In that sense, the first expected contribution is that the software satisfies the intended functionality, i.e., the software provides the ASSC in the specified conditions (SC1). In addition, the software has to protect such functionality to avoid a hazardous action in mixed traffic, e.g., an autonomous machine does not stop because it is not detected or the detection parameters are out of range. Thus, the second contribution is that the software in the ASSC is protected from component malfunctions and malpractices (SC2). SC1 and SC2 can be seen in the rows colored with dark gray in Table 2.

Table 2. Software Safety Goals

ID	Goal
SC1:	The software provides the ASSC in the specified conditions.
SS1	Definition of the intended functionality (i.e., provision of the ASSC).
SG1	The intended functionality (i.e., the provision of the ASSC) is properly designed.
SG2	The intended functionality satisfies defined performance criteria.
SC2:	The functionality is protected from components malfunctions and malpractices.
SS2	Mitigation of system component's failure that contributes to software failure.
SG3	The SCS's communication interface with external components must ensure safe operation.
SG4	The software system provides detection and control of components failure.
SS3	Mitigation of systematic failure.
SG5	Relevant process requirements in compliance with EN ISO 13849:2023 have been followed.

Software contributions SC1 and SC2 are matched with strategies. In particular, the definition of the intended functionality, i.e., the provision of the ASSC (SS1), is the suited strategy for reaching SC1 since the safety control system does not have more responsibilities. Two different strategies were considered appropriate for SC2, i.e., the mitigation of system component failure that contributes to software failure (SS2) and the mitigation of systematic software failure (SS3). SS1, SS2, and SS3 are shown in the rows colored with light gray in Table 2.

Finally, each strategy is mapped to the criteria in Table 1. In particular, criteria 1 and 2 are related to the provision of the expected functionality (SS1), while criteria 3 and 4 refer to mitigation of the system's component failure (SS2). Systematic failure (SS3) can be reached by providing evidence regarding applicable process-related requirements proposed by the standard. SG1 to SG5 are shown in the white rows in Table 2.

4.3 Identification of Software Safety Requirements

Initial brainstorming is done to identify the software safety requirements in alignment with the goals defined in Table 3. In particular, for SG1, which is related to the intended functionality, two requirements have been considered, i.e., SSR1.1 (i.e., monitoring the distance between the two types of vehicles) and SSR1.2 (i.e., provision of the ASSC in case of minimum safety distance violation). SG2 is related to performance criteria, which in this case concerns the response time (i.e., SSR2.1 and SSR2.2.) and availability (SSR2.3) of the safety function. In the case of SG3, two requirements are initially defined, i.e., monitoring and diagnostics of communication failures (SSR3.1), as well as validation of integrity data before processing (SSR3.2). For SG4, it is determined that self-tests shall be performed at startup and during operations to ensure component functionality (SSR4.1). In addition, the ASSC shall be issued in case of controller failure (SSR4.2) or input devices failure (SSR4.3). SG5 can be populated with process-related aspects that result from decisions made during software development. For this case study, the decisions corresponding to verification activities (see Sect. 4.5) were relevant. In particular, the functionality is developed in UPPAAL[1] for model checking and automatically translated to software (SSR5.1) that has to be properly included in a simulation tool (SSR5.2) for further verification. Both the UPPAAL and the simulation tool shall support the investigation of the system under consideration (SSR5.3). This could mean that quality control of such tools has to be provided.

Table 3. Software Safety Requirements

SG. ID	Software Requirement
SG1	**SSR1.1:** The SCS shall monitor the distance between the autonomous machines and the manned vehicles for the specified operating zones of the AOZ
	SSR1.2: The Automated Safety Stop Command (ASSC) shall be issued if an autonomous machine violates the safety distance with respect to a manned vehicle
SG2	**SSR2.1:** The ASSC shall be computed within $\{t_1\}$ milliseconds after a violation of the safety distance is detected
	SSR2.2: The ASSC shall be send within $\{t_2\}$ milliseconds after it is computed
	SSR2.3: The ASSC shall be sent if the periodic heartbeat signal from the MSC stops
SG3	**SSR3.1:** The SCS shall include real-time monitoring and diagnostics to detect communication delays, packet loss, or data corruption
	SSR3.2: The SCS shall validate the integrity of incoming data before processing it to avoid hazards caused by corrupted data
SG4	**SSR4.1:** The SCS shall perform self-tests at startup and during operation to ensure all components are functioning properly
	SSR4.1: The ASSC shall be issued in case of controller failure
	SSR4.2: The ASSC shall be issued if control inputs from the position and check-in/out units are missing or are out of range
SG5	**SSR5.1:** UPPAAL models shall be correctly modeled and translated into software code
	SSR5.2: The software code shall be properly included in the simulation tool
	SSR5.3: Tools (UPPAAL & Simulator) shall support the investigation of the system

---

[1] https://uppaal.org/documentation/.

4.4 Definition of Contracts

Contracts consider the expected behavior described in the software requirements, i.e., the guarantee, as well as the analysis required to identify assumptions. In Table 4, we show contracts for SSR1 (i.e., SC1.1) and SSR2 (i.e., SC2.1).

Table 4. Software Safety Contracts

ID	Contract
SC1.1	**A1.1:** (AM[i].I/O-Status = IN) AND (MM[j].I/O-Status = IN);
	G1.1: implies monitoredDistance(AM[i].position,MM[j].position);
SC1.2	**A1.2:** safetyDistance = {MinimumSafetyDistance}
	G1.2: monitoredDistance ≥ SafetyDistance implies (ASSC = TRUE);

Contract SC1.1 assumes that at least one autonomous machine (AM[i]) and one manned vehicle (MM[j]) are inside the AOZ (i.e., I/O-status = IN, provided by the Check-in/out unit). Such an assumption is essential since the SCS does not react to other configurations, i.e., only autonomous machines or only manned vehicles in the AOZ. This assumption establishes a guarantee regarding monitoring such vehicles to provide the current distance between them based on the vehicle's positions (i.e., AM[i].position and MM[j].position, provided by the localization unit). Contract SC1.2 assumes a previously defined minimum safety distance to be maintained between these two types of vehicles. The guarantee is that if the minimum safety distance is violated, i.e., monitoredDistance ≥ SafetyDistance) then the ASSC is issued (ASSC = TRUE).

4.5 Identification of Verification Techniques and Evidence Provision

In our case study, evidence regarding verification shall be aligned with PL d (see Sect. 2.1). In particular, expert reviews are suitable for work products that require manual analysis. i.e., specifications. The code also required reviews. Automated techniques are suitable for software unit testing. We decided to provide a higher level of confidence considering model-checking results for the software by using the UPPAAL model checker. Model checking also allows describing data and control flow in software, which is mandatory for PL d. Simulations are considered relevant at the component and system levels, so complete functionality is probed. A simulation is one of the extended functional techniques suitable for functionality with PL d. As we provide arguments at design time, we can also check the compliance of the process plans. This aspect is out of the scope of this paper, but examples of techniques for providing such compliance checking can be seen in our previous work (see, for example, [3]).

4.6 Composing the Safety Case Arguments

The top-level argument for the safety case is presented in Fig. 5. It starts with the main goal G1, i.e., the software is sufficiently safe in a given context. Sufficiently safe is a general assumption corresponding to the expected software contributions (SC1 and SC2) to safety presented in Table 2. The context C1 corresponds to the scope of the software function identified in Fig. 3. The arguments develop over the considerations of strategies SS1 to SS3 presented in Table 2, which result in the five away goals SG1 to SG5, also presented in Table 2.

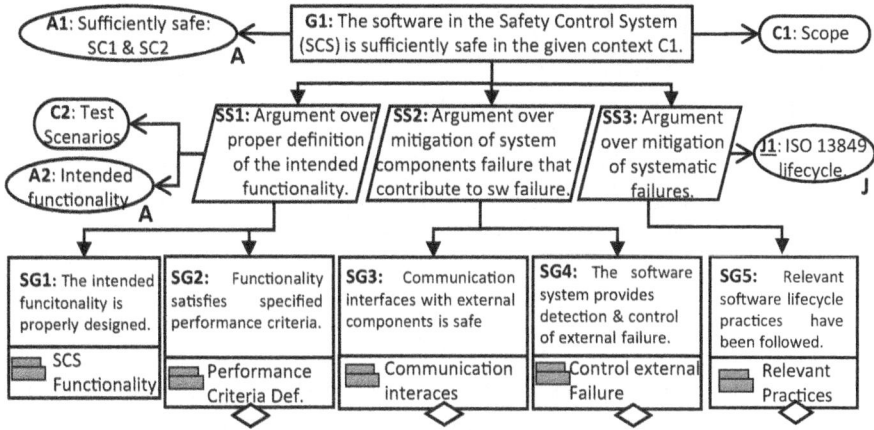

Fig. 5. Top Level Argument

Figure 6 presents a supporting argument for the away goal SG1. Three strategies are identified to support this goal. The first strategy (S1.1) is the proper definition of individual requirements for SG1, which is then developed further with two goals, i.e., SG1.1 and SG1.2, corresponding to the software requirements SSR1.1 and SSR1.2 (see Table 3). Those goals are further augmented with the contracts corresponding to each requirement i.e., SC1.1 and SC1.2 (see Table 4). The second strategy (S1.2) regards the sufficiency of requirements SSR1.1 and SSR1.2 in the implementation of the goal SG1. This argumentation branch reaches the final stages by showing two evidence elements, E1.1 (i.e., expert review) and E1.2 (i.e., simulations results). Finally, the third strategy (S1.3) is the integration of SG1 with other software goals. This branch also reaches a final stage by providing evidence E1.3 (i.e., simulation results). As E1.3 includes the whole integration, it can also be a validation test. Thus, extra evidence related to the user acceptance test results can also be added to strengthen the argument. The strategy S1.3. is the same for the away goals SG1 to SG4 developed at the top-level argument (see Fig. 5). Therefore, this strategy can instead be located in a specific branch related to integration testing at such an argumentation level.

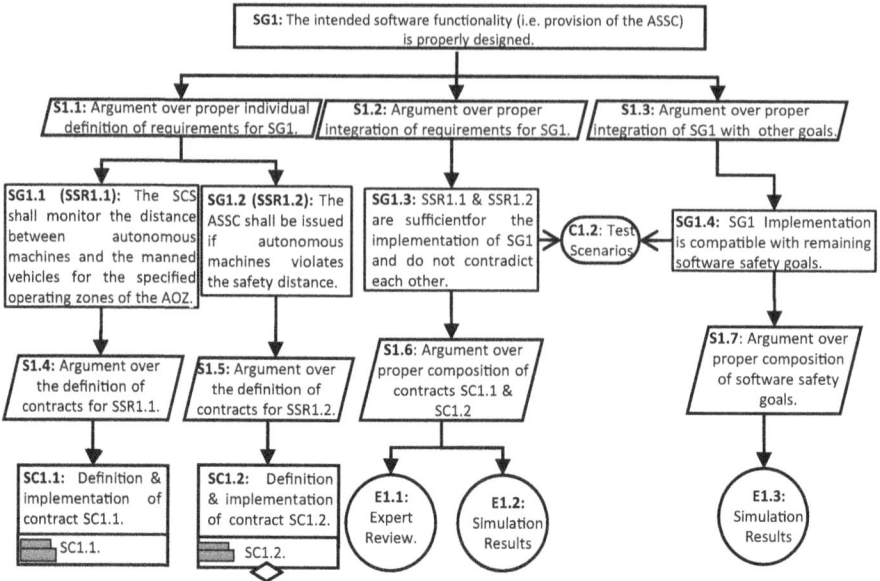

Fig. 6. Argumentation Structure of SG1.

Finally, in Fig. 7, we present the structure for the argument supporting SC1.1 (see Table 4), which corresponds to the definition and implementation of the safety contract linked to requirement SSR1.1. In this case, the argument starts by the contract assumption A1.1 (i.e., (AM[i].I/O-Status=IN) AND (MM[j].I/O-Status=IN)), which means that at least one machine of each type is inside the AOZ. Such an assumption supports the expected contract guarantee, G1.1 i.e., distance monitoring between AM and MM. This argument is extended with strategies regarding contract definition (i.e., S1.1.1), which is

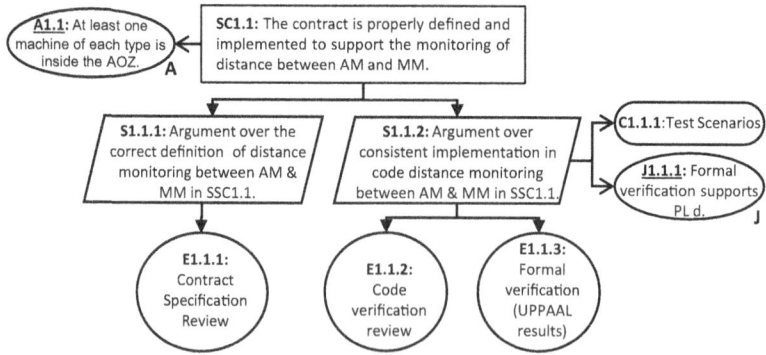

Fig. 7. Argumentation Structure of SC1.1.

manually reviewed (E1.1.1), and consistent implementation, which is also manually reviewed (E1.1.2) as well as formally verified (E1.1.3). The formal verification is done against the test scenarios (i.e., C1.1.1) and based on the justification that, according to the standard, formal verification supports PL d (i.e., J1.1.1.).

5 Discussion

It is challenging to provide a reasonable argument about the safety of software systems. The reason is that predicting quantifiable failure rates (commonly done for hardware) is difficult for software, which behaves uniquely. Thus, we need a strategy that helps us to think about what can go wrong in a specific context. This is especially important when the software function is safety-related and should not fail, as it occurs with software functions allocated to SRP/CS. The accepted practice is based on the qualified confidence given as performance levels (PLs) provided by industry standards. However, it could be challenging to select the appropriate standard, and once selected, its guidance could be difficult to interpret and use in a safety argument. In particular, the guidance for software construction in standards is often process-oriented. However, a careful view of such recommended practices can provide ideas of how to justify safety by considering the software as a product without leaving aside the required processes.

In Sect. 3, we presented a process where the guidance provided by the standard EN ISO 13849:2023 is used to reason about the software functionality. In particular, we departed from the safety-related software specification's structure as described in the standard (see Table 1) to align the safety case with the prescribed practices. From this guidance, we created the software safety goals and used them as safety case modules independent of each other. Such independence allows the creation of new arguments as the software process evolves without changing the initial safety case structure, a feature that can be also aligned with agile practices. The modules also provide readability to the software safety case since it can grow without looking excessively complex.

The structure of the resulting safety assurance case also permits practitioners to conduct brainstorming sessions regarding relevant software functionalities in the SRP/CS, as well as the implications in terms of supporting software functions to protect such functionality. For example, in our case study (see Sect. 4), we considered performance criteria related to response time and availability. However, further iterations can consider other relevant criteria, e.g., resource consumption. Cybersecurity implications also have a place in this argumentation since those practices aim at protecting software functionality. The last argumentation structure presented in Fig. 7 shows the information on the software unit that is traceable to the software safety requirements and component integration in Fig. 6 up to the claim in goal G1 in Fig. 5. In this way, we present a journey covering all the software development lifecycle steps. Thus, this process can be applied to concrete scenarios to create convincing stories that make the safety assurance argument communicable and comprehensible to all stakeholders.

6 Related Work

GSN has been used in the creation of safety argumentation patterns for software. In particular, Weaver [28] presents a framework for articulating software safety arguments based on evidence categorization. Argument patterns for COTS (Commercial-Off-The-Shelf) software are also proposed in Ye's work [29]. We have a similar view as these two approaches regarding evidence provision about the mitigation of software contribution to system-level hazards. However, these works are standards agnostic, which we consider essential in our approach. The Pegasus framework presents safety argumentation related to automated driving systems [19], which is interesting but difficult to apply in the argumentation related to SRP/CS, which is, in essence, less complicated than the required in autonomy systems (i.e., SRP/CS functionality is very reduced and punctual). Ayoub et al. [1] propose the common characteristics map with results in process-oriented evidence. It emphasizes evidence from verification activities considering the techniques, tools, human expertise, and artifacts produced. However, it does not consider the specific software description as we do in our work. General discussions regarding argument sufficiency for software safety are presented in [6,12,13] among others. However, we aim at proposing a context-specific argumentation process that can support practitioners in the machinery context.

Creating convincing arguments for demonstrating machinery safety requires experience in the field and argumentation skills. However, few published general works in this field are available for supporting practitioners. For example, in the work of Gallina et al., [9], there is a knowledge management strategy for handling process artifacts in compliance with the MD, which can be useful for creating evidence artifacts at such a level. In [10], a list of elements that can be used as evidence for creating a safety assurance case is provided. However, these two works do not present structures that support the argument development. In [27], an overall safety assurance argument in GSN is sketched. However, it lacks the level of detail required for a complete argumentation structure. In addition, it only relates to a specific operation (i.e., transportation), which is difficult to extrapolate to machinery aspects beyond such operation. The works presented in [4,17] cover more argumentation details (including the criticality levels prescribed in standards) but do not cover software at the control level, which is our main focus. Methodologies for designing SRP/CS following EN ISO 13849 are provided in [22,23]. However, none of the previous works present explicit argumentation structures covering SRP/CS software in line with the standards harmonized with the MD.

7 Conclusions and Future Work

SAMACS is a process aimed at helping practitioners compose the safety assurance case arguments required to support machinery integration approval in the European context. In particular, SAMACS supports the creation of a safety case for the software included in SRP/CS by adopting current practices in software safety argumentation, i.e., GSN structure and contracts, and the guidance provided by the standard EN ISO 13849:2023. This standard is relevant since it provides accepted guidance for conformance with the Machinery Directive and is expected to be considered for conformance with the Machinery Regulation.

Future work includes providing more case studies to assess the effectiveness of this work's methodological steps in practice. We will also present the process and the resulting safety cases to practitioners and safety assessors to evaluate their perceptions regarding the argument's comprehensiveness and level of coverage. In addition, tools for supporting our process are planned to be investigated.

References

1. Ayoub, A., Kim, B.G., Lee, I., Sokolsky, O.: A systematic approach to justifying sufficient confidence in software safety arguments. In: Ortmeier, F., Daniel, P. (eds.) SAFECOMP 2012. LNCS, vol. 7612, pp. 305–316. Springer, Heidelberg (2012). https://doi.org/10.1007/978-3-642-33678-2_26
2. Bate, I., Hawkins, R., McDermid, J.: A contract-based approach to designing safe systems. In: 8th Workshop on Safety-critical Systems and Software (2003)
3. Castellanos Ardila, J.P., Gallina, B., Governatori, G.: Compliance-aware engineering process plans: the case of space software engineering processes. In: Artificial Intelligence and Law, pp. 1–41 (2021)
4. Castellanos Ardila, J.P., Punekkat, S., Hansson, H., Grante, C.: Arguing operational safety for mixed traffic in underground mining. In: 18th Annual System of Systems Engineering Conference (2023)
5. Castellanos Ardila, J.P., Punekkat, S., Fattouh, A., Hansson, H.: A context-specific operational design domain for underground mining (ODD-UM). In: European Conference on Software Process Improvement, pp. 161–176. Springer, Heidelberg (2022). https://doi.org/10.1007/978-3-031-15559-8_12
6. Chechik, M., Salay, R., Viger, T., Kokaly, S., Rahimi, M.: Software assurance in an uncertain world. In: Hähnle, R., van der Aalst, W. (eds.) FASE 2019. LNCS, vol. 11424, pp. 3–21. Springer, Cham (2019). https://doi.org/10.1007/978-3-030-16722-6_1
7. Cheng, J., Goodrum, M., Metoyer, R., Cleland, J.: How do practitioners perceive assurance cases in safety-critical software systems? In: Workshop on Cooperative and Human Aspects of Software Engineering, pp. 57–60 (2018)
8. Europen Parliament and the Council: *Regulation (EU) 2023/1230* (2023)
9. Gallina, B., Olesen, T.Y., Parajdi, E., Aarup, M.: A knowledge management strategy for seamless compliance with the machinery regulation. In: European Conference on Software Process Improvement, pp. 220–234. Springer, Heidelberg (2023). https://doi.org/10.1007/978-3-031-42307-9_17
10. Global Mining Guidelines Group: Systems Safety for Autonomous Mining (2021)
11. Graydon, P., Bate, I.: The nature and content of safety contracts: challenges and suggestions for a way forward. In: 20th Pacific Rim International Symposium on Dependable Computing, pp. 135–144. IEEE (2014)
12. Habli, I., Hawkins, R., Kelly, T.: Software safety: relating software assurance and software integrity. Int. J. Crit. Comput.-Based Syst. **1**(4), 364–383 (2010)
13. Hawkins, R., Kelly, T.: Software safety assurance-what is sufficient? In: 4th IET International Conference on Systems Safety 2009. Incorporating the SaRS Annual Conference, pp. 1–6. IET (2009)

14. ISO/IEC JTC 1/SC 7: ISO/IEC/IEEE 15026:2019. Systems and software engineering - Systems and software assurance (2019)
15. ISO/TC 199: ISO 12100:2010. Safety of machinery - General Principles for design - Risk Assessment and Risk Reduction (2010)
16. ISO/TC 199: EN ISO 13849-1:2023. Safety of machinery - Safety-related parts of control systems - Part 1: General principles for design (2023)
17. Javed, M.A., Muram, F.U., Hansson, H., Punnekkat, S., Thane, H.: Towards dynamic safety assurance for Industry 4.0. J. Syst. Arch. (2021)
18. Kelly, T.P.: Arguing safety: a systematic approach to managing safety cases. Ph.D. thesis, University of York (1999)
19. Maus, A.: Pegasus safety argumentation (2018). https://www.pegasusprojekt.de/files/tmpl/pdf/PEGASUS%20Safety%20Argumentation.pdf
20. McDermid, J.A.: Software safety: where's the evidence? In: 6th Australian Workshop on Safety Critical Systems and Software, pp. 1–6 (2001)
21. Meyer, B.: Applying design by contract. Computer **25**(10), 40–51 (1992)
22. Porras, A., Romero, J.A.: A new methodology for facilitating the design of safety-related parts of control systems in machines according to ISO 13849:2006 standard. Reliabil. Eng. Syst. Saf. **174**, 60–70 (2018)
23. Söderberg, A., Hedberg, J., Folkesson, P., Jacobson, J.: Safety-related Machine Control Systems using standard EN ISO 13849-1 (2018)
24. Söderberg, A., Johansson, R.: Safety contract-based design of software components. In: International Symposium on Software Reliability Engineering (2013)
25. The Assurance Case Working Group (ACWG): GSN Community Standard. Version 3 (2021)
26. The Council of the European Parliament: Machinery - Directive 2006/42/EC (2006)
27. Volvo Technology AB - Advanced Technology & Research: Automated Safe and Efficient Transport System - VINNOVA Project- Ref: 2015-00612 (2015). https://www.vinnova.se/en/p/automated-safe-and-efficient-transport-system/
28. Weaver, R.A.: The safety of software: constructing and assuring arguments. Ph.D. thesis (2003)
29. Ye, F.: Justifying the use of COTS Components within safety critical applications. Ph.D. thesis, Citeseer (2005)

Sound Non-interference Analysis for C/C++

Daniel Kästner[✉], Laurent Mauborgne, Sebastian Hahn, Stephan Wilhelm, Jörg Herter, Christoph Cullmann, and Christian Ferdinand

AbsInt GmbH. Science Park 1, 66123 Saarbrücken, Germany
Kaestner@absint.com

Abstract. Homologation of vehicles in markets that are subject to U.S. regulations requires, as a part of the certification documentation, an analysis of all input and output signals that influence the control or diagnosis of any emissions-related component or system. On-board diagnostic (OBD) systems are required for every component and system that can cause increases in emissions. In this article we present a sound non-interference analysis at the C/C++ code level, that allows to automatically determine which output signals can be influenced by a given input signal, or vice versa, and which can demonstrate the independence between selected input and output signals. The analysis is based on integrating a generic taint analysis framework into a sound static runtime error analyzer that can take the impact of runtime errors on data and control flow into account. Soundness of the underlying analysis is an essential property since it provides full data and control coverage; in particular it can guarantee that all data and function pointer values have been taken into account. Our approach can be applied to signal flow analysis, to demonstrate freedom of interference between software components at the source code level, to show compliance between source code and software architecture, and to satisfy cybersecurity requirements. We will outline the underlying theoretical concepts, present our taint propagation algorithm for non-interference analysis, and report on practical experience on industry-grade example projects.

Keywords: functional safety · cybersecurity · static analysis · taint analysis · abstract interpretation · signal flow analysis · non-interference analysis · freedom of interference

1 Introduction

The homologation of vehicles in markets that are subject to U.S. regulations requires demonstrating compliance to the legislation of the U.S authorities, which includes the regulations of the California Air Resource Board (CARB). In the last years, CARB has been adopting increasingly stringent criteria for emissions standards for vehicles and engines [4]. On-board diagnostic (OBD) systems are required for every component and system that can cause increases in emissions [29]. Legislation amendments currently proposed by CARB aim at enhancing monitoring requirements and including new data tracking and reporting requirements for emission-related systems. In particular, CARB requires, as a part of the certification documentation, an analysis of all input and output signals that influence the control or diagnosis of any emissions-related component or system [31]. To achieve this goal, a deep understanding of the propagation of signals within and between the vehicle's Electronic Control Units (ECUs) is required, and for this, automatic tool support is essential.

A prerequisite is to determine the data and control flow of the software, which is a fundamental verification obligation of functional safety norms. As an example, ISO 26262 and DO-178C demand that a correct relationship exists between the components of the software architecture, and that the source code matches the data flow and control flow defined in the software architecture. Unintended data and control flow must be avoided. Furthermore, freedom of interference between software components in mixed-criticality software (cf. Sec. 7.4.9 and Annex D of [11] and other safety norms) has to be demonstrated.

In general, the data and control flow of the software can be determined by semantical static analysis. Semantics-based methods can be further grouped into unsound and sound approaches. Abstract interpretation is a formal method for *sound* semantics-based static program analysis which provides assurance that there are no false negatives with respect to the classes of defects under consideration. For data and control flow analysis the soundness of the analysis ensures that all possible variable values and all possible control paths are taken into account, including all potential targets of data and function pointers. Furthermore, runtime errors (invalid pointer accesses, data races, etc.) need to be taken into account, since they can induce errors in the control and data flow.

In this article we propose a comprehensive methodology for static non-interference analysis that can be tailored to signal flow analysis, to demonstrating freedom of interference, and to various cybersecurity properties. The methodology is based on a generic taint analysis framework which is integrated into the sound runtime error analyzer Astrée and which itself is sound with respect to the specified interference properties. Most previous work on taint analysis aims at interference analysis, i.e., interferences are detected, but the absence of interferences cannot be shown. Our main contribution is a scalable non-interference analysis which covers all control flow mechanisms of C/C++, considers concurrent multi-threaded execution and scales for industry-size software projects. The core idea of our signal flow analysis is to automatically taint all relevant input signals with unique hues, compute a sound taint propagation via data and control flow, and record all taint hues arriving at the specified output signals. Absence of the taint hue for input signal X at a particular output signal Y means independence of output signal Y from input signal X. We will outline the basic concept, illustrate design choices for capturing different levels of interferences, and report on experiments on industry-grade example projects.

2 Sound Static Source Code Analysis

The term static analysis is used to describe a variety of program analysis techniques with the common property that the results are obtained *statically*, i.e., by investigating the source code, without executing the program. Purely syntactical static methods can be applied to check syntactical coding rules as contained in all relevant coding guidelines, including MISRA C/C++ [20,24], or SEI CERT C/C++ [26]. A deeper understanding of the code such as knowledge about variable values, pointer targets, etc. requires semantical static analysis. It can be applied to check semantical coding rules which are also contained in the coding guidelines mentioned above, or to identify semantical code defects. *Unsound* analyzers may choose to reduce complexity by not taking certain program effects or certain execution scenarios into account. A *sound* analyzer is not allowed to do this; all potential program executions must be accounted for.

Abstract interpretation is a formal method for sound semantics-based static program analysis [7]. It supports formal correctness proofs: it can be proved that an analysis will terminate and that it is sound, i.e., that it computes an overapproximation of the concrete semantics. Imprecisions can occur, but it can be shown that they will always occur on the safe side. Abstract interpretation-based static analyzers provide full control and data coverage and allow conclusions to be drawn that are valid for all program runs with all inputs. Nowadays, abstract interpretation-based static analyzers that can detect stack overflows and violations of timing constraints [27] and that can prove the absence of runtime errors and data races [8,16], are widely used for developing and verifying safety-critical software [13].

Runtime Error Analysis
At the source code level, the data and control flow of a program might be accidentally affected by unintended behavior, including unspecified and undefined behaviors of the programming language. Hence, a sound analysis of data and control flow – i.e., an analysis which does not miss any potential data and control flow – must be implemented on top of a runtime error analysis designed to report undefined/unspecified behaviors of the programming language used.

In runtime error analysis, *soundness* means that the analyzer never omits to signal an error that can appear in some execution environment. If no potential error is signaled, definitely no runtime error can occur: there are no false negatives. When a *sound* analyzer does not report a division by zero in a/b, this is a proof that b can never be 0. If a potential error is reported, the analyzer cannot exclude that there is a concrete program execution triggering the error. If there is no such execution, this is a *false alarm* (false positive).

Throughout this article, we will focus on the Astrée analyzer as an example of sound static runtime error analyzer [16,23]. Astrée's main purpose is to report program defects caused by unspecified and undefined behaviors in C/C++ programs. The reported code defects include integer/floating-point division by zero, out-of-bounds array indexing, erroneous pointer manipulation and dereferencing (e.g., buffer overflows, null pointer dereferencing, dangling pointers), accesses to uninitialized variables, and further sequential programming defects. In addition, Astrée's sound thread interleaving semantics enables it to also report concurrency defects, such as data races, lock/unlock problems, and deadlocks. Hence, Astrée not only determines the data and control flow within one thread of control, but can also capture interferences between different threads and their effects on the data and control flow within those threads.

Practical experience on avionics and automotive industry applications are given in [16,19,22]. They show that industry-sized programs of millions of lines of code can be analyzed in acceptable time with high precision for runtime errors and data races.

3 Data and Control Flow Errors

Undefined or unspecified behaviors of the programming language might have an effect on data or control flow – an example is a division by 0, which can cause the program to stop with a trap, obviously causing an unexpected change in control flow. The C standard explicitly lists undefined and unspecified behaviors in Annex J [12]. Defect classes originating from undefined and unspecified behaviors that specifically affect memory safety or control flow behavior of the program can be grouped in data flow errors and control flow errors.

Relevant data flow errors are out-of-bounds array accesses, invalid pointer dereference and manipulation, invalid dynamic memory allocation, memory leaks, uninitialized variable accesses, data races, writes to constant memory, and pointer aliasing. Relevant control flow errors are non-returning functions, incompatible function calls, deadlocks, recursions, infinite loops, lock/unlock problems, C++ exceptions, and pure virtual function calls.

It is apparent that these defects may invalidate any assumptions about the data and control flow behavior of the program, and therefore, must also be considered for data and control flow analysis. Hence sound static runtime error analysis can be seen as prerequisite for data and control coupling analysis and non-interference analysis. As emphasized in Sect. 2 the defect classes are not limited to sequential program execution but also include program defects induced by concurrent thread execution. All listed defect classes are reported by Astrée ; it also provides an explicit list of the undefined and unspecified behaviors from [12] along with their coverage by the analyzer.

4 Data and Control Flow Analysis

Classical global data and control flow analysis determines the variable accesses and function invocations throughout program execution. It is centered on concepts included in the programming language, as compared to data and control *coupling* analysis that focuses on software components which are not expressed by programming language constructs. The purpose of data and control coupling analysis [25] is to determine the effective data and control flow between software components which might be desired or undesired, depending on the properties of the software architecture. Data and control flow analysis constitutes the required basis for data and control coupling analysis [15], and for the non-interference analysis presented in this article.

In its basic data and control flow analysis module, Astrée tracks accesses to global variables, static variables, and local variables whose accesses are made outside of the frame in which the local variables are defined (e.g., because their address is passed into a called function). All data and function pointers are automatically resolved. The soundness of the analysis ensures that all potential targets of data and function pointers are taken into account.

5 Taint Analysis

Taint analysis was first introduced as a dynamic analysis technique (e.g., in PERL), to try to find out at runtime which part of a code could be affected by some inputs. The original technique consisted in flipping normally unused bits, that would be copied around by operations and assignments. The same idea can be extended to static analysis by enhancing the concrete semantics of programs with tainting, the formal equivalent of the unused flipped bit in the dynamic approach. In the context of abstract interpretation, it is easy to abstract this extra information in an efficient and sound way, using dedicated abstract domains. Conceptually, taint analysis consists in discovering data dependencies using the notion of *taint propagation*. Taint propagation can be formalized using a non-standard semantics of programs, where an imaginary taint is associated to some input values. Considering a standard semantics using a successor relation between program states, and considering that a program state is a map from memory locations (variables, program counter, etc.) to values in $\mathcal{V}$, the *tainted* semantics relates tainted states, which are maps from the same memory locations to $\mathcal{V} \times \{\text{taint}, \text{notaint}\}$, and such that if we project on $\mathcal{V}$ we get the same relation as with the standard semantics.

To define what happens to the *taint* part of the tainted value, one must define a *taint policy*. The taint policy specifies:

- **Taint sources** which are a subset of input values or variables such that in any state, the values associated with that input values or variables are always tainted.
- **Taint propagation** describes how the tainting gets propagated. Typical propagation is through assignment, but more complex propagation can take more control flow into account, and may not propagate the taint through all arithmetic or pointer operations.
- **Taint alarms** specify when a tainted value must be reported. A convenient way uses expressions and the use of taint propagation inside these expressions.

In addition, it is common to have the following aspects (which could be defined by a combination of the previous elements):

- **Taint cleaning** is an alternative to taint propagation, describing all the operations that do not propagate the taint. In this case, all assignments not containing the taint cleaning will propagate the taint.

– **Taint sinks** are sets of memory locations. They specify alarm conditions (when normal taint propagation would assign tainted values to these memory locations), and amend the taint propagation, cleaning the taints of all values stored at these memory locations.

A sound taint analyzer will compute an over-approximation of the memory locations that may be mapped to a tainted value during program execution. The soundness requirement ensures that no taint alarm will be missed by the analyzer.

In our implementation for taint analysis, we implemented a first abstraction, which is to store tainting at the byte level. Once this is done, most C constructs are easy to analyze, following our already existing analyses, since Astrée already determines which bytes are read in any C expression and which bytes may be written. The only complication to consider is when trying to implement complex taint propagation, like control flow tainting. We'll discuss that in more details in Sect. 5.1. C++ input programs are processed by Astrée using the *clang* compiler frontend to transform them into Astrée's C-like intermediate representation. The taint analysis is performed on this intermediate representation. Hence, there are no C++-specific constructs that complicate the analysis.

5.1 Modeling Interference

Flow analysis can be more precisely described using the notion of interference: considering a program with variables s and d, and a set of starting environments $\mathcal{E}$, s *interferes* with d in $\mathcal{E}$ iff there are two starting environments E_1 and E_2 in $\mathcal{E}$, such that E_1 and E_2 are equal on all variables except s, and after program execution, the values of d are different.

Interference is a general notion, used in many aspects of security, such as integrity, confidentiality or compartmentation. It is a difficult notion to analyze, as one cannot prove or disprove interference by monitoring a single execution trace. It is possible to over-approximate the notion of interference using tainting, but defining taint propagation rules that are sound with respect to interference requires special care, and the knowledge (or an over-approximation) of the program execution environments. On the concrete semantics, it requires reasoning with sets of execution traces, instead of the usual one execution trace at a time.

When a program contains only direct assignments, taint must be propagated through each assignment. To be perfectly precise, one also needs to remove the tainting from the source when it can only take one value in all the starting environments. It gets a bit harder when the right-hand side of the assignments can be any arithmetic expression. At least soundness is easy: it is sufficient to taint an expression as soon as one sub-expression is tainted. Again, it is possible to be more precise by removing all sub-expressions of an expression that may only evaluate to one value, whatever the valid environments. Such optimization will allow the analysis to consider that expressions like 0 * e is not tainted even if e is tainted.

Designing a sound taint propagation policy, with respect to interference, gets more difficult when the program contains data driven control flow, such as conditionals, indirect function calls, or lazy logical operators. The easiest sound propagation rule is to taint all assignments dominated by a tainted data driven control flow. It is possible to be more precise by restricting to the dominance frontier of the branches of the control flow, but that requires some extra computation, especially in the presence of gotos. Instead, we choose a more precise approach that follows the actual reachable code for a given set of start environments: when a control flow can have only one reachable output, meaning that the outcome does not depend on the expression (e.g. the program always takes the true branch of an if), we do not add tainting to the following assignments. When there is more than one possible outcome, each outcome is enriched with a control taint environment for that control flow. Such control taint environment for a given control flow point is removed as soon as all the flows are merged (e.g. after an if-then-else). Note that this is only possible because we define taints based on sets of execution traces. Also, this precise taint propagation only makes sense for a sound analysis, where we can prove that a given branch is unreachable, and it is one of the main contributions of our approach on non-interference analysis.

Sources, Sinks and Multiple Taint Hues. To fully describe the taint policy we also need a way to specify taint sources and taint alarms. Taint sources in Astrée are described using the directive __ASTREE_taint_add. This is a bit more general than the usual taint source policy, as it may be used anywhere in the program code, allowing the tainting of variables even outside of the start environment. The semantic meaning of such directive is with respect to sub-programs starting at the point where the directive is. We consider one such sub-program for each context that may reach that point. If more than one variable is tainted, the interference semantics is the join of the interference for each individual variable and each program point. To enable alarms about taint flows, Astrée provides __ASTREE_taint_alarm directives, that evaluate expressions and raise an alarm if the evaluation of the expression is tainted. Astrée also provides __ASTREE_taint_sink directives to specify variables that act as taint sinks. Note that in that case, interference properties are only soundly approximated up to the first time the variable is assigned a tainted value and the corresponding alarm has been raised.

In addition, in order to be able to track more than one source interference at a time, Astrée allows the introduction of different hues of tainting, each hue described by a unique string. The extension of the taint semantics described above to multiple taints is straightforward. The global semantics is sound for each taint hue in isolation.

Soundness of the non-interference analysis results from the composition of sound abstractions [6]: we prove our tainting propagation policy to be sound with respect to non-interference, and then soundly over-approximate that policy with Astrée. In cases where the Astrée analysis introduces imprecision, it is possible for the end-user to recover by introducing directives to clear the tainting (directive __ASTREE_taint_remove) when the end-user is sure no interference can happen at a given point in the program.

A direct consequence of having a sound non-interference analysis is that no interference can be missed by the analysis, so if the program analysis shows no alarm, then it is proven that there is no interference from the taint sources to the taint sinks or to the expressions in __ASTREE_taint_alarm.

5.2 Signal Flow Analysis

The idea of signal flow analysis is to analyze the flow from input signals to output signals through a complex software. By identifying input signals as taint sources and output signals as taint sinks, taint analysis computes an over-approximation of the possible propagation or *flow* of an input signal. Hence, taint analysis is a good fit for signal flow analysis. The following depicts a small signal-flow example.

```
__ASTREE_taint_add((input; "INPUT"));
__ASTREE_taint_sink((output; all));
auto a = input;        // (1)
f(&b, a); // f: *b <- a (2)
if (b) ...             // (3)
output = b + ...;      // (4)
```

The input variable is tainted by hue "INPUT" where arbitrary strings allow to have a speaking signal naming. The output variable is declared as sink with all specifying that all possible signal influences are relevant for the output. Alternatively to all, a list of only a relevant subset of input signals can specified.

The taint analysis now tracks signal propagation in simple direct cases like (1), indirect flow via pointers as in (2), as well as through complex computation exemplified by (4). Whenever the output taint sink is reached by a taint, an alarm is reported of the kind:

```
ALARM (D) taint_sink: tainting 4 byte(s) at
offset 0 in variable output with hue "INPUT"
```

Optionally, the influence of signals on control flow decisions can be reported such as in (3):

```
ALARM (D) taint_sink: tainting control flow with hue "INPUT"
```

The taint directives can be included as externally-stored analysis configuration, so no actual code modifications are required.

As the taint analysis builds upon a full-fledged semantic analysis of the software under analysis, the influence of application configuration parameters on the signal flow is taken into account. Signal flow analysis can be performed for a specific application configuration, avoiding false signal flows not present in that specific configuration. Alternatively, for a generic configuration, the influence of application parameters on the output signal can be determined by adding them to the set of taint sources.

If Astrée does not report an influence of a tainted input signal on an output signal declared as taint sink, independence has been proven. If an unexpected influence is reported, Astrée helps users understand the signal flow and either improve the code, or increase the analysis precision [18].

Determining whether an input signal actually influences an output signal is an undecidable problem. The soundness of Astrée's taint analysis ensures that no possible signal flow is missed, but Astrée might report false signal flows. Signal flows in complex software can be very subtle, so reviewing the potential signal flows and determining whether it is an actual flow or a false positive can be challenging, in particular, when taints are propagated via (data or function) pointer accesses.

To support this task, Astrée provides a visualization of signal flows by means of a *taint graph*. Nodes in the graph are either variables or control flow decisions through which a taint is propagated. Edges model the actual propagation and are cross-linked with the corresponding source code location. The graph can additionally be filtered for a specific taint hue and/or a specific output signal, as well as to only show the *shortest path* from input to output.

Figure 1 shows an example graph with the linked source code. The dashed line is drawn to represent user interaction between the graph and the actual source code.

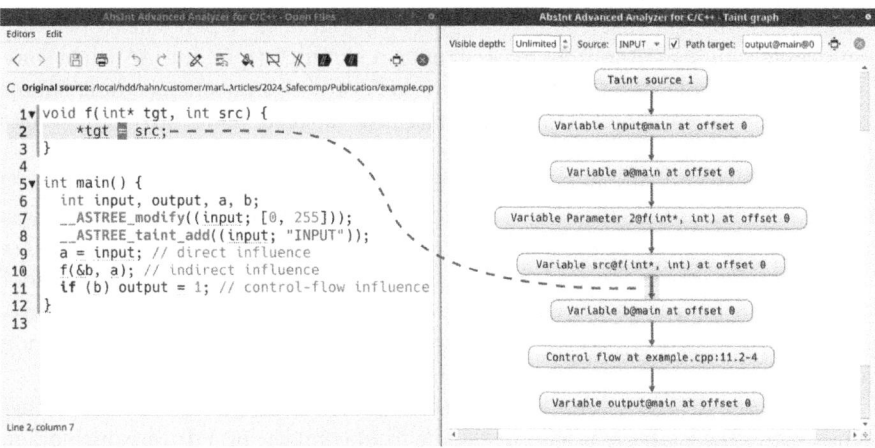

Fig. 1. Astrée's taint graph showing the shortest path from the input to the output signal.

Since the set of relevant input and output signals is application-specific, Astrée cannot automatically generate the required directives by itself. They can either be put in the source code, as shown in the examples, or specified externally to the source code as Astrée annotations (cf. [17]). In the latter case, they can also be automatically generated by a CI/CD environment from an appropriate specification of input and output signals. To confirm correctness of the tainting directives, they need to be explicitly reviewed. To facilitate this, they are listed in a compulsory section (*semantic hypotheses*) of all Astrée reports, along with other directives expressing assumptions about the environment (e.g., absolute addresses).

5.3 Freedom of Interference Between Software Components

Since there is no established understanding of the granularity of "component", a specification mechanism is needed that allows users to specify their concept of software components and the "interesting" data or control flow between them. For data and control coupling analysis, the full power of special-purpose architecture description languages (ADL) is not needed (cf. AADL [9], ArchiMate [2], SysML [30], or UML-based architecture specifications), since the analysis operates at the source code level and focuses on component interferences. A simple starting point is to define a software component as a collection of all variables and functions defined in a set of source files. This can be provided by (conceptual) annotations of source files or functions:

```
"f1.c" insert : __ASTREE_attributes((component("trusted")));
"f2.c" insert : __ASTREE_attributes((component("non-trusted")));
```

This schema can be easily extended to more complex component definitions, e.g., based on individual functions, all files in a sub-directory, etc.

Based on this software component specification, data dependencies between components can be computed by assigning a distinct taint hue to each component, and use all global and local variables of a component as taint source with the component hue. To be automatically notified about all components where these values flow, all reads in each component are declared as taint sinks for all other components hues. The required __ASTREE_taint_add and __ASTREE_taint_sink directives are generated automatically. The result is a sound overapproximation of the data coupling between the specified software components. To cover control coupling, Astrée provides an option to consider values used in guards (for conditional statements, while loops or switch statements) and in function pointer dereferences as taint sinks. No user interaction is required, except providing the component specification.

This approach also allows to account for authorized interactions in a fine-grained way. One example is that data flow from component Y to X is forbidden, unless the access is made by specific gateway functions. Such interactions can be modeled by taint-cleaning operations which remove the taint in those gateway functions (sanitization points), reducing the number of legitimate interferences that need to be examined.

In summary, the software component tainting analysis as described above detects every potential data and control interference between the components and, hence, is able to demonstrate freedom of interference between the components at the source code level. In cases where Astrée reports dependences between components, its data and control flow views help users understand the causes and, if needed, improve the code. On the other hand, the number of dependences reported provides a reliable measure for cohesion and coupling between the components.

6 Experimental Results

The taint analysis presented in the preceding sections is integrated into the sound runtime error analysis of Astrée, which makes it possible to perform the analysis without significant runtime or memory overhead, compared to the same Astrée analysis without tainting.

In the following we will compare the performance of an Astrée analysis with and without tainting on two sequential projects (*AVI* and *AUC*), and on six automotive AUTOSAR projects at the full integration level (*AUA1–AUA6*). Project *AVI* is from the avionic domain and *AUC* is a component-level automotive project. The AUTOSAR projects use a concurrent setup, i.e., , all thread interleavings between tasks and ISRs (Interrupt Service Routines) are taken into account, and concurrency defects like data races or deadlocks are reported. In *AUC* we manually added taint directives on selected global variables representing inputs. In the other projects we partitioned the code in software components and ran an automatic component tainting analysis. The results are comparable, since the same tainting mechanism is used in all these cases. Each input resp. each component is mapped to its own taint hue. Note that one Astrée analysis run determines the interferences of all modeled inputs or components at the same time. To provide additional insight into the scaling with respect to the number of taint hues in concurrent analysis projects, we created two versions of projects *AUA5* and *AUA6*, where each source file has been declared as its own component.

Table 1 shows the characteristics of the projects and gives the results obtained without and with taint analysis. Column S of Tab. 1 gives the size of the projects in millions of physical lines of code[1] after preprocessing, column N_H indicates the number of taint hues. Column T and M show the analysis time and memory consumption in their original configuration. Column T_t and M_t show the analysis time resp. memory consumption with taint analysis. The increase in analysis time and memory consumption associated with taint analysis is given in column ΔT resp. ΔM.

---

[1] I.e., comment lines and empty lines are not counted.

Table 1. Analysis results with/without taint analysis

Project	$S[MLOC]$	T	$M\ [GB]$	N_H	T_t	$M_t\ [GB]$	ΔT	ΔM
AVI	0.51	11m 30s	5.48	3129	11m 52s	5.86	3,19%	6.93 %
AUC	0.028	25s	0.63	83	25s	0.63	0%	0 %
AUA1	4.57	2h 2m	15.35	13	2h 8m	16.12	4.92%	5.02 %
AUA2	4.13	6h 3m	21.29	23	6h 44m	22.06	11.29 %	3.62 %
AUA3	5.38	4h 22m	19.53	22	4h 51m	21.07	11.07 %	7.89 %
AUA4	5.03	3h 8m	31.80	29	5h 52m	38.83	87.23 %	22.11 %
AUA5	3.24	3h 21m	37.61	5	3h 45m	39.66	11.94 %	5.45 %
AUA5'				818	6h 18m	76.52	88.06 %	103.46 %
AUA6	2.86	2h 6m	10.89	20	2h 14m	12.01	6.35 %	10.28 %
AUA6'				380	2h 16m	16.0	7.94 %	46.92 %

The results show that the overhead of taint analysis on sequential projects is very low. In spite of the large number of 3129 taint hues in project *AVI*, the increase in analysis time is 3.19% and the increase in memory consumption is 6.93%. On the component level, analysis analysis time and memory consumption are the same with and without tainting (83 hues).

On the AUTOSAR projects which consist of multiple tasks and ISRs analyzed with all potential thread interleavings, the overhead of tainting is higher since the taint hues need to be propagated between concurrent threads, which increases complexity. In most projects, the increase in analysis time is between 4.92% and 11,94%, with the exception of *AUA4* where the increase is 87.23%, and with the exception of the two configurations *AUA5'* and *AUA6'* (cf. later). The increase in memory consumption is between 3.62% and 22.11%. So in general the overhead is moderate, but depending on the software structure can also be more significant.

The configurations *AUA5'* and *AUA6'* have been created with a large number of taints. Compared to the configuration without taint analysis, analysis time increases by 88.06% and 7.94%, resp., and memory consumption by 103.46% and 46.92%.

Comparing the configuration with different numbers of taint hues, in the case of *AUA5'*, increasing the number of taint hues from 5 to 818 (factor of 163.3) causes analysis time to increase by 68% and memory consumption by 92.94%. In configuration *AUA6'* the number of taint hues is increased from 20 to 380 (factor of 19), which leads to an increase in analysis time by 1.49% and an increase in memory consumption by 33.22%. This comparison shows that both analysis time and memory consumption increase but scale well with increasing numbers of taint hues.

7 Related Work

The focus of this article is *non-interference* analysis, which is a much more challenging topic than interference analysis and can only be addressed within a sound analysis framework. In general, interference at the source code level can be caused by data and control dependencies. Dependencies in the code can also be discovered by dynamic methods and unsound analyzers. However, with dynamic methods and unsound static analysis there may be traces which are not covered, so by design, these methods cannot reason about absence of dependencies. Our approach soundly reports all potential interferences and, hence, can demonstrate freedom of interference.

The first tools to help detecting interferences were dynamic tools, mostly using tainting. Such tools, as far as we know, are not used on an industrial scale, since the code instrumentation has a performance and memory cost, and the analysis requires many runs to have a minimal coverage. The cost of instrumented runs can be mitigated with limitations on the traces (like in [5]) but this only introduces even more missed interferences. Other tools help validating existing codes, sometime based on type systems, but they require highly trained specialists and so far only analyze small codes (like [3]).

Most successful tools have been designed for cybersecurity purposes and use unsound static analysis. They usually might miss effects of indirect function calls, only support one taint hue and are rather imprecise, leading to many false interferences. We cannot cite here all dynamic or unsound approaches using tainting, as they are too many and do not fully cope with true non-interference.

From a theoretical point of view, non-interference was studied in the abstract interpretation framework by [10], leading to a sound formal definition. The link with tainting of low-level languages with gotos was proposed by [1], but they lack the set of traces based semantics. Instead, they rely on so-called execution point graphs, and the definition of non-interference depends on some initial abstraction, which leads to soundness issues, compared to the concrete notion. More recently, [28] used tainting to soundly show non-interference in the tool Julia, but their only control flow taint was function calls (which we also support). One recent approach by [21] is also based on abstract interpretation. It is parametric in the tainting policy (including the taint propagation, with hue transformations in the way we used for Spectre [14]). This allows for a fast, precise and sound analysis, but the article only deals with assignments and function calls, and does not consider any other data-controlled flow (e.g., conditionals or lazy logical operators), nor asynchronous programs.

8 Conclusion

In this article we have presented a sound non-interference analysis, created by integrating a user-configurable taint analysis framework into the sound runtime error analyzer Astrée. The analysis can be applied to perform signal flow analysis at the C/C++ code level and automatically determines which output signals can be influenced by a given input signal, or vice versa, and which can demonstrate the independence between selected input and output signals. It can also be applied to demonstrate source-level freedom of interference between software components and to analyze cybersecurity properties such as interactions between trusted and non-trusted software components. One analysis run simultaneously determines the interferences of all modeled inputs or components at the same time. The soundness of the underlying analysis is an essential property since it can guarantee that all data and function pointer values have been taken into account, such that no critical data and control flow will be missed. Experimental results demonstrate that the overhead of the tainting propagation algorithm is feasible for practical use, even on large-scale projects. The analysis can be applied to determine the dependences of input and output signals for emission control systems, which is expected to become a prerequisite for vehicle homologation in the near future.

References

1. Aldous, P., Might, M.: Static analysis of non-interference in expressive low-level languages. In: Blazy, S., Jensen, T. (eds.) SAS 2015. LNCS, vol. 9291, pp. 1–17. Springer, Heidelberg (2015). https://doi.org/10.1007/978-3-662-48288-9_1
2. The ArchiMate Enterprise Architecture Modeling Language. https://www.opengroup.org/archimate-forum/archimate-overview [2021]
3. Barthe, G., et al.: JACK — a tool for validation of security and behaviour of java applications. In: de Boer, F.S., Bonsangue, M.M., Graf, S., de Roever, W.-P. (eds.) FMCO 2006. LNCS, vol. 4709, pp. 152–174. Springer, Heidelberg (2007). https://doi.org/10.1007/978-3-540-74792-5_7
4. California Code of Regulations (CCR).: 13 CA ADC §1968.2. Malfunction and diagnostic system requirements – 2004 and subsequent model-year passenger cars, light-duty trucks, and medium-duty vehicles and engines (2022)
5. Clang 19.0.0 git documentation. DataFlowSanitizer design document. https://clang.llvm.org/docs/DataFlowSanitizerDesign.html [2024]
6. Cousot, P.: Semantic foundations of program analysis. In S. Muchnick and N. Jones, editors, Program Flow Analysis: Theory and Applications, chapter 10, pp. 303–342. Prentice-Hall (1981)
7. Cousot, P., Cousot, R.: Abstract interpretation: a unified lattice model for static analysis of programs by construction or approximation of fixpoints. In Proceedings of POPL 1977, pp. 238–252. ACM Press (1977)
8. Delmas, D., Souyris, J.: ASTRÉE: from research to industry. In: Proceedings of the 14th International Static Analysis Symposium (SAS2007), number 4634 in LNCS, pp. 437–451 (2007)
9. Feiler, P., Gluch, D., Hudak, J.: Technical Note CMU/SEI-2006-TN-011. The Architecture Analysis & Design Language (AADL): An Introduction. Technical report, Software Engineering Institute, Carnegie Mellon University (2006)
10. Giacobazzi, R., Mastroeni, I.: Abstract non-interference: parameterizing non-interference by abstract interpretation. In: Jones, N.D., Leroy, X., editors, Proceedings of the 31st ACM SIGPLAN-SIGACT Symposium on Principles of Programming Languages, POPL 2004, Venice, Italy, January 14-16, 2004, pp. 186–197. ACM (2004)

11. ISO 26262. Road vehicles – Functional safety (2018)
12. ISO/IEC JTC1/SC22/WG14 working group. ISO/IEC 9899:2018 information technology – programming languages – c. Technical Report N2310, ISO & IEC (2018)
13. Kästner, D.: Applying abstract interpretation to demonstrate functional safety. In: Boulanger, J.-L. (ed.) Formal Methods Applied to Industrial Complex Systems. ISTE/Wiley, London, UK (2014)
14. Kästner, D., Mauborgne, L., Ferdinand, C., Theiling, H.: Detecting spectre vulnerabilities by sound static analysis. In: Anne Coull, R.F., Chan, S., editor, The Fourth International Conference on Cyber-Technologies and Cyber-Systems (CYBER 2019), vol 4 of IARIA Conferences, pp. 29–37. IARIA XPS Press (2019)
15. Kästner, D., Mauborgne, L., Wilhelm, S., Mallon, C., Ferdinand, C.: Static Data and Control Coupling Analysis. In: 11th Embedded Real Time Systems European Congress (ERTS2022), Toulouse, France (2022)
16. Kästner, D., et al.: Finding All Potential Runtime Errors and Data Races in Automotive Software. In SAE World Congress 2017. SAE International (2017)
17. Kästner, D., Pohland, J.: Program analysis on evolving software. In: Roy, M. (ed.) CARS 2015 - Critical Automotive applications: Robustness & Safety. France, Paris (2015)
18. Kästner, D., Wilhelm, S., Mallon, C., Schank, S., Ferdinand, C., Mauborgne, L.: Automatic sound static analysis for integration verification of AUTOSAR software. In: WCX SAE World Congress Experience, SAE International (2023)
19. Kästner, D., et al.: Analyze This! Sound Static Analysis for Integration Verification of Large-Scale Automotive Software. In Proceedings of the SAE World Congress 2019 (SAE Technical Paper). SAE International (2019)
20. Limited, M.: MISRA C++:2008 Guidelines for the use of the C++ language in critical systems (2008)
21. Logozzo, F., Mohamed, I.: How to make taint analysis precise. In: Arceri, V., Cortesi, A., Ferrara, P., Olliaro, M. (eds) Challenges of Software Verification. Intelligent Systems Reference Library, vol 238. Springer, Singapore (2023). https://doi.org/10.1007/978-981-19-9601-6_3
22. Miné, A., Delmas, D.: Towards an industrial use of sound static analysis for the verification of concurrent embedded avionics software. In Proceedings of the 15th International Conference on Embedded Software (EMSOFT'15), pp. 65–74. IEEE CS Press (2015)
23. Miné, A., et al.: Taking static analysis to the next level: proving the absence of runtime errors and data races with astrée. In: 8th European Congress on Embedded Real Time Software and Systems (ERTS 2016), Toulouse, France (2016)
24. MISRA (Motor Industry Software Reliability Association) Working Group.: MISRA-C:2012 Guidelines for the use of the C language in critical systems. MISRA Limited (2013)
25. Radio Technical Commission for Aeronautics. RTCA DO-178C. Software Considerations in Airborne Systems and Equipment Certification (2011)
26. Software Engineering Institute SEI – CERT Division. SEI CERT C Coding Standard – Rules for Developing Safe, Reliable, and Secure Systems. Carnegie Mellon University (2016)
27. Souyris, J., Le Pavec, E., Himbert, G., Jégu, V., Borios, G., Heckmann, R.: Computing the worst case execution time of an avionics program by abstract interpretation. In: Proceedings of the 5th International Workshop on Worst-Case Execution Time (WCET) Analysis, pp.21–24 (2005)

28. Spoto, F., et al.: Static identification of injection attacks in java. ACM Trans. Program. Lang. Syst. **41**(3), 1–58 (2019)
29. State of California – Air Resources Board. Public hearing to consider the proposed revisions to the on-board diagnostic system requirements and associated enforcement provisions for passenger cars, light-duty trucks, medium-duty vehicles and engines, and heavy-duty engines. staff report: Initial statement of reasons. https://ww2.arb.ca.gov/sites/default/files/barcu/regact/2021/obd2021/isor.pdf [Retrieved: Jan 2024], 2021
30. OMG Systems Modeling Language (OMG SysML^{TM}) Version 1.6. https://www.omg.org/spec/SysML/1.6/PDF [Retrieved: Jan 2021]
31. Van Gilder, J.F.: Carb mandated obd compliance reporting update. WCX SAE World Congress Experience WCX 2023, https://www.sae-itc.com/binaries/content/assets/itc/content/hrcs/2023-wcx-carbmandatedobdsignalflowanalysisupdate.pdf [retrieved: Jan. 2024], 2023

Autonomous Systems

A Dynamic Assurance Framework for an Autonomous Survey Drone

Philippa Ryan[✉], Sepeedeh Shahbeigi, Jie Zou, Ioannis Stefanakos, and John Molloy

Department of Computer Science, University of York, Heslington YO10 5DD, UK
{philippa.ryan,sepeedeh.shahbeigi,jie.zou,ioannis.stefanakos,
john.molloy}@york.ac.uk

Abstract. Typical practice for software safety assurance requires the generation of large amounts of assurance data, which can be complex and very expensive to maintain. This assurance data is often presented in the form of an assurance case or safety case, which justifies that the software is considered acceptably safe for use in a given context. Many modern systems are also difficult to assure without being very conservative about worst case performance, particularly when using technology such as multi-core processors and GPU accelerators. This is exacerbated when they are deployed in dynamically changing environments, and means resources can be under utilised. In this paper we present a framework for dynamic risk assessment and assurance of a highly configurable autonomous unmanned drone. We show how the use of continually updated confidence metrics, combined with dialectic arguments, can support a more agile dynamic assurance case approach. We examine two example monitors in detail, and link a battery charge monitor to the assurance case using a dialectic approach to communicate the impact and meaning of confidence shortfalls. We comment on our findings and link to other related work.

Keywords: Safety · Certification · Dynamic assurance

1 Introduction

Typical practice for software safety assurance requires the generation of large amounts of static assurance evidence, which can be complex and expensive to maintain. For example, software requirements matrices for millions of lines of code, with traceability to system safety requirements and testing reports, timing analysis etc., all of which requires review and sign off. This evidence is often presented in the form of a safety case, which justifies that the software is considered acceptably safe for use in a given context. Production of the safety case is a manual, labour intensive process, and it is difficult to maintain and update. This approach, typical of certification practice in many domains, is not feasible for highly reconfigurable complex systems so new methods of assurance must be established. Dynamic assurance is one element of this, monitoring and mitigating for undesirable events and establishing confidence where there is uncertainty in the initial safety case.

It can be very hard to predict the behaviour of complex software systems (e.g., with multi-core processors). Assurance is needed of the performance (e.g., against real-time requirements), of minimum interference (e.g., where network and memory are shared), and overall behaviour (to review for emergent behaviour and ensure system safety requirements still hold). A small change in configuration can lead to unexpected effects and loss of overall safety and operational objectives. Hence, an overly pessimistic approach is taken, reducing resource utilisation and functionality with limited fixed configurations. When using a small survey drone as per our case study, battery life is also extremely limited and whilst being conservative may increase confidence in our ability to meet safety requirements (e.g., enough power to reach a designated safe location), it reduces our ability to take full advantage of the drones capability. Additionally, fixed configurations reduce our ability to adapt to failure conditions or changes in the environment. Therefore, we want to establish a method which monitors areas of known uncertainty which impact safety, and take action when they occur. Ideally, we want to link this to a safety case to assist in certification.

The concept of *dynamic assurance* [9,13] is a proposed solution. In a dynamic assurance approach run-time monitors are utilised to assess past, current and predicted performance which can be compared with assurance objectives and claims within the safety case, and provide updates to evidence supporting those claims. Additionally, we can set triggers to assess when re-configuration may be necessary, e.g., due to failure to meet safety objectives or due to them changing. In order to do this, we must be able to formalise the assurance objectives and how to monitor them. We must also make transparent any shortfall or change in assurance. (We note that not all assurance properties or artefacts may be amenable to formalisation.)

In this paper, we examine confidence and uncertainty metrics for a battery power monitor and a worst case timing prediction monitor for an autonomous survey drone example. We show how these can trigger a configuration process, which trades-off different properties. We link these to elements of an assurance case, monitoring for unacceptable levels of uncertainty and using a dialectic approach to visualise the meaning of these. We discuss how challenging this can be, for even relatively simple monitoring components and systems.

This paper is structured as follows. We first discuss related literature in this area (Sect. 2) to set the scene and present the challenges. Then we describe our case study in more detail and introduce the different monitors, and our approach for reconfiguration (Sect. 3). Then we look at the assurance case for the battery monitor, showing a software interface linking confidence data to continuously update the case (Sect. 4). Finally, we present our conclusions and future directions in (Sect. 5).

2 Related Literature

There have been many papers considering dynamic assurance cases, assurance confidence, visualisation and automatic maintenance and generation of safety cases. We describe some of the most recent, noting that this is an active area of research.

In [13] Hawkins and Ryan present a method for developing run-time monitors using dialectic arguments, based on a systematic review of the assurance case. Our paper takes those principles a step further, with worked examples of monitors and confidence metrics for an Unmanned Aerial Vehicle (UAV).

In [9] the authors describe their vision for dynamic assurance cases (using a UAV as an example) but stop short of describing specific formalisations. They note the importance of confidence in assurance data, which is also considered in our work. In [2] the authors look at dynamic assurance of autonomous systems specifically for Machine Learning (ML). They note the need for specific supporting infrastructure for monitoring, and provide a high level example. They also note the differences between dynamic and static assurance artefacts (both are required for certification).

Graydon and Holloway explore many methods of combining confidence over a case, concluding that these can deliver implausible results [10]. Bloomfield and Rushby [5] also consider aggregation and formalisation of confidence over an assurance case. They emphasise the need to consider challenges to claims in the case, as well as communicate the overall risk to reviewers. Implicit in their work is that the case is relatively static, whereas our approach tackles dynamic assurance. In our paper, we also emphasise the need for clear communication of risk, further noting the challenges of how to do this when risk is changing over time.

In [14] the authors examine dynamic assurance in the particular domain of Industry 4.0, using safety contracts to model assumption/guarantees relationships on different components, such as LiDAR. Monitors provide updates on breaches of these contracts. Their aim is to improve confidence in evidence from simulations by comparisons with operational experience. Our work adds to this by the use of the dialectic approach to clearly visualise the impact of confidence shortfalls.

In [6,8,22,26,27] a number of different approaches to auto-generation of assurance cases are presented. Fang et al. [26,27] use design artefacts such as hazard logs to generate the assurance case, and improve maintainability. They also use formal proofs of correctness to assess the validity of the case. Radu et al. [6] develop a case incrementally over time for a self adaptive system, again using formal proofs to analyse properties at run-time. The DesCert project [22] provides a database of evidence artefacts to support maintenance and generation of safety cases, but does not consider dynamic risk impact. Denney et al. [8] automatically generate arguments from small fragments, and use formal proofs of correctness. Again, they discuss the importance of visualisation and risk communication.

Finally, Koopman discusses [17] the complexity of dynamic environments for automated road vehicles. Koopman also discusses Safety Performance Indicators for autonomous vehicles in [16] and the need to link these to the assurance case. An illustration of the complexity of metrics need to monitor safety performance for pre-crash scenarios is presented by Sharath et al. in [23].

In addition to these studies focusing on dynamic assurance, numerous studies have tried to advance dynamic risk management methodologies, offering promising potential for integration with dynamic assurance cases. For example, authors in [3] propose utilizing a dynamic Fault Tree Analysis (FTA) to evaluate the dynamic risk probability of three failure scenarios in a UAV operation. In [20] authors propose a Situation-Aware Dynamic Risk Assessment (SINADRA) framework which utilizes Bayesian networks to synthesize a probabilistic run time risk monitor.

3 Case Study and Monitor Examples

To illustrate our dynamic assurance framework, we utilize an autonomous survey drone as a case study[1]. The drone is tasked with gathering information on other aerial vehicles and objects of interest and communicating this information to a remote location. To maintain focus on the assurance case and ensure conciseness, we only consider two core functional components to support the mission:

- **Battery Monitor:** This component keeps track of the current power levels, consumption of power, and helps predict future usage requirements. Additionally, a Health Monitor (HM) function reports uncertainties associated with these parameters. We describe this component in more detail in 3.1.
- **Perception Component:** This component encompasses both hardware (Camera) and software (perception tasks such as data preprocessing and an object detection algorithm such as YOLO [15]). Environmental information undergoes preprocessing before being processed by YOLO to classify objects of interest. The resultant data can be streamed to a remote station or stored for subsequent review. The performance of this component can vary across different levels.

The autonomous survey drone's mission consists of takeoff, cruise/loiter (where the survey is undertaken), and landing phases. There are two Top Mission Goals (TMG). The first is primarily about the quality of detection of objects of interest and the other is concerned with safety. In an ideal conditions we have **TMG1** and **TM2**, if necessary (e.g., due to limited battery power) we reconfigure to support a degraded mode with **TMG1'** and **TMG2'**:

- Ideal: **TMG1**: Situational awareness is maintained such that known vs unknown objects in nearby airspace can be determined within 30 s.

---

[1] The battery monitor code, reconfiguration approach and a demonstration of the dynamic assurance case are available at https://github.com/uoy-research/CODAF.

- Ideal: **TMG2**:Drone must fly safely to end point.
- Degraded: **TMG1'**: Situational awareness is maintained such that known vs unknown objects in nearby airspace can be determined within 60 s
- Degraded: **TMG2'**: Drone must fly safely to recovery point

The high level architecture of these tasks is shown in Fig. 1a for the ideal scenario (**TMG1** and **TMG2**) where there have been no hardware failures or changes to the software configuration on the drone. Each of these components have their own dedicated computer processor (with local communications). However, there may be a number of different reasons why we need to reconfigure the software, such as failure of a CPU, or too little battery power to return home if we continue processing at full image resolution, and/or have encountered heavy wind conditions. An example reconfiguration for **TMG1'** and **TMG2'** is shown in Fig. 1b, in which the perception functionality is reduced, and the battery monitor has been moved to a different computer processor.

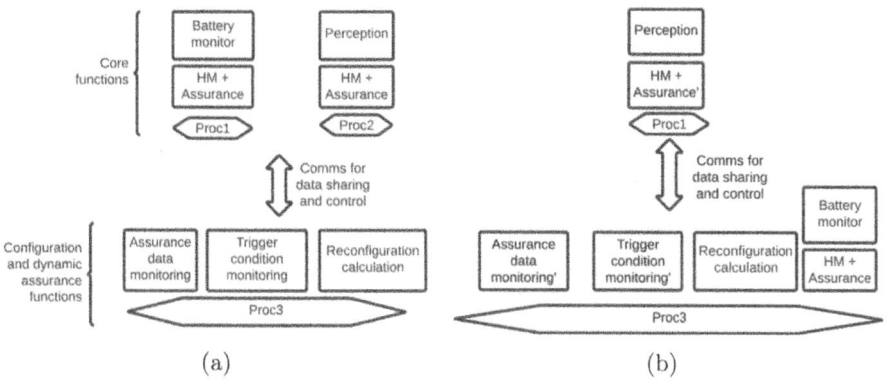

Fig. 1. Different software configurations for the drone

Also shown in Fig. 1 are a number of necessary supporting configuration and dynamic assurance components. These include Health Monitoring (HM) and assurance monitoring functions, both for individual components and also for the system as a whole. Additionally, there is a component which keeps track of whether triggering conditions have been reached which require reconfiguration.

If a triggering condition has been breached we may need to reconfigure the system. For anything other than a trivial system, we will not be able to determine all possible valid configurations prior to flight. Therefore, we need another component to calculate valid configurations dynamically, based on trade-offs of different properties including: schedulability, overall mission objectives, assurance confidence, and time taken to reconfigure. Our proposed approach for this is presented in Sect. 3.3.

3.1 Monitoring Battery State and Required Power Consumption

A critical aspect affecting both safety and capability to fulfill the mission goal involves the real-time monitoring of available power through State-of-Charge (SoC) estimation of the battery. The SoC is defined as the ratio of the available capacity at time t, $Q(t)$, and the maximum possible charge that can be stored in a battery, i.e., the rated capacity Q_{rated}. Manufacturers provide the rated capacity value for each battery, representing the maximum charge the battery can hold. SoC at time t depends on the power consumption of the drone components from the beginning of battery operation t_0 until time t. SoC based on capacity and power consumption is formulated as:

$$\text{SoC}(t) = \frac{Q(t)}{Q_{\text{rated}}} = \text{SoC}(t_0) - \frac{1}{Q_{\text{rated}}} \int_{t_0}^{t} P(t') \, dt' \qquad (1)$$

where:

- $\text{SoC}(t)$ is the State of Charge at time t,
- $Q(t)$ is the available capacity at time t,
- Q_{rated} is the rated capacity provided by the manufacturer,
- $\text{SoC}(t_0)$ is the initial SoC at time t_0,
- $\frac{1}{Q_{rated}} \int_{t_0}^{t} P(t') \, dt'$ is the power consumption from time t_0 to t normalized by Q_{rated}.

Addressing the formidable challenges posed by the high nonlinearity and time-varying characteristics of batteries, estimating the SoC has proven to be a complex task. Acknowledging the inherent uncertainty in estimated values becomes paramount in facilitating well-informed decision-making for the successful fulfilment of mission requirements. Consequently, it is imperative for the assurance case, that the HM report the confidence associated with the estimated SoC. Over the decades, extensive research has been dedicated to developing methods that enhance the accuracy of estimating this critical parameter. Adaptive filters, known for their robustness and ability to correct errors in measurements, are commonly used as the estimation method. These methods involve predict and update steps. The predict step anticipates the future state of the parameter using the battery model and input parameters, while the update step calculates a filter gain using the measurements (battery's current, voltage, and temperature readings) to refine the predicted state. The updated state is then fed back to the predict step as the initial state. Further to the state, these filters also update uncertainty measures around the estimates in this recursive process.

Another critical parameter affecting the reconfiguration and assurance case, is the required power to complete the mission (and return home) with the current configuration and its associated power consumption. In this study, we assume that the mission consists of fixed, known waypoints. Therefore, by averaging the current power consumption per second, and estimating the remaining time of the mission it is possible to predict whether there is enough capacity to carry the drone to the end of mission. If it is defined that the current capacity is insufficient a change in configuration (e.g. camera downgraded mode) request is triggered.

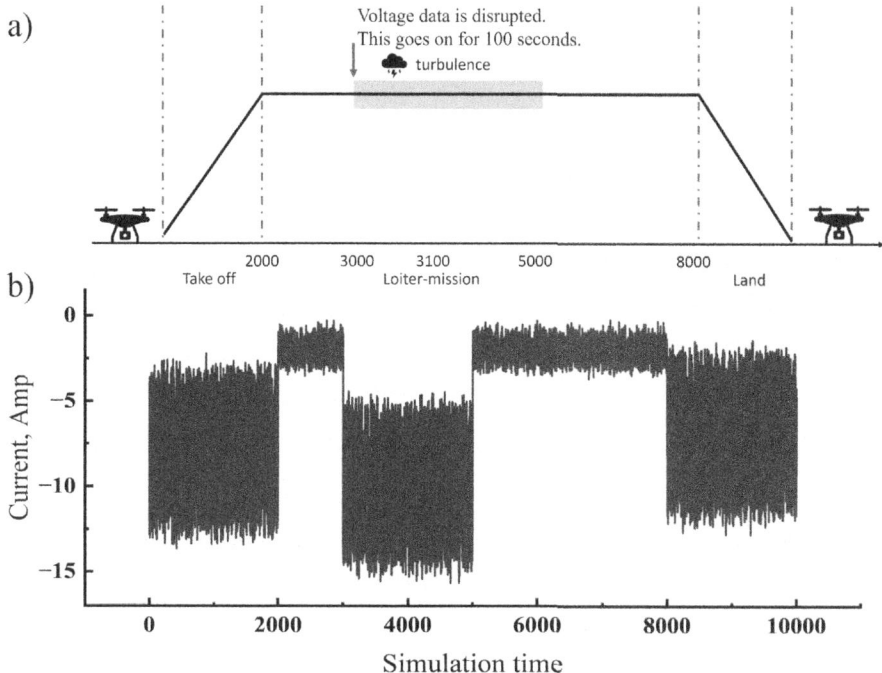

Fig. 2. Mission description: a) Various phases of the mission including when there is turbulence and voltage reading disruption, b) Current draw for each phase.

The battery used in this case study is a simplified circuit model consisting of 6 lithium-ion batteries connected in series. Each cell has a voltage of 3.7 v resulting in a pack voltage of approximately 22 v. The total battery pack capacity is around 30000 mAh, facilitating an extended mission duration. The heat transfer between battery cells is not considered in this study; However, the pack dissipates heat by transferring it to the environment. The environment temperature is assumed to be constant at $10°C$. The thermal mass of the pack is 2100 J/k.

For SoC estimation, an adaptive Unscented Kalman Filter (UKF) [24] is employed. The adaptive UKF allows for increased flexibility in accommodating slight differences between the actual and modeled system (battery), and it is generally more stable compared to conventional UKFs. The drone starts the mission with SoC=0.75. The operation and power consumption variation throughout various mission phases is demonstrated in Fig. 2. At some point in the cruise phase, voltage readings become unavailable as soon as a turbulence state is occurred. This lasts for 100 s and during this time, the UKF uses the last available voltage reading and continues the estimation. However, this causes an overestimation in the SoC until the end of the mission.

Further to the disruption in voltage readings and the turbulence, there are other reasons that cause the change in the uncertainty in the SoC estimation. For

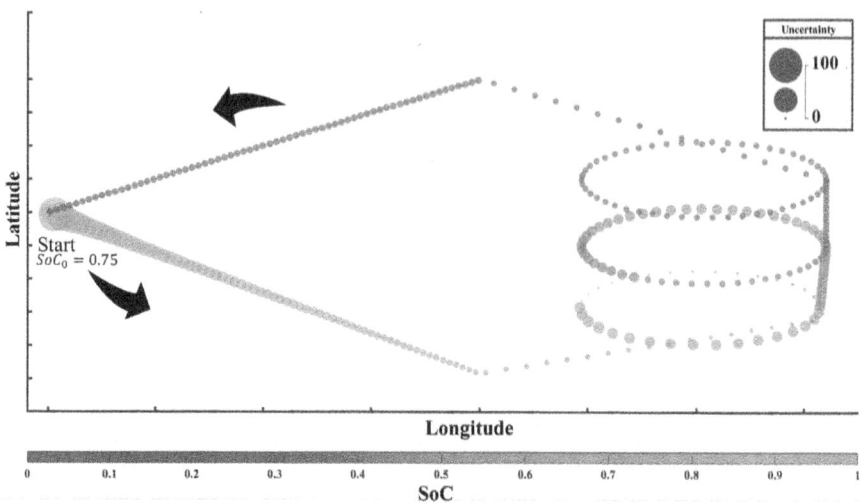

Fig. 3. Changes in battery State-of-Charge (SoC) and its uncertainty during operation. Each circle color shows the remaining SoC and its size shows the uncertainty around it.

example, at the start of the mission the UKF is in a transient phase and there is very high uncertainty in its estimate. Additionally, changes in the operation phase (e.g. starting to land) causes an increase in the uncertainty in the SoC estimate. Thus, not only does the SoC value play a pivotal role in risk assessment, but also the uncertainty surrounding its value holds equal significance. For the case study in this work, the change in SoC and its uncertainty is shown in Fig. 3.

3.2 Monitoring and Predicting Task Scheduling and Executing Times

In this section we describe our monitor for task scheduling and execution times. This provides a contrasting example of a monitor, to help illustrate the variety required, and different ways in which assurance confidence can change over time.

When using advanced multi-core heterogeneous architectures anticipating the impact of collisions among concurrently executing tasks on the execution time is difficult. Software tasks may share the same hardware resource (e.g., CPU core or caches, etc.) and attempt to access it simultaneously. In contrast to conventional single-core platforms, the complexity of heterogeneous architectures in advanced hardware, means constructing highly accurate models is impossible due to limited accessible information [1] (with some components working as "black boxes," e.g., Nvidia GPU [25]). Coupled with substantial workloads (over millions of lines of code), conventional static methods are impractical for timing analysis, as they depend on precise system models [7,11]. Therefore, in this work, statistical analysis techniques are employed to understand the combinations of resources that introduce variabilities in execution times.

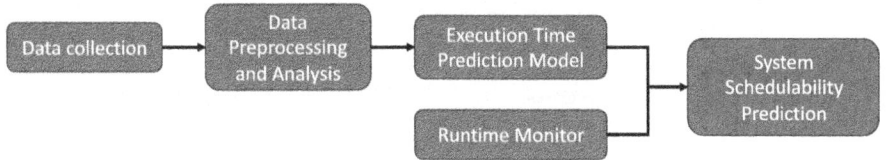

Fig. 4. Timing monitor workflow

To realize the survey mission, tasks in different configuration modes are identified during the design phase. Predicting their execution times is based on understanding their characteristics and the interference among them. Tasks can be classified into various types (e.g., CPU-bound tasks, Streaming Data tasks, etc.), each exhibiting unique characteristics that impact different system components and introduce varying levels of interference with concurrently executed tasks [21]. For instance, Streaming Data Workloads often exert high pressure on the Last Level Cache (LLC), while CPU-Bound Workloads lead to high CPU utilization.

The interference between different types of tasks can result in varying levels of execution time increases. To construct an accurate execution time prediction model, it is essential to collect system performance data during both the individual execution of tasks, to profile their characteristics, and their execution alongside other tasks to understand the sources of interference and the differences in impact levels.

In this work, we use relatively simple system performance data (e.g., CPU utilization (U_{cpu}), memory consumption (M), and power consumption (P)) to provide a relatively simplified execution time prediction, which can be further refined with more detailed data (e.g., L1 Cache misses, Instructions, etc.). For example, an increase in CPU utilization of the core (ΔU_{cpu}) where the analyzed task (τ) is deployed can imply that the execution time of the task will increase. If memory consumption also increases (ΔM), the impact level could be severe. Based on the collected data, we can formulate an equation to describe the relationship between the severity of the execution time increase ($\Delta Exe(\tau)$) and the monitored system performance data, as shown below. Where a, b, c and d are coefficients.

$$\Delta Exe(\tau) = a \times \Delta U_{cpu}^2 + b \times \Delta P^2 + c \times \Delta M + d \qquad (2)$$

At runtime, it is essential to monitor the CPU utilization, power, and memory consumption. By applying the fitted equation, the execution times of tasks can be predicted. However, considering the risks associated with underestimating execution times, we incorporate a 10% redundancy. This approach could be significantly less pessimistic than traditional worst-case execution time estimation methods. Using these adjusted execution times, we can assess the system's schedulability. The detailed timing analysis procedure is beyond the scope of this paper.

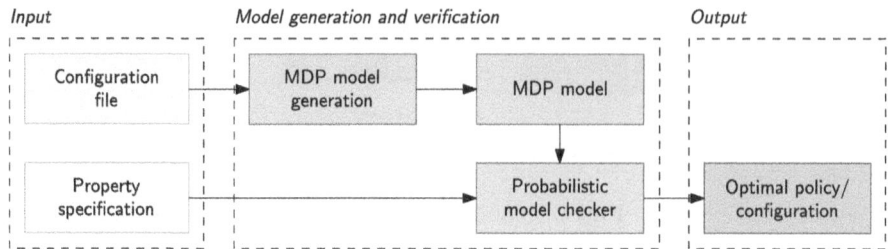

Fig. 5. High-level diagram of the approach for verification of (re)configurations

The monitor workflow is summarised in Fig. 4. Input data is gathered during and prior to operation, including measurements of execution time, CPU utilisation, memory consumption, and power consumption (*Data Collecion*). Output is the likelihood of the system being schedulable and confidence in the prediction results (*System Schedulability Prediction*). Data pre-processing requires an understanding of task interaction for the monitored data. The interaction impacts on resource contention and when this changes it impacts our confidence in our prediction. Interaction is examined through pairwise comparison of the collected data types listed in the previous paragraph. This helps us target which elements to monitor.

Our execution time prediction model is based on the strongly correlated data. Our confidence in that prediction uses measured vs original model data. Where our prediction is close to the measured timing values our confidence is high and vice versa. Our model can be improved over multiple missions as well as on the current one as we gather more and more measurements and combinations of task interactions. The timing predictions provide evidence for our safety case. As our confidence increases or decreases in the predictions, our understanding of how well the case is supported increases.

3.3 Reconfigurations Calculation

Finally, we describe briefly how the reconfigurations are calculated. As seen in the high-level diagram of Fig. 5, we enable the selection of optimal system reconfigurations through the application of formal modelling and verification [18]. The verification process starts with an input *configuration file* that includes system parameters such as the number of available configurations and their associated properties, i.e., execution time, battery consumption and assurance confidence level. The second input file (*property specification*) contains the mission requirements formally expressed in Probabilistic Computation Tree Logic (PCTL) [12].

Using the information from the configuration file, we synthesise a Markov Decision Process (MDP) model representing the system and the earlier specified configuration options. The MDP model is defined in the high-level modelling language of the probabilistic model checker PRISM [19]. The synthesised MDP model, along with the expressed in PCTL property file, are then supplied as input

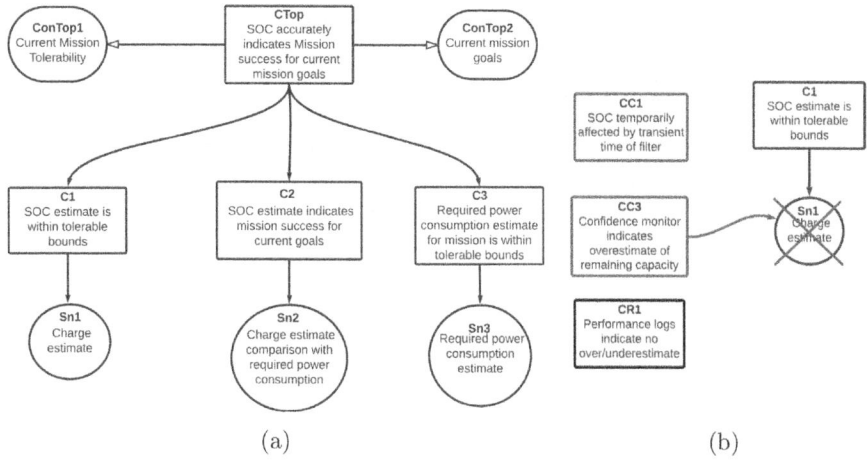

Fig. 6. Simplified argument structure and example challenges: a) battery monitor argument, b) example challenges to Sn1.

into PRISM to verify whether the model satisfies these properties. Based on the outcome of the verification, an optimal policy/configuration is determined, i.e., if there is a need for system reconfiguration and which of the available options satisfy the mission requirements, while at the same time maintaining a trade-off between execution time, battery consumption, etc.

4 Dynamic Assurance Case Approach

In this section, we present how we combine the monitors described in Sect. 3.

4.1 Battery Monitor Assurance Case

We have used the Goal Structuring Notation (GSN) with dialectic argument extensions to present and develop our safety case [4,13]. A simplified version of the argument structure for the battery monitor is shown in Fig. 6a. Our top claim (*CTop*) is that the SOC accurately indicates mission success for the current mission goals (these are defined in *ConTop2*, e.g. **TMG1** and **TMG2**). These goals may change during the flight, in which case this impacts on the sub-goals and evidence. Additional context is in *ConTop1*, and this refers to tolerability. For example, we may set a percentage limit on how precise our charge estimates need to be, such as at least 5% more power than estimated should be available.

Our next level of claims then refer to the current SOC estimate (*C1, Sn1*), the estimated required power consumption (*C3,Sn3*), and a simple comparison of these (*C2,Sn2*).

There are a number of different challenges (red claims) which undermine our confidence in the evidence *Sn1*, some of which are shown in Fig. 6b. At the start of each flight, we know that our estimate may not be accurate (*CC1*), and following a failure in the voltage sensor we know that there may be an overestimate (see Sect. 3.1) of remaining charge. We may have a rebuttal of these challenges, based on our understanding of the performance, either from this mission or earlier ones. A key issue to note is that although our confidence may be low, this doesn't mean that we should immediately trigger a reconfiguration of the system. If we are aware that we are in a known state of uncertainty which will correct itself, then we should also monitor the trend in uncertainty for a set period of time in order to see if the situation recovers (*CR1*). The combination of confidences of various evidence artifacts affects the goals in higher levels. However, these effects are very complex and application-specific and beyond the scope of the current research.

4.2 Discussion

Each assurance case fragment has different confidence metrics associated with claims, challenges, rebuttals and evidence. For example, for the battery monitor we provide a measure of *uncertainty* around the SoC estimate, which when high indicates low confidence. For this purpose, we needed to develop a detailed failure model for the battery, its associated sensors and SoC estimation method. Some of the failures are expressed as challenges(the red claims in Fig. 6b) for our case. However, some of the failures have limited impact on the SoC estimation (or risk) so are not needed within the case.

We developed a similar assurance fragment for the timing monitor in 3.2. Its challenges, and how they change in validity over the period of the mission, are different in nature to the battery monitor. They relate to interactions between tasks, which are much harder to predict for multiple tasks. For example, contention for a particular computing resource will increase when there are multiple tasks with similar priorities. In the fragment, associated confidence metrics were based on common statistical measures of standard error and confidence intervals[2].

Another issue is understanding the relative importance of each challenge, in context, which we propose should be related to risk. We can assume that ultimately a risk holder will need to be able to make decisions about, for example, continuing the mission or ensuring that if the drone loses power it lands on an unpopulated area. The risk holder may not be expected to understand the significance of low, unclear or high confidence metrics in isolation. For this reason we argue that the dialectic approach is a valuable method for providing clarity on importance. For example, we may have low confidence in the SOC due to an overestimate or underestimate. The impact of these is quite different - an overestimate may increase risk that we cannot achieve **TMG2**, thus we fail to return home and the drone crashes. An underestimate may mean we move to the degraded mode with **TMG1'** unnecessarily, and do not identify another object in time to prevent collision. On the other hand, an underestimate in the

---

[2] We note that the appropriateness of the confidence metrics must also be justified on a case by case basis.

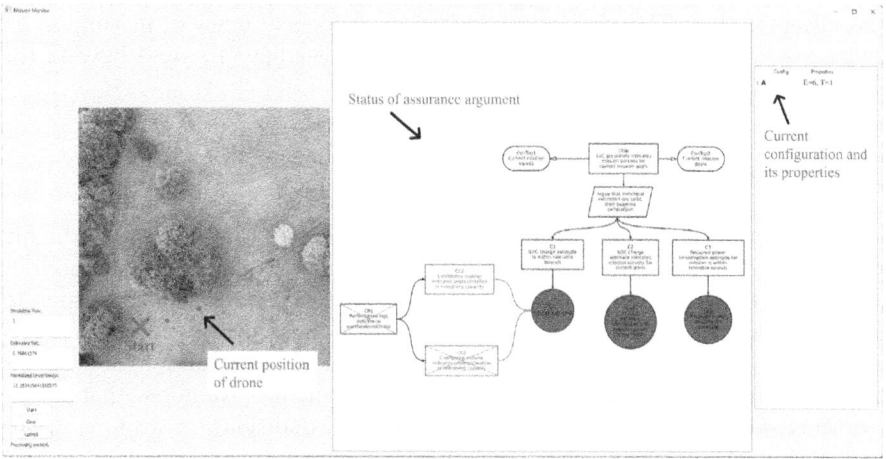

Fig. 7. Mission monitor display with its components, from left to right: time and battery estimate information, assurance argument, and reconfiguration data.

processing resource needed to run tasks may mean the tasks overrun and **TMG1** cannot be maintained. The severity and immediacy of the risk will impact how the system or risk holder should respond to confidence shortfalls.

We discovered that, even for this small example, we need to consider relationships between the confidence measures, and again this can be very context specific. For example, confidence in $Sn2$ (charge comparison) is highly dependent on that for $Sn1$ and $Sn3$. For the timing monitor, confidence in predictions is interrelated by the pairwise comparisons of different performance values. We believe that the scalability of this approach in practice will be difficult, and monitors and relationships must be considered from the earliest stages of design.

4.3 Online Monitoring Interface

Based on our understanding of this system, and need for contextual understanding of risk, we have developed an initial user display demonstrated for an instance of mission in Fig. 7.

In this instant of the mission, SoC value is estimated to be enough to complete the mission. Furthermore, there is high confidence around the SoC and required power consumption (i.e. the predicted power required to complete the mission). Therefore, the solutions $Sn1$, $Sn2$, and $Sn3$ are in dark green colors. The very low confidence around counter claim 3 ($CC3$) means it is defeated and is shown in the interface as crossed out. However, $CC2$'s low confidence has not passed the threshold and it is only shown in red to reflect this. Due to this information, performance logs claim is not considered trustworthy enough to defeat the related counter claims anymore and they are crossed out. Throughout the mission, uncertainties around predicted values are reflected in the assurance argument component of the interface to enable real-time risk decision-making.

The effect of these changes in confidence on evidence artifacts in support of higher level goals is very complex, potentially misleading [10], and beyond the scope of this paper.

5 Conclusions

In this paper we have presented a dynamic assurance framework for a highly configurable UAV. We have described confidence metrics for a battery monitor and timing monitor, and how we can use these to trigger reconfigurations to maintain safety performance. We argued that the meaning of confidence shortfalls must be communicated to risk holders, and showed an example of how this might be achieved via a dialectic assurance case. Further, we noted that reconfiguration should only be triggered under specific conditions, to avoid spurious changes or overly conservative responses.

For the next stages of our research we will extend the approach to consider other types of evidence such as certification data, e.g., test reports, as our confidence in these may also be undermined by reconfiguring the system where there are hardware dependencies. Additionally, we will be exploring in more detail the inter-dependencies between confidence measures, and their impact on the case. We have applied the approach to a UAV, but will consider how it can be extended to other types of systems and domains. Key challenges will be in scalability and how to generalise the approach.

References

1. Alves, E.E., Bhatt, D., Hall, B., Driscoll, K., Murugesan, A., Rushby, J.: Considerations in assuring safety of increasingly autonomous systems. Tech. rep. (2018)
2. Asaadi, E., Denney, E., Menzies, J., Pai, G.J., Petroff, D.: Dynamic assurance cases: a pathway to trusted autonomy. Computer **53**(12), 35–46 (2020)
3. Aslansefat, K., et al.: Safedrones: real-time reliability evaluation of UAVs using executable digital dependable identities. In: International Symposium on Model-Based Safety and Assessment, pp. 252–266. Springer (2022). https://doi.org/10.1007/978-3-031-15842-1_18
4. Assurance Case Working Group, S.C.S.C.: GSN Community Standard Version 3 (2021)
5. Bloomfield, R., Rushby, J.: Assessing Confidence with Assurance 2.0 (2023)
6. Calinescu, R., Weyns, D., Gerasimou, S., Iftikhar, M.U., Habli, I., Kelly, T.: Engineering trustworthy self-adaptive software with dynamic assurance cases. IEEE Trans. Software Eng. **44**(11), 1039–1069 (2018)
7. Dai, X., Zhao, S., Lesage, B., Bate, I.: Using digital twins in the development of complex dependable real-time embedded systems. In: International Symposium on Leveraging Applications of Formal Methods, pp. 37–53. Springer (2022). https://doi.org/10.1007/978-3-031-19762-8_4
8. Denney, E., Pai, G.: Automating the assembly of aviation safety cases. IEEE Trans. Reliab. **63**(4), 830–849 (2014)

9. Denney, E., Pai, G., Habli, I.: Dynamic safety cases for through-life safety assurance. In: 2015 IEEE/ACM 37th IEEE International Conference on Software Engineering, vol. 2, pp. 587–590 (2015)
10. Graydon, P.J., Holloway, C.M.: An investigation of proposed techniques for quantifying confidence in assurance arguments. Saf. Sci. **92**, 53–65 (2017). https://doi.org/10.1016/j.ssci.2016.09.014
11. Griffin, D., Lesage, B., Bate, I., Soboczenski, F., Davis, R.I.: Forecast-based interference: modelling multicore interference from observable factors. In: Proceedings of the 25th International Conference on Real-Time Networks and Systems, pp. 198–207 (2017)
12. Hansson, H., Jonsson, B.: A logic for reasoning about time and reliability. Formal Aspects Comput. pp. 512–535 (1994)
13. Hawkins, R., Ryan Conmy, P.: Identifying run-time monitoring requirements for autonomous systems through the analysis of safety arguments. In: Computer Safety, Reliability, and Security: 42nd International Conference, pp. 11–24 (2023)
14. Javed, M.A., Muram, F.U., Hansson, H., Punnekkat, S., Thane, H.: Towards dynamic safety assurance for industry 4.0. J. Syst. Archit. **114**(C) (2021)
15. Jiang, P., Ergu, D., Liu, F., Cai, Y., Ma, B.: A review of YOLO algorithm developments. Procedia Comput. Sci. **199**, 1066–1073 (2022)
16. Koopman, P.: Safety Performance Indicator (SPI) Metrics. https://safeautonomy.blogspot.com/2020/12/safety-performance-indicator-spi.html (2020)
17. Koopman, P., Fratrik, F.: How Many Operational Design Domains, Objects, and Events? In: SafeAI@AAAI (2019)
18. Kwiatkowska, M., Norman, G., Parker, D.: Stochastic model checking. In: Formal Methods for Performance Evaluation, pp. 220–270 (2007)
19. Kwiatkowska, M., Norman, G., Parker, D.: PRISM 4.0: Verification of probabilistic real-time systems. In: Computer Aided Verification, pp. 585–591 (2011)
20. Reich, J., Trapp, M.: Sinadra: Towards a framework for assurable situation-aware dynamic risk assessment of autonomous vehicles. In: 2020 16th European Dependable Computing Conference (EDCC), pp. 47–50 (2020). https://doi.org/10.1109/EDCC51268.2020.00017
21. Sen, R., Ramachandra, K.: Characterizing resource sensitivity of database workloads. In: 2018 IEEE International Symposium on High Performance Computer Architecture (HPCA), pp. 657–669. IEEE (2018)
22. Shankar, N., et al.: DesCert: Design for Certification (2022)
23. Sharath, M.N., Mehran, B.: A literature review of performance metrics of automated driving systems for on-road vehicles. Front. Future Transp.**2** (2021)
24. Sun, F., Hu, X., Zou, Y., Li, S.: Adaptive unscented kalman filtering for state of charge estimation of a lithium-ion battery for electric vehicles. Energy **36**(5), 3531–3540 (2011)
25. Wagle, R., Tong, Z., Sites, R.L., Anderson, J.H.: Want predictable GPU execution? Beware SMIs! In: Proceedings of the 29th IEEE International Conference on Parallel and Distributed Systems (2023)
26. Yan, F., Foster, S., Habli, I.: Safety case generation by model-based engineering State of the art and a proposal. In: 11th International Conference on Performance, Safety and Robustness in Complex Systems and Applications, pp. 4–7 (2021)
27. Yan, F., Foster, S., Habli, I., Wei, R.: Model-based generation of hazard-driven arguments and formal verification evidence for assurance cases. In: Proceedings of the 10th International Conference on Model-Driven Engineering and Software Development - Volume 1: MODELSWARD, pp. 252–263 (2022)

Redefining Safety for Autonomous Vehicles

Philip Koopman[1] and William Widen[2]

[1] Carnegie Mellon University, Pittsburgh, PA, USA
koopman@cmu.edu
[2] University of Miami – School of Law, Miami, FL, USA
wwiden@law.miami.edu

Abstract. Existing definitions and associated conceptual frameworks for computer-based system safety should be revisited in light of real-world experiences from deploying autonomous vehicles. Current terminology used by industry safety standards emphasizes mitigation of risk from specifically identified hazards, and carries assumptions based on human-supervised vehicle operation. Operation without a human driver dramatically increases the scope of safety concerns, especially due to operation in an open world environment, a requirement to self-enforce operational limits, participation in an ad hoc sociotechnical system of systems, and a requirement to conform to both legal and ethical constraints. Existing standards and terminology only partially address these new challenges. We propose updated definitions for core system safety concepts that encompass these additional considerations as a starting point for evolving safety approaches to address these additional safety challenges. These results might additionally inform framing safety terminology for other autonomous system applications.

Keywords: Automated vehicles · autonomous vehicles · safety terminology · safety case · safety engineering · risk

1 Introduction

The proliferation of critical systems that employ machine learning-based technology requires a fresh look at what we mean for a computer-based system to be "safe". It is time to ask whether our current definitions of safety are, themselves, fit for purpose. Existing definitional frameworks can be stretched further, but are showing significant signs of wear. Moreover, the mindset and processes informed by those legacy definitions are themselves eroding under the complexity of this new technology.

An update to the definition of safety and related terms can promote a more robust view of safety. We ground our proposed updates in incidents already seen in real-world driverless Autonomous Vehicle (AV) deployments as well as incidents involving human-supervised driving automation features, identifying broader underlying themes. We propose an updated set of definitions based on a need to evolve terminology used in current relevant industry consensus standards: ISO 26262, ISO 21448, and UL 4600.

As an initial example of the types of issues that are arising, there have been numerous reports of robotaxis in San Franciso interfering with emergency responder operations.

This often does not involve overtly dangerous vehicle motion such as a crash, but rather might involve a vehicle immobilizing itself in a bid to mitigate the risk of a crash in a situation it cannot handle safely. However, such safety-motivated behavior can increase societal risk by impeding the progress of emergency response vehicles. This is not the type of loss event primarily contemplated by current safety terminology.

We use examples of recent vehicle automation incidents to identify shortcomings of current safety concepts and terminology. We then propose adjusted definitions for core safety terms. While we focus on terminology and vehicle automation in this paper, this is the start of a much larger discussion regarding the safety of autonomous systems for a variety of application domains. The nature of that new technology demands a major shift and expansion of safety perspective to deal with these challenges.

Section 2 of this paper summarizes current definitions for safety and related concepts, using international standards relevant to automated vehicle safety. In subsequent sections, we cover why those definitions struggle under the demands placed upon them by autonomous systems, and identify core gaps illustrated by real-world incidents. We then propose updated key definitions to address the gaps while not breaking existing safety engineering practices, staying as close to existing definitions as is practicable.

Out of scope for this paper are: organizational safety philosophy, how to best achieve a defined safety goal, and how to integrate engineering terms into law. We leave for other discussions important topics such as safety improvement approaches and safety management systems. Similarly, hazard identification and analysis techniques remain applicable but beyond our scope. We also acknowledge that there is extensive work on safety and risk in the legal domain which informs our work. However, we do not seek to propose a redefinition of legal terminology—a topic for another paper.

2 Existing Safety Definitions

At its heart, our work is an inquiry into changing the nature of what "safety" should mean when developing a computer-based safety-critical system – especially one that must operate within a sociotechnical framework that goes beyond the bounds of both traditional functional safety and human-supervised system safety.

This work builds on our initial work finding that acceptable autonomous vehicle safety requires more than a net risk "better than human driver" approach [14]. In that work, we identify complicating factors including: risk subsidy, risk transfer, negligence, responder role safety, the proper role of blame, standards conformance, regulatory requirements, ethical concerns, equity concerns, and per-behavior regulatory approaches to risk acceptability. That work encompasses various contributions from [1, 2, 5, 23], and [26], but does not consider the question of whether the language of risk and safety itself should be revisited to help resolve the issues raised.

We start with an examination of key definitions of safety from autonomous vehicle safety standards. An overarching concept of existing definitions in the automotive domain is that automotive feature safety is often considered to mean an *Absence of Unreasonable Risk (AUR)*, typically as described in a chain of definitions in ISO 26262.

2.1 ISO 26262

ISO 26262:2018 is a functional safety standard that applies to road vehicles, whether autonomous or not [9]. That standard's Part 1 safety definitions are based on layers of risk, hazards, and harm. The core stack of definitions is summarized below, omitting cross-references and supporting notes. Defined terms are italicized:

- **Safety:** absence of *unreasonable risk*
- **Unreasonable risk:** *risk* judged to be unacceptable in a certain context according to valid societal moral concepts
- **Risk:** combination of the probability of occurrence of *harm* and the *severity* of that *harm*
- **Severity:** estimate of the extent of *harm* to one or more individuals that can occur in a potentially *hazardous event*
- **Hazardous event:** combination of a *hazard* and an *operational situation*
- **Hazard:** potential source of *harm* caused by *malfunctioning behavior* of the *item*
- **Harm:** physical injury or damage of persons
- **Malfunctioning behavior:** *failure* or unintended behavior of an *item* with respect to its design intent
- **Operational situation:** scenario that can occur during a vehicle's life
- **Safety case:** argument that *functional safety* is achieved for *items* or *elements*, and satisfied by evidence compiled from *work products* of activities during development.

We stop at this point without diving deeper into other defined terms. For our purposes, an "item" is an autonomous vehicle including its runtime support infrastructure, although it could be interpreted in other ways according to the standard. A failure is related to an abnormal condition, including both random and systematic faults, caused by component malfunctions, software defects, and any other sources.

As a functional safety standard, the emphasis is on avoiding harm to people due to malfunctioning behaviors, omitting mention of other types of potential losses such as property damage. The approach implicitly assumes that a system that perfectly implements its design intent (which includes risk mitigation measures) will be safe.

In practice, the automotive industry tends to identify hazards, analyze the risk of each hazard, and perform mitigation to ensure that no individual hazard results in an unreasonable risk. If that criterion is met, the system is deemed safe.

While it might happen in practice, there are no explicit requirements to examine relationships between risks presented by multiple unrelated hazards, to quantify net risk presented by the system as a whole, or to consider risks due to sources other than malfunctioning behaviors that violate design intent. In practice, a vehicle is presumed to be safe when it is released into series production. Any safety defect discovered after release is, in essence, due to a defective development or manufacturing process that might lead to a regulatory recall or other deployment of a manufacturer remedy.

The Automotive Safety Integrity Level (ASIL) approach considers severity, exposure, and controllability to determine the mitigation required for each identified hazard. Addressing controllability for uncrewed vehicles is potentially problematic.

2.2 ISO 21448

ISO 21448:2022 covers Safety of the Intended Function (SOTIF) [10], expanding the definition of safety from ISO 26262 while serving as a complementary standard. Its scope encompasses the driving behaviors of autonomous vehicles.

ISO 21448 does not change the definition of risk, nor of safety, instead adopting the ISO 26262 definitions. It does, however, add a new term covering SOTIF. Again, definitions are listed with supporting material omitted and defined terms italicized:

- **Safety of the intended functionality (SOTIF):** absence of *unreasonable risk* due to *hazards* resulting from *functional insufficiencies* of the *intended function* or its implementation
- **Hazard:** potential source of harm caused by the hazardous behavior at the vehicle level
- **Functional insufficiency:** *insufficiency of specification* or *performance insufficiency*
- **Performance insufficiency:** limitation of the technical capability contributing to a hazardous behavior or inability to prevent or detect and mitigate reasonably foreseeable indirect *misuse* when activated by one or more *triggering conditions*
- **Operational design domain (ODD):** specific conditions under which the system is designed to function

Again we stop short of listing the entire linked set of defined terms.

SOTIF overall deals with two types of functional insufficiencies generally beyond the scope of ISO 26262: (1) incomplete requirements, and (2) technical limitations of the system. A functional insufficiency might produce hazardous behavior in a specific circumstance, known as a triggering condition.

The incomplete requirements aspect is intended to deal with the fact that public roads are complex operational environments. The SOTIF methodology is based on reducing the "unknown hazardous" scenario categories until AUR has been achieved ([10] figures 8 and 10). The possibility of requirements gaps is acknowledged via reference to a potential insufficiency of specification. While this standard contemplates that an iterative experimental approach will be required for hazard analysis and scenario identification, it presumes that complete-enough specifications can be developed to achieve AUR before deployment. Its definition of hazard is modified from that in ISO 26262 to emphasize dangerous vehicle-level behavior.

Technical limitations of any system with external sensors will inevitably require building an internal model of the external world based on limited, incomplete, and noisy information. Not every radar pulse will produce a return. Sensors do not have an infinite range. Some objects are occluded from the point of view of any particular sensor. And various types of noise will introduce uncertainty into any model of the outside world. This standard recognizes that an automated driving system must be designed for AUR despite those potential performance insufficiencies.

Again, the emphasis is on the vehicle's behavior, and a general assumption is that loss events will primarily result from crashes caused by sensor limitations, vehicle behavioral deficiencies, or specification insufficiencies.

The inclusion of the concept of a defined Operational Design Domain (ODD) raises an important point in terms limiting the scope of safety considerations. In classical terminology, safety tends to be limited to some defined ODD scope. For example, the definition of a safety case in DefStan 00–56 is limited to cover "a given application in a given operating environment" [4]. However, the consideration of how operation is limited to that defined scope is often deferred to the judgment of human operators, such as a pilot who is supposed to avoid flying in extreme environmental conditions. With an autonomous system, a limitation of safety to defined conditions is inadequate, because some actor (by default the autonomous system) must also ensure safety by avoiding operating outside the ODD. This might include refusing to start a mission outside the ODD. But it should also require some acceptably safe response to being forcibly ejected from the ODD on short notice due to an unforeseen condition or event. Thus, enforcement of the ODD must be within the scope of a relevant definition of safety.

2.3 ANSI/UL 4600

UL 4600 is a system-level safety standard specific to autonomous vehicles [27]. Its terminology is intended to be compatible with ISO 26262 and ISO 21448, while addressing a broader system-level scope. Definitions were methodically reconciled with safety standards from other domains and are summarized below:

- **Safe:** having an *acceptable* post-mitigation *risk* at the *item* level as defined by the *safety case*
- **Acceptable:** sufficient to achieve the overall *item risk* as determined in the *safety case*
- **Risk:** combination of the probability of occurrence of a *loss* event and the severity of that *loss* event
- **Loss:** a substantive adverse outcome, from damage to property or the environment, to animal injury or death, to human injury or death
- **Safety case:** structured argument, supported by a body of evidence, that provides a compelling, comprehensible, and valid case that a system is safe for a given application in a given environment

As with previously discussed standards, risk is related to the probability and severity of a specific loss event. However, the notion of acceptability of risk has to do with an overall item risk rather than individual risks, including for example a highly recommended approach of total item risk summing (UL 4600 prompt element 6.1.1.3.a). Additionally, the definition of a loss is expanded past harm to humans to include other types of adverse outcomes, such as property damage.

The scope of a safety case is broadened, but its interaction with the definition of the term "safe" gives latitude to the authors of the safety case. It is the responsibility of the creator of the safety case to define what "safe" might mean, and to ensure that the safety case provides a suitable argument showing that the goal of safety has been achieved. The source for the definition of 'safety case' is Def Stan 00–56 [4], which limits the scope of the safety case to predetermined applications in a given operational environment. However, UL 4600 provides extensive lists of prompt elements to encourage robust consideration of potentially exceptional aspects of the ODD.

2.4 Other Safety Definitions

Other automotive safety standards in development that we are aware of plan to adopt ISO 26262 and potentially ISO 21448 terminology in whole or in part.

Another definition of safety widely cited in the automotive industry is Positive Risk Balance (PRB). This is one safety consideration proposed in a BMVI report [1]. That report proposes other restrictions on ethical safety, such as avoiding using personal characteristics to choose victims in a no-win crash scenario. But PRB is typically the criterion singled out for broader automotive industry safety discussions.

Using PRB as a sole criterion is problematic in no small part because of the difficulties in establishing a comparable baseline for human-driven vs. autonomous vehicles [14]. Nonetheless, it is the primary criterion promoted by both Waymo [29] and Cruise [3] when messaging their safety. PRB is often presumed to yield AUR in such discussions, although examples of AV safety problems discussed in the next section suggest this will not necessarily be the case.

An important regulatory user of the concept of AUR is the US National Highway Traffic Safety Administration (NHTSA). NHTSA pursues enforcement action when "the Agency finds either non-compliance or a defect posing an unreasonable risk to safety" [20]. Their approach tends to have two elements. The first is compliance with the Federal Motor Vehicle Safety Standards (FMVSS) [21], a set of specific tests for specific safety features primarily applicable to conventional vehicle safety rather than automated vehicle functions. FMVSS is necessary but not sufficient for safety.

The second NHTSA criterion is that they consider a safety defect to be a specific behavior, design defect, or other issue that they can associate with a pattern of loss events. NHTSA can launch investigations and require safety recalls. Typical NHTSA recalls and investigations involve a failure to conform to FMVSS or, less often, patterns of mishaps and incidents that are more difficult to link to a specific technical defect. NHTSA decisions have not historically considered net PRB at the vehicle level.

3 Examples of AV Safety Problems

In this section we consider some of the many real-world incidents and mishaps suffered by robotaxis to provide grounding for identifying gaps in current safety terminology. Examples are drawn from robotaxis and other vehicle automation incidents. Whether a safety supervisor human driver was present for an incident is irrelevant—a safety issue that happened with a human safety driver might also happen without one.

Pedestrian Dragging. A pedestrian was hit by another vehicle and thrown into the path of a robotaxi [12]. The robotaxi braked aggressively but struck the pedestrian. Arguably the robotaxi could have driven more defensively to perhaps avoid that initial strike. Regardless, after having stopped, the robotaxi lost track of the pedestrian and decided to pull to the side of the road, dragging the pedestrian under the vehicle, ending up with the pedestrian pinned almost entirely under the rear of the vehicle. The robotaxi company involved attempted to portray this as an unforeseeable freak event. Nonetheless, moving a vehicle without first ascertaining the location of an injured pedestrian who has just been struck by that same vehicle is highly problematic.

Crash with a Firetruck. A robotaxi entered an intersection as permitted by a green light but then collided with a fire truck, resulting in a passenger injury [16]. The fire truck had emergency annunciators active (siren, lights, horn) and was proceeding through a red light in cross traffic while responding to an emergency call. The robotaxi failed to yield to that emergency vehicle as required by road rules.

Hitting a Bus. A robotaxi became confused when following a long bus with a mid-body articulation pivot [24]. The robotaxi tracked the front half of the bus and ignored the back half. It then crashed into the back half of the bus because the tracking system had decided to ignore the detected back half in favor of the front half.

Interfering with Emergency Responders. City of San Francisco emergency responders reported at least 55 incidents of robotaxi interference with their operations [7], later increased to 74 incidents reported by the fire department [8]. While no incidents were definitively shown to result in harm to a person, they presented risk by delaying emergency responders and requiring attention from emergency response personnel that would be better spent tending to the actual emergency.

Encroachment on Closed Roads. There have been numerous incidents of encroachment on closed roads and emergency scenes which presented potential hazards to vehicles, vehicle occupants, and other road users. Examples include: dragging downed power lines and emergency scene yellow barrier tape down the street [24], and driving through a construction zone only to get mired in wet concrete [12].

Mass Strandings. Numerous mass strandings of vehicles have occurred, under a variety of circumstances. One that got particular attention was attributed to a loss of communications due to cellular phone system overload by a concert event in a proximate geographic region – even though that event did not involve the street on which the stranding took place [18]. This raises questions as to what would happen in a communication disruption or traffic control device power outage caused by a natural disaster such as an earthquake or other common cause infrastructure failure.

Child Debarking a School Bus. A vehicle with driving automation activated struck and injured a child debarking from a school bus [17].

Failure to Stop at Stop Signs (Rolling Stops). A NHTSA recall for safety defects was implemented for a driving automation system that was programmed to roll through stops at speeds up to 5.6 mph, in violation of traffic laws [28].

Emergency Responder Injuries and Fatalities. An investigation and initial recall were conducted for a pattern of collisions with emergency response vehicles that, over time, involved at least 14 crashes, 15 injuries, and 1 fatality [22]. An eventual recall was not for a specific reproducible behavioral defect, but rather a pattern of losses with a common theme related to lack of ODD enforcement combined with inadequate enforcement of human driver attention on the road [11].

Collisions with Stopped and Crossing Vehicles. Related to the emergency responder crash investigation are numerous reports of collisions with stopped vehicles that have been reported for a particular driving automation system. That includes multiple fatalities in scenarios involving under-running a crossing heavy truck (e.g. [25]). While the manufacturer says that the design intent is for the driver to manually avoid such driving situations and such collisions, crashes keep accumulating nonetheless.

Elevated Collisions in a Vulnerable Community. A concerning 11 out of 74 San Francisco Fire Department reported robotaxi incidents, including a collision with a fire truck, occurred in the vicinity of the geographically small Tenderloin district [8]. This district is renowned for being disadvantaged and having a historically at-risk community. For likely related reasons, it is one of the most active emergency response locations in the US. Nonetheless, robotaxi companies have seen fit to continue testing in that area, presumably because of its location in downtown San Francisco.

Attracting Passengers Away from Mass Transit. Even if robotaxis were as safe as human-driven vehicles, human-driven vehicles are far more dangerous than mass transit [19]. A widespread adoption of robotaxis could degrade safety by shifting mass transit passenger-miles to robotaxi passenger-miles [31]. Loss of mass transit patrons could additionally erode the funding and viability of safer modes of transportation, increasing net fatalities totaled across all modes of transportation.

We acknowledge that human drivers can and do make all of the types of mistakes recounted above. But we are interested in understanding the true scope of safety, which should apply to both human drivers and vehicles with automated driving features.

Many of these situations did not involve actual harm to a person. But all were considered safety issues by at least some relevant stakeholders due to the potential for direct or indirect harm. In many of the above cases it is easy to blame some actor other than the driving automation capability for responsibility. But casting blame to dodge changing the status quo is unlikely to prevent future mishaps.

4 What is Missing from Safety Definitions

Based on these observed failures and a general understanding of autonomous vehicle safety requirements, we identify four general characteristics of autonomous vehicles that profoundly affect safety engineering: operating in an open world environment, self-enforcing operational limits, deployment in an ad hoc sociotechnical system, and addressing external constraints such as legal limitations. Dealing with all four of these areas has historically been allocated to the human vehicle driver. However, the whole point of having an autonomous vehicle is to no longer need that human driver, imposing these additional requirements on technical systems instead. (As an interim measure, remote operations teams might assist with some of these issues. But scalable deployment requires minimizing the need for such remote human operator intervention).

4.1 Open World Environment

Autonomous vehicles are engineered systems that must operate within a framework of uncertainty and incomplete training for all of the objects and events they will encounter while operating on public roads at scale.

The SOTIF approach of ISO 21448 is largely intended to address this issue. However, that standard's approach – and the practical approach of many developers – is to assume that enough possible scenarios, objects, and events will have been identified and mitigated before deployment to result in net AUR. This might not be possible in a practical system if the distribution frequency of encountering hazards is heavy-tailed, involving a large number of individual hazards that each have a very low frequency of arrival [13]. While there are possible technical approaches to ensure acceptable safety despite a heavy-tailed hazard situation, the probability of substantive risk from such hazards remaining after deployment cannot be ignored. ISO 21448 accommodates an iterative improvement approach to deal with newly emergent hazards, but still presumes that AUR will be achieved at initial release.

UL 4600 has more comprehensive mechanisms that recognize a substantive degree of uncertainty might be present at initial deployment, meaning AUR might be expected, but that expectation might itself have a substantial degree of uncertainty. UL 4600 requires the use of Safety Performance Indicators and field engineering feedback to manage a continuous improvement process to identify and mitigate risks due to a changing, open world environment as well as encounters with unforeseen heavy-tail events.

A comprehensive definition of safety should contemplate two issues that will be a reality for autonomous vehicles for the foreseeable future: (1) proactive management of inevitable requirements gaps in deployed systems, and (2) support for continual updates over the vehicle's lifecycle to mitigate emergent hazards and risks due to environmental and other changes.

4.2 Self-enforcement of Operational Limitations

A common approach to reaching a situation outside the operational limits of an AV is some sort of safety shutdown or abnormal mission termination. For an AV this might mean pulling to a safe stopping location, or even stopping in the middle of a travel lane in favorable conditions. While executing a reasonable safety stop can be complex, recognizing that the AV has exceeded its operational limitations is even more challenging.

Machine learning-based technology has a fundamental challenge with recognizing that it has encountered a meaningful data dimension that is relevant to safety but has not been captured as a data feature of some sort during its training. For example, if there are too few people dressed in yellow clothing in a training data set, a yellow-garbed construction worker might not be recognized as a construction worker directing traffic, or might not even be recognized as a person at all. Or emergency scene yellow tape might be recognized as an insubstantial bit of plastic that poses no collision threat instead of constituting a safety-relevant "keep-out" warning.

A core element of safety will need to be recognizing and responding to situations that exceed the intended nominal operational environment of the system. These might be unforeseen situations such as novel objects and events. But they might also be an unexpected arrival of a foreseen out-of-ODD situation, such as a sudden torrent of rain on a day with a sunny weather forecast. This couples with the open environment issue to make it challenging to self-enforce operational limits inherent to the system that were not contemplated as such by the design team. Current definitions tend to limit safety to being considered within an understood, specified environment. However, autonomous systems must also ensure safety when operating in an under-specified environment, as well as be able to react in some reasonable way to unexpectedly finding themselves outside the environment they were designed to operate within.

4.3 Ad Hoc Systems of Systems

Autonomous systems must operate as a component of societal systems, which are often under-specified and, for the most part, beyond the ability of the autonomous system design team to control. From the point of view of current safety definitions, the design team cannot control which scenarios it will have to deal with other than by, in some cases, establishing operational limitations that the system must then self-enforce.

A common theme of some stakeholder safety concerns is that an autonomous vehicle is acting in a narrowly safe way by not crashing into things, but is causing negative externalities for other road users. Examples include an in-lane stop when the AV is unsure what to do next, missing cues that an apparently open driving lane (intended for construction vehicle use) should not be used by a robotaxi due to the context of "road closed" signs in adjacent lanes, a practical necessity to break some normal road usage practices in exceptional circumstances to provide room for a passing emergency vehicle, elevated risk of crashes due to so-called "phantom" braking, etc.

Some safety concerns have a more subtle context sensitivity. For example, a human driver might reroute due to seeing a huge structure fire a few blocks ahead to avoid becoming ensnarled with the likely on-scene traffic chaos. A robotaxi that proceeds without recognizing the situation will potentially impede emergency response activity.

While one might try to analyze all the hazards present at the system-of-systems level, this is typically beyond the scope and resources of an AV safety engineering effort. We believe it is more practical to instead express mitigation for negative externalities as constraints on permissible behavior. An example rule might be to pull out of travel lanes when a fire truck with active annunciators is in the vicinity regardless of on-board software's estimate of the risk of collision. Another example might be to avoid using a road where a preceding robotaxi has gotten stuck to avoid clustering stuck robotaxis.

4.4 Legal and Ethical Constraints

There are a number of constraints on permissible system behavior that not only are difficult to express as hazards, but might reduce the theoretical net benefits (or possibly even net safety) of a system. A risk-centric approach will struggle with such constraints.

As an example, consider a hypothetical situation in which a driving automation system reduced total fatalities, but increased the rate of fatalities imposed on emergency responders at roadside crash scenes. Or consider a more extreme hypothetical situation in which total road fatalities were reduced by half—but pedestrian fatalities doubled in number, becoming a much bigger fraction of that reduced net fatality rate. Both outcomes will be problematic for some stakeholders, even if net harm is reduced.

A technical system should not be considered acceptably safe by societal stakeholders if it does not satisfy constraints on both individual vehicle behaviors and patterns of behaviors that would, for example, also apply to human drivers in such situations. Consider a vehicle that rolls through stop signs when it determines the intersection is clear. Perhaps, hypothetically, a study might show that this reduces rear-end collisions and therefore improves net safety. Such a system would still likely be seen as presenting unreasonable risk due to automating the violation of traffic laws. It would also present a negligent driving liability exposure if an AV were to hit an undetected pedestrian after ignoring the requirement to come to a full and complete stop at a stop sign.

There are also a number of ethical, equity, and legal concerns that are not obviously tied to malfunctioning vehicle motion control, such as concerns of over-weighting public road testing of immature technology in vulnerable communities [30]. Some such constraints can be converted to functional requirements, but others might better be treated as constraints on design optimization choices [15].

5 A Proposal for More Robust Definitions

5.1 What Needs to be Addressed?

We propose an updated set of core safety definitions to address the issues identified based on examples of vehicle automation incidents, in light of concepts already present in the standards and other sources mentioned in the preceding sections.

Reviewing the example incidents and analysis of gap areas in Sects. 3 and 4 above, we believe that the following aspects of safety need to be addressed more directly at the definitional level. We use keywords and phrases from definitions surveyed in Sect. 2 as tags for each concept.

- **Safe/Safety:** The system must not only mitigate hazards, but also meet externally imposed constraints. Constraints might involve ethical- and equity-based prohibitions on engineering optimizations that might otherwise improve specific aspects of safety [15], and also prohibit unacceptable patterns of risks. We prefer "has **acceptable safety**" to "safe".
- **ODD/Given environment:** The environment cannot be assumed to be fully characterized, nor unchanging over time. Rather, acceptable safety must be assured regardless of the real-world environment, even if the system must self-enforce a risk mitigation response due to exceeding its operational limitations.
- **Given application:** The system needs to enforce potential misuse, such as an operator attempting to engage driving automation on roads with cross-traffic when that might violate a limited access highway-only system design intent restriction.
- **Risk:** The time-worn formulation of risk as a combination of probability and severity might serve for single-dimensional optimization of net harm, especially in situations in which monetary compensation is a morally acceptable plan for mitigation. But it is an overly narrow viewpoint for societal and other constraints such as patterns of harm or violations of constraints not readily reduced to a classical risk value, especially when a purely utilitarian approach might be deemed undesirable.
- **Hazardous event:** There are some risks and constraints that are challenging to evaluate at the hazardous event level because they involve safety tradeoffs at the system-of-systems level and potentially unknown requirements (think of future case law that has not yet been established). It is unreasonable to expect vehicle designers to fully assess, let alone mitigate every single such risk up front before systems are deployed.

Patterns of risk such as risk transfer onto vulnerable populations are rather far removed from individual loss events.
- **Severity:** Harm done by any particular loss event is important, but the severity of any individual event might not capture the importance of that event if it is part of a larger

---

- **Acceptable:** meets all *safety constraints* as shown by a *safety case*
  Note: The phrase "acceptably safe" might be used in some contexts for clarity. While "Safety" is used as a modifier, use of the word "safe" alone should be avoided. Safety constraints encompass to whom the safety must be acceptable.
- **Safety case:** structured argument, supported by a body of evidence, that provides a compelling, comprehensible, and sound argument that *safety engineering* efforts have ensured a system meets a comprehensive set of *safety constraints*
  Note: This emphasizes meeting constraints rather than net risk. A limit to defined operational environments is intentionally excluded, but ODD enforcement might be allocated to human operators in the safety case when appropriate.
- **Safety engineering:** a methodical process of ensuring a system meets all its *safety constraints* throughout its lifecycle, including at least hazard analysis, risk assessment, risk mitigation, validation, and field engineering feedback
  Note: Requires safety engineering beyond brute force test validation. Hazard analysis is broadened to address all safety constraints. Explicitly requires addressing safety over the system's lifecycle.
- **Safety constraint:** a limitation imposed on *risk* or other aspects of the system by stakeholder requirements
  Note: This implicitly requires the identification of stakeholders who might be affected by losses, and makes it more straightforward to view safety as a multi-dimensional constrained optimization problem rather than a mostly one-dimensional pure risk optimization problem [15]. Safety constraints might include: AUR, PRB, limits on individual risks, limits on net risk, exposure limits for specified types of risk patterns, and issues that are difficult to trace to pure risk.
- **Risk:** combination of the probability of occurrence of a *loss*, or pattern of *losses*, and the importance to stakeholders of the associated consequences
  Note: Consequence (severity) might be an overriding concern regardless of probability. Net importance can be non-linearly related to individual losses if forming a pattern. Correlated loss events, inequitable loss patterns, and loss patterns involving a failure to mitigate emergent loss trends are in-scope.
- **Loss:** an adverse outcome, including damage to the system itself, negative societal externalities, damage to property, damage to the environment, injury or death to animals, and injury or death to people
  Note: This is broader in scope than some other typical definitions of loss or harm. Some types of loss might be assigned very low severity in some application domains. Allocation of blame does not affect whether a loss occurred.

Fig. 1. Proposed definitions.

pattern of unethical or inequitable outcomes – even if that event would otherwise be considered low severity in terms of utilitarian personal harm.
- **Malfunctioning:** Incidents can occur not just because a vehicle has displayed dangerous motion, but also because in an attempt to improve tactical safety it inflicts potential damage at the system-of-systems level. Immobilized robotaxis blocking emergency vehicles after a safety shutdown are a poster child of this issue.
- **Harm:** Harm must be considered to go beyond personal injury or fatalities (as is already the case in some definitions). Even indirectly caused harm to property might legitimately be seen as a safety issue for some domains.

5.2 Proposed Safety-Related Definitions

We propose a set of new definitions for core safety terminology in Fig. 1 above. While specific proposals for each standard's particular use of terms are beyond the scope of this paper, we believe that if these terms were to be adopted by UL 4600 the changes imposed on the remainder of the document would guide a beneficial evolution to the scope of coverage of safety cases conforming to that standard. Other standards such as ISO 26262 might still appropriately use more restrictive terms, but should align terminology so as not to preclude conformance if these more expansive definitions are used by design teams instead.

At a high level, the approach suggested here considers the concept of "safety" not as an optimization process to reduce risk, but rather as the satisfaction of a set of safety constraints. One such constraint will typically be sufficient risk mitigation in keeping with more traditional safety engineering approaches. However, other constraints together with an increased scope for the concept of a loss event can address legal and ethical issues. The open world environment issue is addressed by explicitly including lifecycle considerations in the definition of safety engineering, and requiring safety engineering to be tied to a safety case. The self-enforcement of operational limitations issue is addressed via removing the phrase "for a given environment" and avoiding reference to an ODD in the definition of the safety case. The ad hoc system of systems issue is addressed via including the notion of stakeholders beyond the system designer in the definitions of risk and safety constraint.

6 Conclusions

While there is a continual stream of new safety considerations in the evolution of the area of safety engineering, we believe that the advent of autonomous systems represents a watershed moment. Moreover, the concerns motivating the proposed changes are not just theoretical, but have actually played out on US public roads. This is just the beginning. It is time for the safety community to revisit what safety really means.

One might consider addressing the concerns we raise via aggressive reinterpretation of existing terminology. However, any compliance-centric users of a standard are strongly incentivized to interpret definitions in the way most favorable to a low-cost compliance effort, undermining any potential educational efforts to promote reinterpretation to expand definitional scope. Additionally, a standard should say what it means, and not be dependent on non-normative, independent interpretational guidance.

While the standards mentioned in this paper were all written in good faith and have served honorably, it is time to update their definitions of safety to address the concerns raised by real-world experiences with the increasing complexities of automation technology. To be clear, we consider these proposed definitions to be the start of a discussion within the safety community rather than a finished conclusive outcome.

While we have used autonomous vehicles as a motivating example, similar issues will arise across a broad spectrum of safety-critical systems, and these definition proposals might inform autonomous system safety more broadly. What has changed is the lack of a closely attentive human operator to address issues such as enforcing operational limits, or to potentially serve as a moral crumple zone [6] to shield the system from blame for unmitigated equipment failures. An additional factor to consider is the relationship between the definitions for safety and cybersecurity terminology, especially when security failures can compromise safety.

Broadening the scope of safety will be an improvement for all systems, and becomes increasingly important as technology continues to insinuate itself into the fabric of everyday society. The more we automate beyond a practical ability for humans to exercise effective oversight, the more pressing these issues will become.

Thanks to Dr. Mallory Graydon and the anonymous reviewers for their comments.

References

1. BMVI. Ethics Commission: Automated and Connected Driving Report, Federal Ministry of Transport and Digital Infrastructure, Germany (2017). https://perma.cc/6UBX-KH5G
2. CDEI. Responsible Innovation in Self-Driving Vehicles. Policy Paper, Center for Data Ethics and Innovation (2022). https://www.gov.uk/government/publications/responsible-innovation-in-self-driving-vehicles
3. Cruise. Safety (2024). https://getcruise.com/safety/
4. UK Ministry of Defence. Defence Standard 00-56 Issue 7 (Part 1): Safety Management Requirements for Defence Systems. p. 26 (2017). https://s3-eu-west-1.amazonaws.com/s3.spanglefish.com/s/22631/documents/safety-specifications/def-stan-00-056-pt1-iss7-28feb17.pdf
5. European Commission. Directorate-General for Research and Innovation, Ethics of connected and automated vehicles – Recommendations on road safety, privacy, fairness, explainability and responsibility, Publications Office (2020). https://data.europa.eu/doi/10.2777/035239
6. Elish, M.: Moral crumple zones: cautionary tales in human-robot interaction. Engag. Sci. Technol. Soc. **5** (2019). https://bit.ly/47MmxpO
7. Eskenazi, J., Jarrett, W.: Explore: See the 55 reports—so far—of robot cars interfering with SF fire dept. Mission Local (2023). https://missionlocal.org/2023/08/cruise-waymo-autonomous-vehicle-robot-taxi-driverless-car-reports-san-francisco/
8. Farivar, C.: A mystery around a robotaxi, the fire department, and a death in San Francisco. Forbes (2023). https://bit.ly/490Awtg
9. INT'L ORG. FOR STANDARDIZATION [ISO]. ISO 26262-1:2018 Road vehicles—Functional safety (2018). https://www.iso.org/standard/68383.html
10. INT'L ORG. FOR STANDARDIZATION [ISO]. ISO 21448:2022 Road vehicles—Safety of the intended functionality (2022). https://www.iso.org/standard/77490.html

11. Hawkins, A.: Tesla's huge recall will make it harder—but not impossible—to misuse Autopilot. The Verge (2023). https://bit.ly/424ZHsx
12. Kerr, D.: Driverless car startup Cruise's no good, terrible year. Nat. Pub. Radio (2023). https://bit.ly/3O9Ffkr
13. Koopman, P.: The heavy tail safety ceiling. In: Automated and Connected Vehicle Systems Testing Symposium (2018). https://bit.ly/47N0zmP
14. Koopman, P., Widen, W.: Breaking the tyranny of net risk metrics for automated vehicle safety. Saf.-Crit. Syst. eJ. (2024). https://scsc.uk/scsc-191
15. Koopman, P., Widen, W.: Ethical design and testing of automated driving features. IEEE Design Test (2024). https://doi.org/10.1109/MDAT.2023.3281733
16. Korosec, K.: Cruise robotaxi involved in a crash with fire truck, one passenger injured. TechCrunch (2023). https://techcrunch.com/2023/08/18/cruise-robotaxi-involved-in-a-crash-with-fire-truck-one-passenger-injured/
17. Krisher, T.: US probes crash involving Tesla that hit student leaving bus. Associated Press (2023). https://apnews.com/article/tesla-school-bus-student-hurt-firetruck-d282a5dd63874f22f5e1a6fc8168801b
18. Mitchell, R.: San Francisco's North Beach streets clogged as long line of Cruise robotaxis come to a standstill. L.A. Times (2023). https://www.latimes.com/california/story/2023-08-12/cruise-robotaxis-come-to-a-standstill
19. National Safety Council. Deaths by Transportation Mode (Undated). https://injuryfacts.nsc.org/home-and-community/safety-topics/deaths-by-transportation-mode/. Accessed 22 Jan 2024
20. NHTSA. Understanding NHTSA's Regulator Tools (Undated). https://www.nhtsa.gov/sites/nhtsa.gov/files/documents/understanding_nhtsas_current_regulatory_tools-tag.pdf. Accessed 27 May 2024
21. NHTSA. Laws and Regulations (Undated). https://www.nhtsa.gov/laws-regulations. Accessed 27 May 2024
22. NHTSA, ODI Resume EA 22-002 (2022). https://static.nhtsa.gov/odi/inv/2022/INOA-EA22002-3184.PDF
23. SASWG. Safety Assurance Objectives for Autonomous Systems, The Safety of Autonomous Systems Working Group (2022). https://scsc.uk/r153B:1?t=1
24. Stumpf, R.: Cruise autonomous robotaxis rear-ends bus, get tangled in power lines. The Drive (2023). https://www.thedrive.com/news/cruise-autonomous-robotaxis-rear-ends-bus-get-tangled-in-power-lines
25. Thadani, T., et al.: The final 11 seconds of a fatal Tesla Autopilot crash. Wash. Post (2023). https://bit.ly/4b9H76s
26. Law Commission of England and Wales, and Scottish Law Commission. Automated Vehicles: joint report, HC 1068 SG/2022/15 (2022). http://bit.ly/lawcom2022
27. UNDERWRITERS LAB'YS. ANSI/UL 4600 Standard for Safety for the Evaluation of Autonomous Products, 3rd edn. (2023)
28. Vincent, J.: Tesla to disable 'rolling stop' feature after NHTSA says it can 'increase the risk of a crash. The Verge (2022). https://bit.ly/48FeAUI
29. Waymo. Safety (2024). https://waymo.com/safety/
30. Widen, W.: Highly Automated vehicles & discrimination against low-income persons. Univ. N. C. J. Law Tech. **24**, 115 (2022). https://papers.ssrn.com/sol3/papers.cfm?abstract_id=4016783
31. Zipper, D.: San Francisco has a problem with Robotaxis. The Atlantic (2023). https://bit.ly/3O8Es3i

Author Index

A
Acquaviva, Andrea 169
Alemzadeh, Homa 33
Annable, Nicholas 134
Askarpour, Mehrnoosh 134

B
Backeman, Peter 251
Barbierato, Luca 169
Barchi, Francesco 169
Bartolini, Andrea 169
Bonomi, Silvia 200

C
Cappai, Stefano 200
Castellanos-Ardila, Julieth Patricia 251
Chiang, Thomas 134
Coppa, Emilio 200
Cullmann, Christoph 267

D
D'Argenio, Pedro R. 235
Dai, Yifan 184
de Andrés, David 3
Denney, Ewen 51
Durling, Michael 84

F
Fan, Wenjun 184
Ferdinand, Christian 267

G
Garagiola, Nazareno 235
Gracia-Morán, Joaquín 3

H
Hahn, Sebastian 267
Hansson, Hans 251
Hermanns, Holger 235
Herter, Jörg 267

I
Iliasov, Alexei 68

K
Kästner, Daniel 267
Kodama, Hideaki 150
Kokaly, Sahar 134
Koopman, Philip 119, 300

L
Laibinis, Linas 68
Lawford, Mark 134
Lim, Eng Gee 184
Lisitsa, Alexei 184
Liu, Jia 184
Lopuhaä-Zwakenberg, Milan 218
Lu, Yiyang 33

M
Matsuno, Yutaka 150
Mauborgne, Laurent 267
Meng, Baoluo 84
Moitra, Abha 84
Molloy, John 285
Munaro, Tiziano 18
Muntean, Irina 18
Musa, Alberto 169

O
Okada, Manabu 150
Ota, Hiroshi 150

P
Pai, Ganesh 51
Paige, Richard F. 134
Parisi, Emanuele 169
Patti, Edoardo 169
Paul, Saswata 84
Pretschner, Alexander 18
Punnekkat, Sasikumar 251

© The Editor(s) (if applicable) and The Author(s), under exclusive license
to Springer Nature Switzerland AG 2024
A. Ceccarelli et al. (Eds.): SAFECOMP 2024, LNCS 14988, pp. 315–316, 2024.
https://doi.org/10.1007/978-3-031-68606-1

R
Romanovsky, Alexander 68
Ruiz, Juan Carlos 3
Ryan, Philippa 285

S
Saiz-Adalid, Luis-J. 3
Schmedding, Anna 33
Schoepp, Ulrich 100
Schowitz, Philip 33
Sethu, Ramesh 134
Shahbeigi, Sepeedeh 285
Siu, Kit 84
Smirni, Evgenia 33
Soltani, Reza 218
Sorokin, Lev 100
Stefanakos, Ioannis 285
Stoelinga, Mariëlle 218

T
Takai, Toshinori 150
Taylor, Dominic 68
Tsuchiya, Tomoyuki 150

V
Volante, Franco 169

W
Wassyng, Alan 134
Widen, William 300
Wilhelm, Stephan 267

Y
Yang, Lishan 33

Z
Zhou, Xugui 33
Zou, Jie 285

SPRINGER NATURE

GPSR Compliance

The European Union's (EU) General Product Safety Regulation (GPSR) is a set of rules that requires consumer products to be safe and our obligations to ensure this.

If you have any concerns about our products, you can contact us on ProductSafety@springernature.com

In case Publisher is established outside the EU, the EU authorized representative is:

Springer Nature Customer Service Center GmbH
Europaplatz 3
69115 Heidelberg, Germany

The manufacturer's authorised representative in the EU is Springer Nature Customer Service Centre GmbH, Europaplatz 3, 69115 Heidelberg, Germany. If you have any concerns regarding our products, please contact ProductSafety@springernature.com

Printed and bound by CPI Group (UK) Ltd, Croydon, CR0 4YY

25/03/2026

02078187-0017